ION
INTERACTIONS
IN
ENERGY
TRANSFER
BIOMEMBRANES

ION INTERACTIONS IN ENERGY TRANSFER BIOMEMBRANES

EDITED BY
G. C. PAPAGEORGIOU
Greek Atomic Energy Commission
Athens, Greece

J. BARBER
Imperial College of Science and Technology
London, England

and
S. PAPA
University of Bari
Bari, Italy

PLENUM PRESS • NEW YORK AND LONDON

Library of Congress Cataloging in Publication Data

Main entry under title:

Ion interactions in energy transfer biomembranes.

"Proceedings of an international workshop organized jointly by UNESCO Working Group IV of the European Expert Committee of Biomaterials and Biotechnology and the National Research Center, Demokritos, held April 8–12, 1985, in Athens, Greece" — T.p. verso.
    Includes bibliographical references and index.
    1. Membranes (Biology) — Congresses. 2. Bioenergetics — Congresses. 3. Ion pumps — Congresses. 4. Ion channels — Congresses. 5. Electron transport — Congresses. I. Papageorgiou, George C. II. Barber, J. (James), 1940-   . III. Papa, S. IV. Unesco. European Expert Committee of Biomaterials and Biotechnology. Working group IV. V. Kentron Pyrénikon Ereunón Démokritos (Athens, Greece)
QH601.I66   1986   574.87'5   85-31707
ISBN 978-1-4684-8412-0   ISBN 978-1-4684-8410-6 (eBook)
DOI 10.1007/978-1-4684-8410-6

Proceedings of an international workshop organized jointly by UNESCO
Working Group IV of the European Expert Committee on Biomaterials and
Biotechnology and the National Research Center, Demokritos,
held April 8–12, 1985, in Athens, Greece

PREFACE

The Expert Committee on Biomaterials and Biotechnology for the
European and the North American Region was founded by the General
Assembly of UNESCO at its 21st Session, in 1981. The Committee comprises
a Coordinating Group and four Working Groups, defined in the following
scientific areas:

    Group I   : Proteins: source, structure and function.
    Group II   : Nucleic acids: the hereditary materials.
    Group III : Immune materials and mechanisms.
    Group IV   : Membranes and transport in biosystems.

In fulfilment of one of the objectives of the Committee, which have
been adopted by the General Assembly of UNESCO in 1981, namely the
intensification of the exchange of scientific information on biomaterials
and biotechnology, Working Group IV organized an international workshop
on Ion Interactions in Energy Transport Systems, which was convened in
Athens, Greece, from 8 to 12 April, 1985. Scientific papers presented at
that workshop make up the chapters presented in this volume.

The present volume focusses on natural and artificial membranes that
are involved in energy transduction. Several chapters are devoted to
membranes and membrane components that convert and utilize light, such as
the thylakoid membrane of oxygenic photosynthetic organisms (eukaryotic
and prokaryotic), the chromatophore membrane of nonoxygenic
photosynthetic bacteria and the purple membrane of the halophilic
bacteria. Other systems examined include the mitochondrial membranes and
their adenine nucleotide carrier, the plasma membrane of animal cells,
and lipid bilayer vesicles, either reconstituted or not, with enzymes.

The most widespread mechanisms of biological energy transduction are
those associated with the photosynthetic and the respiratory processes,
which involve flow of energy among the following forms:

Light

$\searrow$ Redox   Electrochemical   Phosphate bond
          potential $\rightarrow$ potential $\rightarrow$ potential

Metabolites

Intensive research, in the recent years, has succeeded in
elucidating the basic mechanisms by which these transformations occur.
Electrochemical potential differences established between membrane
compartments having low permeability to ions play a central role in the
transduction of redox energy to phosphate bond energy and have been
studied and reported in depth. What distinguishes, however, the present
volume from other publications dealing with these subjects is an effort
to view energy transduction from the perspective of the electrostatic

properties of membrane-solution interfaces. In the physiological pH range, biological membranes carry an overall negative electric charge on their surfaces, which is due to the preponderance of ionized acidic groups over basic groups, although positively charged domains and components are known to exist. The electrostatic parameters of the membrane-solution interface, namely the surface and space charge densities, the electrostatic force field, and the electrostatic potential are influenced by the electrolyte structure of the aqueous media in contact with the membrane, and particularly by the cations present. In this way, through the electrostatic properties of the membrane-solution interfaces, electrolytes control several structural and physicochemical manifestations of membranes, such as binding of extrinsic proteins, membrane-membrane adhesion, surface area concentrations of ions, and the magnitude of electrochemical potential differences across membranes. The aim of the workshop was that of gathering together experts from various countries to report and debate their more recent studies in this particular area of biochemistry and biophysics. Judging from the contributions delivered at the workshop and published in this book, the meeting served as an unique opportunity for concerted and critical discussions of various aspects of ionic interactions with energy coupling membranes and related topics.

The workshop on Ion Interactions in Energy Transfer Systems and the volume on Ion Interactions in Energy Transfer Biomembranes would not have been possible without the support of UNESCO, IUPAB, IUB-IUPAB Bioenergetics Group and the National Research Center Demokritos of Greece. To all these organizations, and to their officials we express our gratitude.

September, 1985

G. Papageorgiou
J. Barber
S. Papa

CONTENTS

ION REQUIREMENTS OF PHOTOSYNTHETIC ELECTRON TRANSPORT

THE ELECTROSTATIC AND ELECTROKINETIC PROPERTIES OF BIOLOGICAL

MEMBRANES: NEW THEORETICAL MODELS AND EXPERIMENTAL RESULTS

Stuart McLaughlin

Dept. of Physiology & Biophysics
HSC, SUNY
Stony Brook, NY 11794

INTRODUCTION

Gouy (1) and Chapman (2) formulated the concept of the diffuse double layer by combining the Poisson and Boltzmann equations to describe the profile of the electrostatic potential in an aqueous phase adjacent to a charged surface. Stern (3) modified this concept by introducing the possibility that ions could bind to the surface. The Gouy-Chapman-Stern theory of the diffuse double layer describes adequately the electrostatic potential adjacent to a phospholipid bilayer: as the theory requires, the charges on a negative phospholipid are located in a plane at the bilayer surface.

Figure 1A illustrates the distribution of ions at a given instant in time near a negatively charged surface immersed in an aqueous solution (4). The negative charges produce an electric field that attracts counterions, ions of the opposite sign to the charge on the membrane, and repels coions, ions of the same sign to the charge on the membrane. The magnitude of the electrostatic potential is predicted to decrease with distance from the membrane as illustrated in Fig. 1B. If we assume that 23% of the lipids bear a single net negative charge and that the remaining lipids are net neutral, the average charge density is about 1 electronic charge/3 nm^2 because a phospholipid occupies an area of about 0.7 nm^2. If the concentration of salt in the bulk aqueous phase is 0.1 M, the theory predicts that the potential at the aqueous side of the membrane-solution interface is -60 mV, as illustrated in Fig. 1B. When the surface potential is not too high, the magnitude of the potential decreases in an approximately exponential manner with distance from the membrane. As indicated in Fig. 1B, the distance at which the potential falls to 1/e its value at the surface of the membrane is called the Debye length. The Debye length is predicted to be about 1 nm when the bulk concentration of monovalent ions is 0.1 M and about 10 nm when the concentration of monovalent ions is 0.001 M. Recent experiments by Israelachvili and his collaborators in Australia and by Parsegian, Rand, Lis and their collaborators in America confirm these predictions. The concentration of ions at any distance away from the membrane may be calculated from the potential illustrated in Fig. 1B via the Boltzmann relation. These concentrations are illustrated in Fig. 1C. If the concentration of monovalent ions in the bulk aqueous phase is 0.1 M and the surface potential is -60 mV, the Boltzmann relation predicts that the concentration of monovalent cations at the surface of the membrane is 1 M and that the concentration of monovalent anions is 0.01 M.

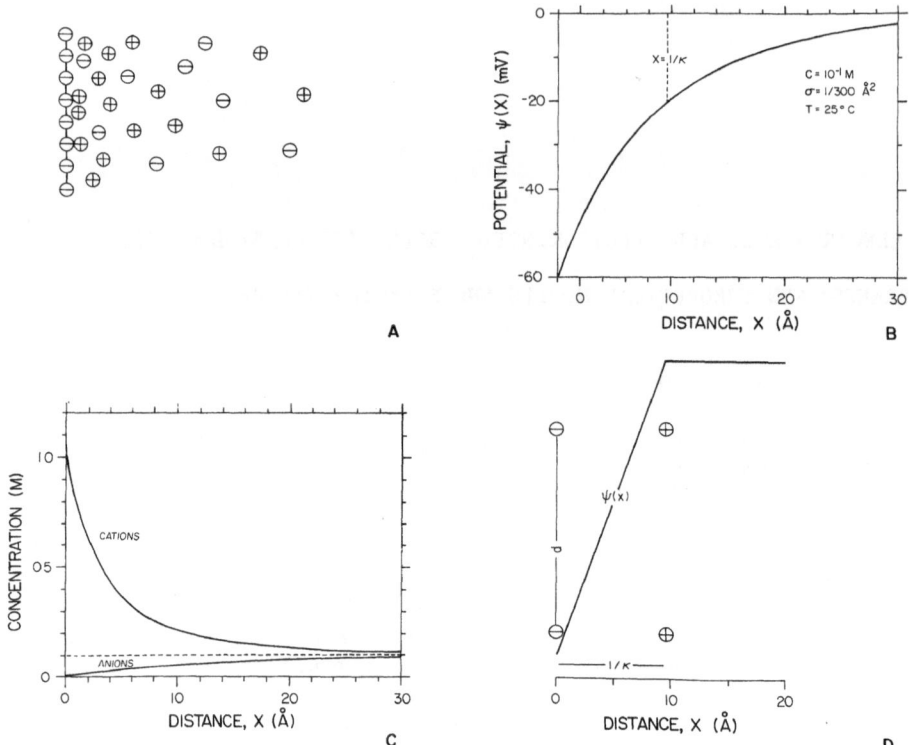

Fig. 1. (A) Schematic diagram of the distribution of ions near a negatively charged membrane. (B) The potential profile predicted by the Gouy-Chapman theory of the diffuse double layer when 20% of the lipids in the membrane bear a net negative charge. (C) The concentration of anions and cations adjacent to the membrane, as predicted by the Gouy-Chapman theory. (D) A parallel plate capacitor model of the diffuse double layer. We assume that the counterions are located a distance $1/\kappa$, the Debye length from the membrane. The average distance between the charges on the surface of the membrane is $d = 1.8$ nm. See text for details. The temperature was 25°C. From reference (4).

A simple analogy illustrates two important features of the diffuse double layer theory. We first note that the membrane plus any volume of fluid that extends for more than a few Debye lengths must be electroneutral. The excess number of counterions in the diffuse double layer must, therefore, be exactly equal to the number of charges on the membrane. As a crude approximation, we can consider all these counterions placed at an average distance from the membrane (Fig. 1D); this average distance is the Debye length. We are thus considering the diffuse double layer (Fig. 1B, 1C) to be analogous to a parallel plate capacitor (Fig. 1D). The analogy is only valid for low values of the surface potential. For a capacitor, the potential falls in a linear manner with distance between the plates, as illustrated in Fig. 1D. The electric field predicted by the capacitor analogy is only identical to the field predicted by the theory of the diffuse double layer at the membrane solution-interface (compare Figs. 1B and 1D), but the analogy does illustrate how the surface potential depends on the charge density and the salt concentration.

In terms of the model illustrated in Fig. 1D, increasing the charge on the membrane will increase the potential at the surface of the membrane.

The potential is predicted to increase linearly with the charge density. For low potentials the Gouy-Chapman theory also predicts the magnitude of the surface potential increases linearly with the charge density.

When the concentration of ions in the bulk phase is reduced, the Debye length or average distance of the counterions from the membrane increases. In terms of the model presented in Fig. 1D, this is equivalent to moving the capacitor plates farther apart. Thus an increase in the Debye length will produce an increase in the magnitude of the surface potential for a given charge density. This is also the behaviour predicted by the theory of the diffuse double layer. When the surface potential is large the Gouy-Chapman theory predicts that the surface potential will increase by 58 mV for a ten-fold decrease in monovalent salt concentration.

RESULTS

The prediction that the surface potential changes by 58 mV when the monovalent salt concentration changes by a factor of ten has been tested experimentally with phospholipid bilayers and monolayers (4). The experimental results agree well with each other and adequately with the theoretical prediction. For example, the recent study by Tocanne et al. (5) demonstrates that the surface potential of a phosphatidylglycerol monolayer changes by 53 rather than 58 mV when the [NaCl] changes by a factor of ten. The surface and zeta potentials of phospholipid monolayers and bilayers change by about 27 mV when the concentration of divalent cations changes by a factor of ten (6, 7), in exact agreement with the theoretical predictions of the Gouy-Chapman-Stern theory.

Two major assumptions implicit in the theory are that the charges on the membrane are smeared uniformly over the interface and that the counterions and coions in the aqueous diffuse double layer are point charges. We recently tested these two assumptions. NMR measurements (7, 8) indicate that discreteness-of charge effects are not of great importance with phospholipid bilayers, provided the charges are at the membrane-solution interface and not buried within the membrane (9). The finite size of the cations in the aqueous phase does not appear to be of significance unless the size approaches the Debye length (10), in agreement with a simple theoretical extension of Gouy-Chapman theory (11).

The Helmholtz-Smoluchowski equation relates the electrophoretic mobility of a phospholipid vesicle to the zeta potential, the potential at the hydrodynamic plane of shear, which is about 0.2 nm from the surface of the membrane in a 0.1 M salt solution (10). However, the combination of the Helmholtz-Smoluchowski and Gouy equations cannot describe the electrophoretic mobility of biological membranes. For example, the mobility of a human erythrocyte is about half the value predicted from the classical theory (12). Donath and Pastushenko (13), Wunderlich (14) and Levine et al. (12) describe the mobility of an erythrocyte in terms of structural parameters of the cell surface. These theories differ from the classical treatment in two respects: they assume that independent spherical elements (e.g. sugars) in the glycocalyx generate additional frictional forces and that the charged elements (e.g. sialic acid residues) are either spread over a volume or concentrated in a plane some distance from the surface. The additional frictional drag is predicted to decrease the electrophoretic mobility; the location of charged groups away from the surface is predicted to increase the mobiility and decrease the magnitude of the surface potential. We tested these predictions experimentally with a simple model system.

3

We formed membranes from mixtures of a ganglioside (e.g. G_{M1}, which has four neutral sugars and one charged sialic acid residue in its head group) and the zwitterionic phospholipid phosphatidylcholine (PC). When the head group of G_{M1} is extended maximally, the charge is about 1 nm from the surface and the head group protrudes about 2.5 nm from the surface. These distance are comparable to the Debye length in a 0.1 M NaCl solution. Thus PC:ganglioside vesicles should be useful for testing the new theories. The results we obtained (15) with PC:G_{M1} vesicles are illustrated in Figure 2.

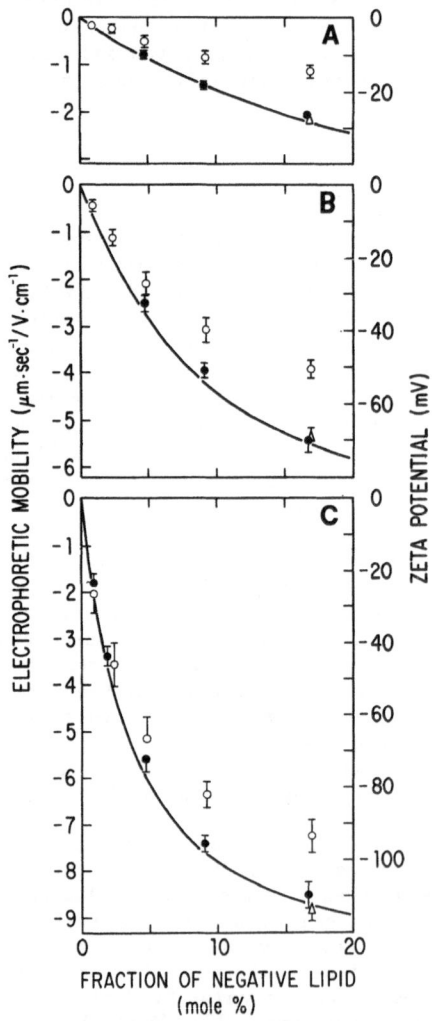

Fig. 2. Electrophoretic mobility and ζ potential of multilamellar vesicles formed from mixtures of the zwitterionic lipid phosphatidylcholine and the negative lipid G_{M1} (open circles), PS (filled circles), or PG (triangles). The curves illustrate the predictions of the Gouy-Chapman-Stern theory if the intrinsic association constant of Na with the negative lipid is 1 M^{-1}. The error bars illustrate the standard deviations of measurements on at least 20 vesicles. The aqueous solutions contained (A) 0.1 M NaCl, (B) 0.01 M NaCl and (C) 0.001 M NACl buffered to pH 7.5 at 25°C with 10^{-2}, 10^{-3}, and 10^{-4} M MOPS, respectively. From reference (15).

Note that in a 0.1 M NaCl solution, where the Debye length is about 1 nm, the electrophoretic mobility of the PC:G_{M1} vesicles is about half the mobility of PC:phosphatidylglycerol (PG) or PC:phosphatidylserine (PS) vesicles of equivalent composition. The mobility of the PC:PG and PC:PS vesicles can be described by the combination of the Gouy-Chapman-Stern and Helmholtz-Smoluchowski equations (curves, Fig. 2). We also studied membranes containing the divalent gangioside G_{D1a} and trivalent ganglioside G_{T1}. We could describe these results with a combination of the the the non-linear Poisson-Boltzmann and Navier-Stokes equations by assuming that the sugar moieties exert hydrodynamic drag and that the sialic acid groups are located at some distance from the vesicle surface. The theoretical predictions depend strongly on the thickness of the head group region and the location of the charges but are essentially independent of the location of the shear plane and the Stokes radius of the monosaccharide residues. We obtained a reasonable fit to the mobility data by assuming that the ganglioside head groups are 2.5 nm thick and that the fixed charges on the gangliosides are located in a plane 1.0 nm from the surface of the bilayer. We tested the latter assumption by measuring the surface potentials of PC:ganglioside bilayers using three techniques: conductance measurements with planar membranes, ^{31}P nuclear magnetic resonance with sonicated vesicles and TNS fluorescence measurements with sonicated vesicles. The results were consistent with our assumptions.

CONCLUSION

The electrostatic potential produced by charges on phospholipids can be well described by the Gouy-Chapman-Stern theory of the diffuse double layer and the electrokinetic properties of phospholipid vesicles can be described by combining this theory with the Helmholtz-Smoluchowski equation. However, the electrostatic and electrokinetic properties of biological membranes are more complex. Several authors have combined the Poisson-Boltzmann and Navier-Stokes equations to describe these properties. Our experimental results support this approach.

REFERENCES

1. Gouy, G. (1910) J. Phys. 9, 457-468.
2. Chapman, D.L. (1913) Phil. Mag. 25, 475-481.
3. Stern, O. (1925) Z. Elektrochem. 30, 508-516.
4. McLaughlin, S. (1977) Curr. Top. Mem. Trans. 9, 71-144.
5. Lakhdar-Ghazal, F., Tichadou, J. and Tocanne, J. (1983) Eur. J. Biochem. 134, 531-537.
6. McLaughlin, S., Mulrine, N., Gresalfi, T., Vaio, G. and A. McLaughlin (1981) J. Gen. Physiol. 77, 445-473.
7. McLaughlin, A., Eng, W., Vaio, G., Wilson, T. and S. McLaughlin (1983) J. Membrane Biol. 76, 183-193.
8. Lau, A., McLaughlin, A. and S. McLaughlin (1981) Biochim. Biophys. Acta 645, 279-292.
9. Andersen, O.S., Feldberg, S., Nakadomari, H., Levy, S. and McLaughlin S. (1978) Biophys. J. 21, 35-70.
10. Alvarez, O., Brodwick, M., Latorre, R., McLaughlin, A. and McLaughlin, S. (1983) Biophys. J. 44, 333-342.
11. Carnie, S. and McLaughlin, S. (1983) Biophys. J. 44, 325-332.
12. Levine, S., Levine, M., Sharp, K.A. and Brooks, D.E. (1983) Biophys. J. 42, 127-135.
13. Donath, E. and Pastushenko, V. (1979) J. Electroanal. Chem. 104, 453-554.
14. Wunderlich, R.W. (1982) J. Colloid Interface Sci. 88, 385-397.
15. McDaniel, R.V., McLaughlin, A., Winiski, A.P., Eisenberg, M. and McLaughlin, S. (1984) Biochemistry 23, 4618-4624.
16. Sharp, K., McDaniel, R.V., McLaughlin, A.C., Winiski, A.P., Brooks, D. and S. McLaughlin (1985) Biophys. J. 47, 47a.

EFFECT OF THE SURFACE POTENTIAL ON MEMBRANE ENZYMES AND TRANSPORT

Lech Wojtczak

Nencki Institute of Experimental Biology
Pasteura 3, 02-093 Warsaw, Poland

The surface charge of biological membranes is produced by dissociable groups of membrane constituents, mainly phospholipids and proteins. The surface potential (ψ_o) is related to the surface charge density (σ) according to the Gouy-Chapman equation which, when simplified for a symmetric electrolyte, is

$$\psi_o = \frac{2RT}{zF} \text{ arc sinh}[\sigma (8RT \epsilon_o \epsilon_r C)^{-\frac{1}{2}}] \qquad (1)$$

where R is the gas constant, T is the absolute temperature, F is the Faraday constant, C denotes the concentration of the electrolyte, and z is its valency, ϵ_o is the permittivity of the vacuum, and ϵ_r is the relative permittivity (dielectric constant) of the medium. Because of electric attraction or repulsion, the concentration of ions in the immediate vicinity of the membrane surface (C_O) is different from that in the bulk solution (C_∞), as described by the Boltzmann distribution

$$C_O = C_\infty \exp(-zF \psi_o /RT) \qquad (2)$$

Since the surface of natural membranes at pH close to neutrality is, in general, negative, the concentration of anions is decreased, whereas that of cations is increased, as compared to bulk concentrations. It has been shown[1,2] for soluble enzymes immobilized on charged solid supports that their kinetics depend on the charge of the support and that of the substrate molecule. The present contribution summarizes studies from the author's laboratory showing that kinetics of enzymes located in biological membranes are also altered by changing the surface potential of these membranes. Evidence is also presented that transport processes through membranes can be affected by the surface potential as well.

The surface potential of natural and artificial membranes can be manipulated by (1) screening by high electrolyte concentrations, (2) changing pH of the medium, (3) ionized amphiphiles, (4) di- and polyvalent ions with high affinity towards membranes, (5) changing phospholipid composition of the membranes, and (6) chemical modification of membrane proteins. These changes of the surface potential can easily be monitored e.g. by using the fluorescent membrane probe, 8-anilino-1-naphthalene sulfonate (ANS).[3,4]

It has been observed[4,5] that enzymes reacting with anionic substrates, like arylsulfatase C (EC 3.1.6.1), glycerol-3-phosphate dehydrogenase (EC 1.1.99.5), and glucose-6-phosphate phosphatase (EC 3.1.3.9), are inhibited by factors making the membrane more negative and activated by those shifting the membrane surface potential to less negative values. Opposite relationships occur for enzymes reacting with cationic substrates, e.g. acetylcholinesterase (EC 3.1.1.7), monoamine oxidase (EC 1.4.3.4), and dimethylaniline oxidase (EC 1.14.13.8).[4-6] In these studies apparent K_m values of the enzymes were altered, whereas, in general, no significant change of the maximum reaction rate at saturating substrate concentration (V_{max}) was observed. This is in agreement with the assumption that mem-

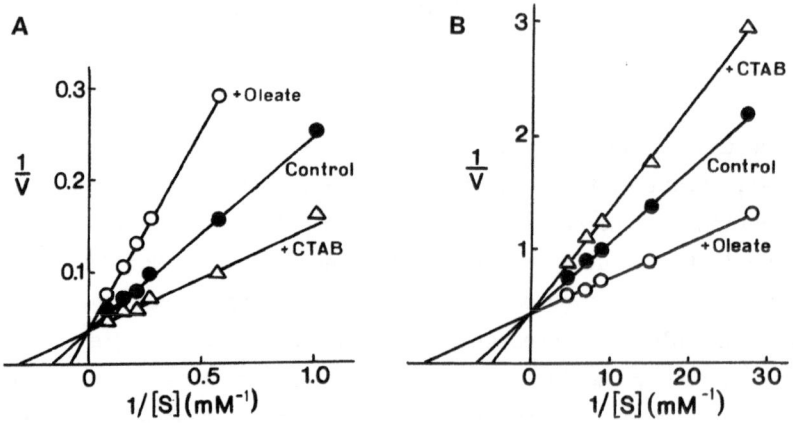

Fig. 1. Effect of ionic amphiphiles on the kinetics of glycerol-3-phosphate dehydrogenase in mitochondria from insect muscles (A) and monoamine oxidase in mitochondria from rat liver (B). Double reciprocal plots. Oleate and cetyltrimethylammonium bromide (CTAB) were added at concentrations of 68 μM and 36-39 μM, respectively. From Wojtczak and Nałęcz.[4]

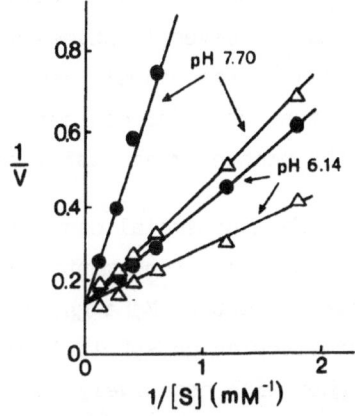

Fig. 2. Effect of pH and cationic sur-
factant cetyltrimethylammonium
bromide (CTAB) on glycerol-3-
phosphate dehydrogenase in
insect muscle mitochondria.
Double reciprocal plots.
●, Control; △, 34 μM CTAB.
From Wojtczak and Nałęcz.[4]

brane-located enzymes sense the local concentration of their substrates at
the membrane surface rather than bulk concentrations. Examples of the
effect of ionic amphiphiles, pH, and phosphorylation of membrane proteins
are shown in Figs. 1, 2, and 3, respectively. These effects disappear when
the membranes become solubilized in nonionic detergents and re-appear on
re-incorporation of the enzymes into liposomes.[4-8] In the latter case apparent K_m also depends on the phospholipid composition of liposomes, being
increased by negatively charged phospholipids for enzymes reacting with
anionic substrates and decreased for those reacting with cationic substrates[7] (Table 1).

Assuming that the change of apparent K_m results only from a change
of the local concentration of the substrate at the membrane surface, the
change of the surface potential can be calculated from the following
equation[4]

$$\Delta\psi_o = \frac{R\,T}{z\,F}\,\ln\frac{K_m''}{K_m'} \tag{3}$$

where K_m' and K_m'' are apparent Michaelis constants before and after the treatment altering the surface potential, respectively. A fairly good agreement of $\Delta\psi_o$ calculated in this way with values obtained by measuring ANS binding has been observed for several enzymes and different ways of changing the surface potential[4-7] (as example see Table 2). This confirms the assumption that the surface potential can control the kinetics of membrane-bound enzymes by the mechanism as postulated.

In a similar way the surface potential can affect the transport of charged molecules through membranes. Fundamental research in this line on model membranes has been carried out by McLaughlin et al.[9] who observed that the permeability of bimolecular phospholipid membranes for the nonactin-K^+ complex was higher with negatively charged phospholipids than with neutral ones. The reverse relationship was found for the permeability for the I_5^- complex. Extensive studies on the effect of the surface potential on inorganic ion transport through the yeast plasma membrane have been made by Borst-Pauwels and coworkers.[10,11]

Numerous studies on mitochondrial transport also point to a role of the surface potential. The inhibitory effect of factors known to increase the negative surface charge of membranes, like fatty acids,[12,13] long-chain acyl-CoA esters,[14-18] anionic detergents,[19] and lipophilic anions,[20] on

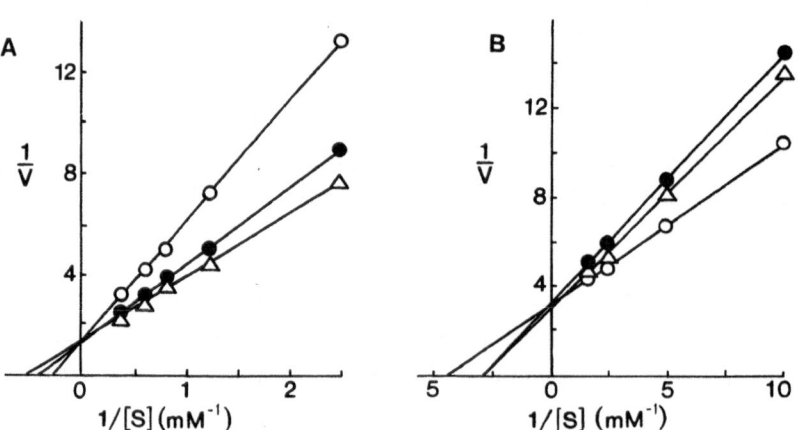

Fig. 3. Effect of phosphorylation and dephosphorylation of membrane proteins on the kinetics of arylsulfatase (A) and dimethylaniline oxidase (B) in rat liver microsomes. Double reciprocal plots. ●, Control microsomes; O, after phosphorylation; △, after subsequent dephosphorylation. From Famulski et al.[5]

Table 1. Effect of phospholipid composition of liposomes on apparent K_m values of incorporated membrane enzymes. Liposomes were prepared from the following phospholipids: phosphatidylcholine (PC), phosphatidylserine (PS), phosphatidylcholine containing 20 mol% dicetylphosphate (PC + DCP), phosphatidylcholine containing 30-40 mol% phosphatidic acid (PC + PA), and phosphatidylcholine containing 30-40 mol% tridecylamine (PC + TDA). From Nałęcz et al.[7]

| Phospholipid composition | Apparent K_m values (mM) of: | |
	arylsulfatase	monoamine oxidase
PC	3.2	0.17
PS		0.14
PC + DCP	5.6	
PC + PA	7.1	0.12
PC + TDA	2.0	0.24

Table 2. Effect of phosphorylation and dephosphorylation of membrane proteins on the surface potential of rat liver microsomes. $\Delta\psi_o$ was calculated from the binding constant for ANS (K_d) and from apparent K_m values of arylsulfatase and dimethylaniline oxidase. From Famulski et al.[5]

| Treatment | $\Delta\psi_o$ (mV) calculated from: | | |
	K_d for ANS	K_m of aryl-sulfatase	K_m of dimethyl-aniline oxidase
Phosphorylation	-10	-11	-10
Dephosphorylation	0	+5	0

carrier-mediated transport of adenine nucleotides and anionic metabolites in mitochondria is well known. Although a specific effect of some of those factors, in particular acyl-CoA, cannot be ruled out, it seems highly likely that their inhibitory effect on anion transport is at least partly due to increasing the negative surface potential. This is indicated, among others,

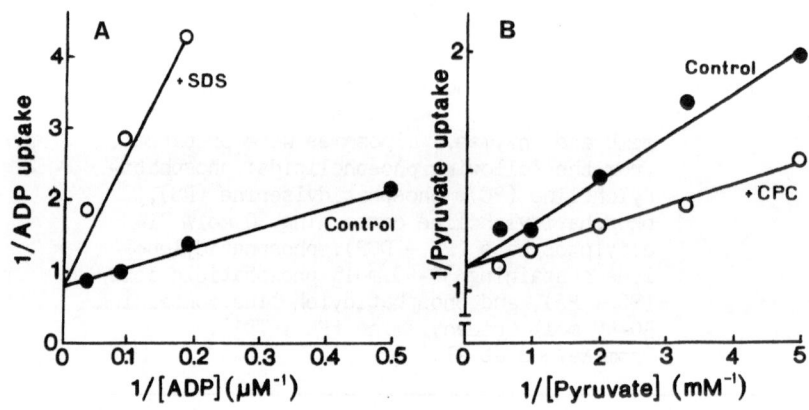

Fig. 4. Effect of ionic amphiphiles on the uptake of ADP (A) and pyruvate (B) by rat liver mitochondria. Double reciprocal plots. SDS, sodium dodecylsulfate, 60 nmol/mg mitochondrial protein; CPC, cetylpyridinium chloride, 12 nmol/mg protein. A, from Duszyński and Wojtczak;[19] B, unpublished, courtesy of Dr. Anna Sterniczuk.

by the fact that the inhibition can be partly overcome by treatments decreasing the negative surface potential, as cationic detergents,[19] Mg^{2+}, and high ionic strength.[21] On the other hand, acyl-CoA esters and free fatty acids greatly increase the permeability of the inner mitochondrial membrane to inorganic monovalent cations.[13,22] Two examples of this effect are illustrated in Fig. 4. It shows that dodecylsulfate, known to increase the surface potential, decreases the transport rate of ADP, whereas cetylpyridinium, a cationic amphiphile, potentiates the transport of pyruvate. In both cases apparent K_m but not V_{max} is changed. Thus, these effects can also be attributed to changes of the local concentrations of the transported ions at the membrane surface.

References

1. L. Goldstein, Y. Levin, and E. Katchalski, A water-insoluble polyanionic derivative of trypsin. II. Effect of the polyelectrolyte carrier on the kinetic bahavior of the bound trypsin, Biochemistry, 3:1913 (1964).
2. E. Katchalski, I. Silman, and R. Goldman, Effect of the microenvironment on the mode of action of immoblilized enzymes, Adv. Enzymol.,34:445 (1971).
3. D. H. Haynes, A fluorescent indicator of ion binding and electrostatic potential on the membrane surface, J. Membrane Biol., 17:341 (1974).

4. L. Wojtczak and M. J. Nałęcz, Surface charge of biological membranes as a possible regulator of membrane-bound enzymes, Eur. J. Biochem., 94:99 (1979).

5. K. S. Famulski, M. J. Nałęcz, and L. Wojtczak, Effect of the phosphorylation of microsomal proteins on the surface potential and enzyme activities, FEBS Lett., 103:260 (1979).

6. K. S. Famulski, M. J. Nałęcz, and L. Wojtczak, Phosphorylation of mitochondrial membrane proteins: Effect of the surface potential on monoamine oxidase, FEBS Lett., 157:124 (1983).

7. M. J. Nałęcz, J. Zborowski, K. S. Famulski, and L. Wojtczak, Effect of phospholipid composition on the surface potential of liposomes and the activity of enzymes incorporated into liposomes, Eur. J. Biochem., 112:75 (1980).

8. L. Wojtczak, K. S. Famulski, M. J. Nałęcz, and J. Zborowski, Influence of the surface potential on the Michaelis constant of membrane-bound enzymes. Effect of membrane solubilization, FEBS Lett., 139:221 (1982).

9. S. G. A. McLaughlin, G. Szabo, and G. Eisenman, Divalent ions and the surface potential of charged phospholipid membranes, J. Gen. Physiol., 58:667 (1971).

10. G. W. F. H. Borst-Pauwels, Ion transport in yeast, Biochim. Biophys. Acta, 650:88 (1981).

11. A. P. R. Theuvenet and G, W. F. H. Borst-Pauwels, Effect of surface potential on Rb^+ uptake in yeast. The effect of pH, Biochim. Biophys. Acta, 734:62 (1983).

12. L. Wojtczak and H. Załuska, The inhibition of translocation of adenine nucleotides through mitochondrial membranes by oleate, Biochem. Biophys. Res. Commun., 28:76 (1967).

13. L. Wojtczak, Effect of long-chain fatty acids and acyl-CoA on mitochondrial permeability, transport and energy-coupling processes, J. Bioenerg. Biomembr., 8:293 (1976).

14. S. V. Pande and M. C. Blanchaer, Reversible inhibition of mitochondrial adenosine diphosphate phosphorylation by long chain acyl coenzyme A esters, J. Biol. Chem., 246:402 (1971).

15. E. Lerner, A. I. Shug, and E. Shrago, Reversible inhibition of adenine nucleotide translocation by long chain fatty acyl coenzyme A esters in liver mitochondria of diabetic and hibernating animals, J. Biol. Chem., 247:1513 (1972).

16. R. A. Harris, B. Farmer, and T. Ozawa, Inhibition of the mitochondrial adenine nucleotide transport system by oleyl CoA, Arch. Biochem. Biophys., 150:199 (1972).

17. M. L. Halperin, B. H. Robinson, and I. B. Fritz, Effect of palmitoyl CoA on citrate and malate transport by rat liver mitochondria, Biochemistry, 11:949 (1972).

18. F. Morel, G. Lauquin, J. Lunardi, J. Duszyński, and P. V. Vignais, An appraisal of the functional significance of the inhibitory effect of long chain acyl-CoA on mitochondrial transports, FEBS Lett., 39:133 (1974).

19. J. Duszyński and L. Wojtczak, Effect of detergents on ADP transport in mitochondria, FEBS Lett., 40:72 (1974).

20. H. Meisner, Inhibition of metabolite anion uptake in mitochondria by tetraphenylboron, Biochim. Biophys. Acta, 318:383 (1973).

21. J. Duszyński and L. Wojtczak, Effect of metal cations on the inhibition of adenine nucleotide translocation by acyl-CoA, FEBS Lett., 50:74 (1975).

22. L. Wojtczak, Effect of fatty acids and acyl-CoA on the permeability of mitochondrial membranes to monovalent cations, FEBS Lett., 44:25 (1974).

REGULATION OF THYLAKOID MEMBRANE STRUCTURE BY SURFACE ELECTRICAL CHARGE

J. Barber

AFRC Photosynthesis Research Group
Department of Pure and Applied Biology
Imperial College of Science and Technology
London SW7, 2BB, U.K.

INTRODUCTION

Macroscopic surfaces are a feature of all biological tissue. Although each type of surface will have its own particular structure and composition all will be electrically charged. The extent and polarity of this surface charge will be dictated by the specific components which make-up the surface and by their pK values. In general biological membranes are negatively charged at neutral pH although regions of net positive charge may exist. Particle electrophoresis, and other techniques, have been extensively used to investigate the surface electrical properties of a wide range of biological systems with the view to identifying the chemical nature of the exposed charged groups and their densities. Such studies are of great importance for understanding a wide range of biophysical, biochemical and cellular processes. For example, surface charge governs the concentration of charged solutes at the membrane/liquid interface; affects transmembrane electrical gradients; regulates pH induced phenomena via their pK values; govern the binding of extrinsic proteins; modify ionic conductivity etc. (see ref.1-3).

In this article I want to focus on one aspect of surface charges, namely their ability to regulate membrane-membrane interaction and control spatial relationships between intrinsic proteins. This subject has implications over a wide biological spectrum ranging from cell fusion and formation of tight and gap junctions to membrane reorganisation in response to hormone and lectin binding, electrical stimulation and protein phosphorylation (4). However, my own experience has come from detailed studies of the organisation and properties of the thylakoid membrane of higher plant chloroplasts. It is this membrane which harbours the components responsible for capturing light and converting it into chemical potential energy. The stabilized products are ATP and NADPH which are derived from the light induced extraction of electrons and protons from water with O_2 being the by-product. Within the intact chloroplast the thylakoid membrane has a complex folded structure giving rise to stacked and unstacked regions thus leading to external membrane surfaces which are either exposed to the aqueous phase or are in tight appression. When isolated, this membrane structure can be manipulated by changing the ionic status of the medium in which it is suspended. Detailed studies of these structural changes show that they can be understood in terms of long range forces of interaction with particular emphasis on the role of the space charge density immediately adjacent to the membrane surface.

Electrophoresis studies with isolated thylakoids have shown them to possess net negative surface charge at neutral pH and to have an isoelectric point at about 4.3 (5,6). The origin of this net charge seems to be carboxyl groups of glutamic and aspartic acid residues of exposed segments of integral membrane proteins with little or no contribution from the head groups of acidic lipids (6). Below pH 4.3 the surface becomes positively charged, the majority of which has been assigned to the guanidino group of exposed arginine residues (7). Although the magnitude of the net negative surface charge density at neutral pH has been estimated many times (see 3), a value of $-2.5 \times 10^{-2} Ccm^{-2}$ is used throughout the theoretical considerations of this paper since this falls within the range of values published.

A GENERAL CONSIDERATION OF SPACE CHARGE DENSITY

The existence of fixed charges on a membrane means that counterions are drawn into a diffuse layer adjacent to its surface. The amount of electrical charge in the diffuse layer will equal the net surface charge density (σ). That is:

$$\sigma = -\int_{x=0}^{x=\infty} \rho_x dx \qquad \text{Eqn.1}$$

where ρ_x = net space charge density of ions in solution in a plane parallel to the membrane surface at distance x. The theoretical details which follow ignore the existence of adsorbed ions and structured water at the immediate membrane-solution interface (Stern layer) since the interactions between macroscopic surfaces which I will be considering, seem only to involve long range forces. Moreover, the treatment presented uses classical electrostatic theory having a number of implicit assumptions including (i) all ions are considered as point charges (ii) the dielectric constant and activity co-efficients within the diffuse layer are the same as in the bulk solution (iii) the surface charges are 'smeared-out' (iv) the surface is planar. The value of ρ_x is given by:

$$\rho_x = \sum_i Z_i FC_{ix} \qquad \text{Eqn.2}$$

where Z_i is the charge carried by the ion i, F is the Faraday and C_{ix} is the concentration of the ion i in the diffuse layer. Assuming an equilibrium then the value of C_{ix} can be related to the bulk concentration of ion i ($C_{i\infty}$) by the Boltzmann equation

$$C_{ix} = C_{i\infty} \exp(-Z_i F\psi_x/RT) \qquad \text{Eqn.3}$$

where ψ_x is the electrostatic potential at distance x from the membrane relative to the potential of the bulk (i.e. $\psi_\infty = 0$). Therefore the space charge density in a plane at distance x from the surface is given by

$$\rho_x = \sum_i Z_i FC_{i\infty} \exp(-Z_i F\psi_x/RT) \qquad \text{Eqn.4}$$

However for considerations of electrostatic screening we normally require to know the integrated space charge density (ρ_x') in the volume between the surface and the plane at x and it is convenient to normalize this quantity against the surface charge density (σ) assuming the latter remains constant.

$$\rho_x' = -\frac{1}{\sigma} \int_{x=0}^{x} \rho_x \, dx \qquad \text{Eqn.5}$$

For a planar charged surface or membrane the Poisson equation is given by:

$$\frac{d^2\psi_x}{dx^2} = -\frac{1}{\varepsilon_r \varepsilon_o}\rho_x \qquad\qquad \text{Eqn.6}$$

where ε_r is the relative permittivity of the solution and ε_o is the permittivity of a vacuum and has the value 8.854×10^{-12} $C^2.J^{-1}.m^{-1}$.

Employing Eqn.6 and recognising that the electric field $(d\psi_x/dx)$ at the surface of the membrane is due to σ so that $\sigma = -\varepsilon_o\varepsilon_r(d\psi_x/dx)_{x=o}$, then Eqn.5 can be rewritten:

$$\rho'_x = 1 - (d\psi_x/dx)/(d\psi_x/dx)_{x=0} \qquad\qquad \text{Eqn.7}$$

Eqn.7 satisfies the condition that as $x \to \infty$, $(d\psi_x/dx) \to 0$ and $\rho'_x \to 1$ and can be integrated analytically even for mixtures of unsymmetrical electrolytes (e.g. $C^+ A^-$, $C^{2+} A^-$). This is done by using a parameter ν as shown in Ref.8 and yields, after suitable manipulation, the expression

$$\rho'_x = 1 - \frac{\{\nu/(\nu^2 - C''_\infty)\}\ \mathrm{sech}^2(\phi_o + \Omega x)}{\{\nu_o/(\nu_o^2 - C''_\infty)\}\ \mathrm{sech}^2(\phi_o)} \qquad\qquad \text{Eqn.8}$$

where

$$\nu = \left[C''_\infty + (C'_\infty + 2C''_\infty)e^{F\overline{\psi}_x/RT} \right]^{\frac{1}{2}}$$

$$\phi_o = \tanh^{-1}(\nu_o/\sqrt{\gamma})$$

$$\gamma = C'_\infty + 3C''_\infty$$

$$\Omega = (F^2\gamma/2RT\varepsilon_o\varepsilon_r)^{\frac{1}{2}}$$

Note $\overline{\psi}_x = [\psi_x]$ and ν_o is value of ν when $\overline{\psi}_x = \overline{\psi}_o$. C'_∞ and C''_∞ are the bulk concentrations of mono- and di-valent cation species respectively.

The value of ψ_x, the electrical potential at any point x, can be obtained by combining Eqns. 4 and 6 and integrating to obtain the non-linear Poisson-Boltzmann relationship.

$$\frac{d\psi_x}{dx} = \mp\left(\frac{2RT}{\varepsilon_r\varepsilon_o}\right)^{\frac{1}{2}}\left[\sum_i C_{i\infty}\left\{\exp\left(\frac{-Z_iF\psi_x}{RT}\right) - 1\right\}\right]^{\frac{1}{2}} \qquad \text{Eqn.9}$$

Eqn. 9 can be combined with the Gauss equation to obtain a value of ψ_o in terms of σ as given in Eqn. 10 and also further integrated to yield a general expression for $\overline{\psi}_x$ (Eqn. 11).

$$\sigma = \mp\left[\{2\varepsilon_r\varepsilon_o RT\sum_i C_{i\infty}\left(\exp\left(\frac{-Z_iF\psi_o}{RT}\right) - 1\right)\}\right]^{\frac{1}{2}} \qquad \text{Eqn.10}$$

$$-\overline{\psi}_x = (RT/F)\ln\{[\gamma\tanh^2(\phi_o + Kx) - C''_\infty]/(C'_\infty + 2C''_\infty)\} \qquad \text{Eqn.11}$$

where $\quad K = (F^2\gamma/2RT\varepsilon_o\varepsilon)^{\frac{1}{2}}$

Usually the surface potential ψ_0 is used as the parameter to discuss various phenomena due to the fixed charges on macroscopic surfaces. In the case of membrane-membrane interaction, however, it is necessary to consider the distribution of diffusible charge within the diffuse layer since it is this quantity which governs the coulombic forces between adjacent surfaces. This is of particular importance when the surface is exposed to media containing mixed electrolytes of different valencies. Under such conditions the relationship between space charge distribution and surface potential can be complex as shown in Fig. 1.

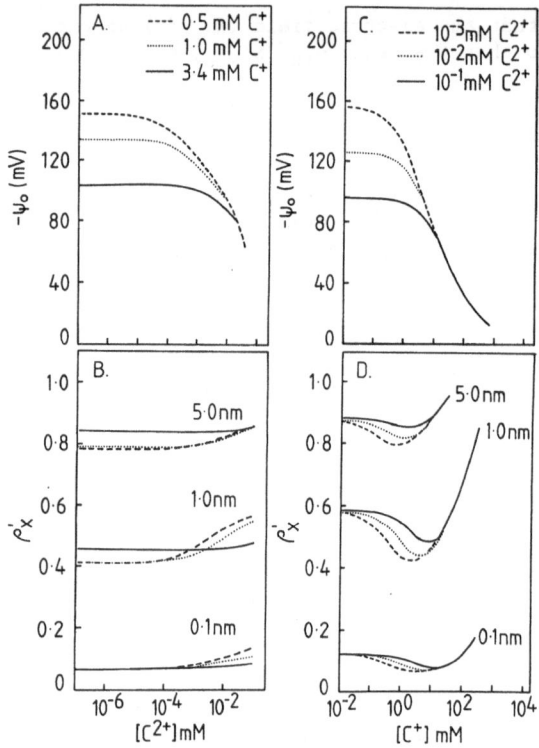

Fig. 1 Computer derived curves showing how changes of the surface potential (ψ_0) and the integrated space charge density (ρ_x') varies under different electrolyte conditions. For the determination of ρ_x' three volumes have been considered per unit area between $x = 0$ and $x = 0.1$ nm, 1.0 nm and 5.0 nm and the calculations have used Eqns. 8 and 10. Fig. (A) and (B) are for conditions when the C^{2+}/A^- concentration is changed with three background C^+/A^- levels and Fig. (C) and (D) are for three background levels of $C2/A^-$ with changing C^+/A^- salts. The surface charge density has been taken as -2.5×10^{-2} Cm^{-2}.

In this figure the integrated space charge density ρ_x' at different distances has been plotted for a membrane possessing a fixed charge density of -2.5×10^{-2} C m^{-2} immersed in media containing C^{2+}/C^- and C^+/C^- electrolytes. Figs. 1A and 1B show profiles of ψ_0 and ρ_x' respectively for conditions where the background monovalent cation concentrations were either 0.5, 1.0 or 3.4 mM and divalent cation level was changed from 10^{-7} to 10^{-1} mM. Although the ρ_x' curves show some interesting 'cross-over' effects the general trend is that an increase in ρ_x' is matched by a decrease in ψ_0. Such anticipated characteristics are not evident when comparing Figs. 1C and 1D. In this case the background electrolyte was divalent cations at 10^{-3}, 10^{-2} or 10^{-1} mM and the monovalent cation level changed from 10^{-2} to 10^3 mM. As would be expected, increasing the level of monovalent cations decreases ψ_0 but the interesting feature is that under these particular electrolyte conditions ρ_x' does not follow ψ_0. It can be seen that with very low monovalent cation levels the predominance of divalent cations in the diffuse layer generates a relatively high value of ρ_x' at any distance out to 5 nm, but particularly at 1 nm, relative to a minimum ρ_x' value achieved when the monovalent cation level is raised.

Clearly Fig. 1 emphasises a very important point that ψ_0 and ρ_x' can respond differently to changes in mixed electrolyte levels. Recognition of this is important for discussing the forces which exist between macroscopic surfaces.

FORCES BETWEEN MACROSCOPIC SURFACES

Non specific forces between two adjacent planar macroscopic surfaces can be divided into long-range (van der Waals attraction and electrostatic repulsion) and short-range (hydration shell repulsion). The repulsive forces due to hydration are not fully understood but decay exponentially within 0.2 nm of the surface (9). Long range attraction is due to the van der Waals dispersion forces resulting from the oscillatory electrical dipoles which exist in all matter due to fluctuations in the relative positions of nuclei and their associated electron clouds. As Eqn. 12 shows this electrodynamic attractive force acting per unit area (F_a) decays with the third power of distance and is a fixed quantity determined by the physical characteristics of the particular surfaces, considered in this case to be two adjacent membranes having thickness d and spaced at distance ℓ (10).

$$F_a = -\frac{A}{6\pi} \left[\frac{1}{\ell^3} - \frac{2}{(\ell+d)^3} + \frac{1}{(\ell+2d)^3} \right] \qquad \text{Eqn. 12}$$

where A is the Hamaker constant for non retarding forces and related to the permittivities of the membrane and aqueous phases (10). For a pure lipid membrane A is about 5.6×10^{-21} J but increases when proteins are also considered to be a constituent of the membrane.

In contrast, the magnitude of the electrostatic force resulting from coulombic repulsion is variable depending on the screening of surface charges by counter ions in the diffuse layer. The parameter which controls the level of screening is ρ_x' which can be be related to the electrostatic repulsive force F_r acting between two planar membrane surfaces by the following expression:

$$F_r = (P_i - P_e) - \tfrac{1}{2} \left[(2\varepsilon_o\varepsilon_i - \varepsilon_o) \{E_x/(1 - \rho_x')\}_I^2 \right.$$

$$\left. - (2\varepsilon_o\varepsilon_e - \varepsilon_o) \{E_x/(1 - \rho')\}_{II}^2 \right] \qquad \text{Eqn. 13}$$

where (P_i, P_e), $(\varepsilon_i, \varepsilon_e)$ are the Kelvin pressures and the dielectric constants, respectively, in the electrolyte medium in contact with the internal (i) and external (e) faces of membrane surfaces. The quantities $\{E_x/(1 - \rho_x')\}_I$, $\{E_x/(1 - \rho_x')\}_{II}$ are the electric fields E_i and E_e, respectively, acting at the internal and external faces of the membrane surfaces. Subscript I denotes the interplate region bounded by the midplane and the internal faces of the two membranes. Subscript II denotes the outer plate region bounded by the external face of each membrane and a plane in the bulk solution corresponding to a zero electric field, and E_x is defined as $E_x = -d\psi_x/dx$. Derivation of Eqn. 13 is given in Ref. 8 but starts with the general expression of Verwey and Overbeek (11).

$$F_r = P_d - P_\infty \qquad \text{Eqn. 14}$$

where P_d is the hydrostatic pressure at the mid-point between the two membrane surfaces and P_∞ is the hydrostatic pressure in the bulk medium. By combining with the non-linear Poisson-Boltzmann equation it is possible to solve Eqn. 13 numerically (see 11) using the general expression

$$F_r = -RT \sum_i C_{i\infty} (C^* + 1) \qquad \text{Eqn. 15}$$

where C* is an integration constant expressed as

$$C^* = -\sum_i C_{i\infty} \exp(-Z_i F\psi_m/RT) / \sum_i C_{i\infty}$$

where $C_{i\infty}$ is the concentration of ith ion with valency Z_i, and ψ_m is the potential at the mid-plane between the two identical membrane surfaces and is calculated from the expression:

$$\sigma^2 = (2RT\varepsilon_o\varepsilon_r) \left[\sum_i C_{i\infty} \exp(-Z_i F\psi_o/RT) - \sum_i C_{i\infty} \exp(-Z_i F\psi_m/RT) \right]$$

$$\text{Eqn.16}$$

The relationship between ψ_m and the distance between the two membrane surfaces (ℓ) is given by:

$$\ell = \frac{2}{(2RT/\varepsilon_o\varepsilon_r)^{\frac{1}{2}}} \int_{\psi_m}^{\psi_o} \frac{d\psi_x}{\left[\sum_i C_{i\infty} \exp(-Z_i F\psi_x/RT) - \sum_i C_{i\infty} \exp(-Z_i F\psi_m/RT) \right]^{\frac{1}{2}}}$$

$$\text{Eqn.17}$$

STACKING AND UNSTACKING OF THYLAKOID MEMBRANES AND ASSOCIATED CHLOROPHYLL FLUORESCENCE CHANGES

The unstacking and restacking of isolated thylakoid membranes in response to changes in the cationic nature of the suspending medium was first reported by Izawa and Good (12) while cation induced chlorophyll fluorescence changes were initially described by Homann (13) and Murata (14). Since then there have been many studies of these salt induced phenomena (15) but perhaps the most important was that of Gross and colleagues (16,17). The basic observations are summarized in Fig. 2.

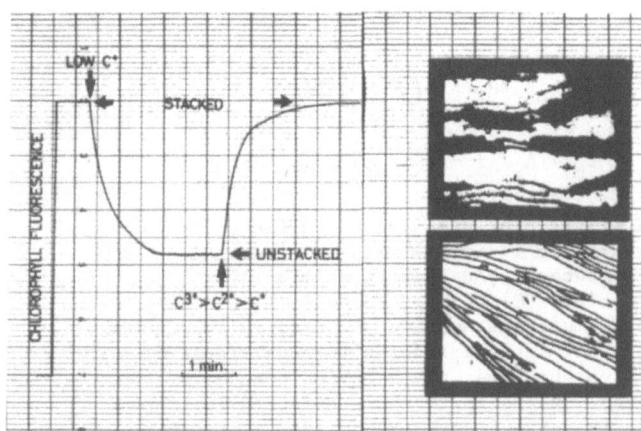

Fig. 2 Shows chlorophyll fluorescence and structural changes which occur when cations are added to un-washed, DCMU treated thyl-akoid membranes initially suspended in an essentially cation free medium (e.g. 100 mM sorbitol, 1 mM Hepes, brought to pH 7.5 with KOH).

When unwashed isolated thylakoids are suspended in extremely low salt containing media (e.g. 100 mM sorbitol, 1 mM Hepes brought to pH 7.5 with KOH), they maintain their stacked configuration and have a high F_m level of chlorophyll fluorescence (F_m is the level of fluorescence when all photo-system two, PS2, traps are closed, e.g. by either very bright light or by addition of 3-(3,-4-dichlorophenyl)-1,1-dimethylurea, DCMU). When low levels of monovalent cations (1 to 30 mM) are introduced into the suspension, the membranes totally unstack and the F_m level is lowered. From this condition the restacking and associated increase in F_m can be accomplished by intro-ducing high levels of monovalents (>100 mM) or low levels of divalents and trivalent cations (<5 mM). The relative order of effectiveness of different valency cations ($C^{3+}>C^{2+}>C^+$) in causing re-stacking and the fluorescence increase indicates that the fundamental process is governed principally by electrostatics, a conclusion reinforced by the lack of dependency of the two phenomena on chemical species within a valency group or on osmotic potential changes. The dependency on valency could, however, imply that either ψ_0 or ρ'_x is the main controlling factor but it is the antagonistic action between low C^+ and high C^+ and between low C^+ and C^{2+} which pin-points ρ'_x as the determining parameter. According to Fig. 1 such a conclusion is justified if isolated thylakoids have divalent cations as their main component within the diffuse layer. Indeed, analyses of unwashed isolated thylakoids and intact chloroplasts, indicate that Mg^{2+} is the dominating cation screening the membrane surface charges (18).

To explore this concept further thylakoids were isolated from pea chloroplasts, washed with EDTA and suspended in various mixed electrolyte solutions. Both stacking and chlorophyll fluorescence were monitored and the results have been published (11). Figs. 3 and 4 show some of the data and compares it with values of the coulombic repulsive force calculated from Eqn. 15.

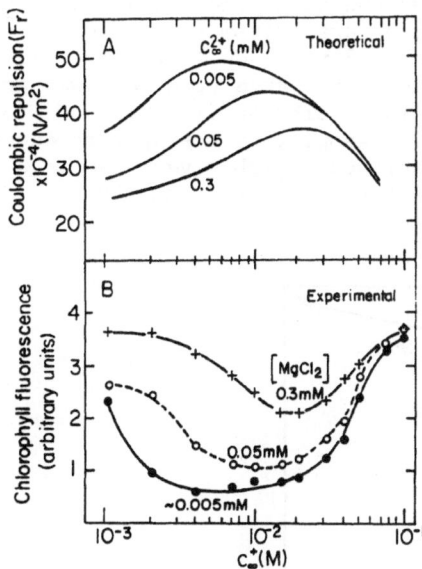 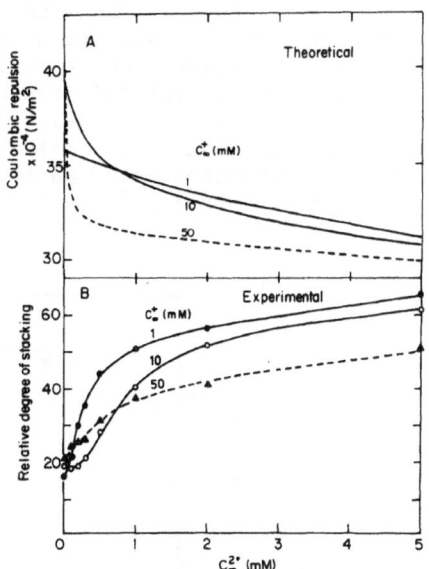

Fig. 3. **A.** Double layer repulsive force at constant surface potential as a function of monovalent cation concentration for three background divalent cation concentrations (C^{2+}_∞) at a membrane separation distance of 2.0 nm. The curves have been calculated using Eqn. 15, taking the surface charge density at infinite membrane separation as -2.5×10^{-2} Cm^{-2}. **B.** Chlorophyll fluorescence from isolated thylakoids treated with 10 μM DCMU and suspended in mixtures of mono- and di-valent cations, $C^{2+}_\infty = Mg^{2+}$ and $C^+_\infty = K^+$, as indicated. Further details in Ref.11.

Fig. 4. **A.** Double layer repulsive force at constant surface potential as a function of divalent cation concentration at three background monovalent cation levels (C^+_∞) as indicated. The curves were calculated using Eqn. 15 for a membrane separation distance of 1.0 nm and assuming a surface charge density at infinite membrane separation as $-2.5 \times 10^{-2} Cm^{-2}$. **B.** Effect of increasing divalent cation concentration ($C^{2+}_\infty = Mg^{2+}$) on thylakoid stacking in the presence of fixed background monovalent cation levels ($C^+_\infty = K^+$) as indicated. Further details in Ref.11.

It can be seen in Fig. 3A that with a constant surface potential and a surface charge density of $-2.5 \times 10^{-2} \text{Cm}^{-2}$ at infinite plate separation, the F_r calculated for an intermembrane distance of 2.0 nm, goes through a maximum when the background level of $C^{2+}A^-$ is fixed at a very low concentration (0.005 mM) and the C^+A^- level increases. This maximum corresponds to a minimum in the yield of chlorophyll fluorescence measured with approximately the same background level of $MgCl_2$ (Fig. 3B). Increasing the background Mg^{2+} concentration in the medium diminishes the minimum in chlorophyll fluorescence and shifts it so as to occur at higher K^+ levels. The same trend is seen in Fig. 3A, where the maximum in the repulsive force, F_r, decreases and shifts along the C^+ axis to higher concentrations. This comparison of calculated F_r values and fluorescence yield changes under identical mixed electrolyte conditions supports the notion that the salt induced phenomena are controlled by the space charge density ρ_x'. To test this concept further the degree of membrane stacking changes was also determined with various mixed electrolytes. The experiments shown in Fig. 4B monitored changes in stacking with increasing $MgCl_2$ concentration with three different background levels of K^+. As Fig. 4B shows, addition of Mg^{2+} caused the isolated thylakoid membranes to stack and that the three curves obtained show different features with some cross-over points. As expected, calculation of F_r, this time for a membrane separation of 1.0 nm, resulted in a decrease in coulombic repulsion with increasing C^{2+} (see Fig. 4A) and that the F_r profiles showed reciprocal features to the stacking curves of Fig. 4B. To satisfactorily mimic the "cross-over" points and the order of interception of the curves on the vertical axis at zero C^{2+}, it was found that the assumption of a constant surface potential gave the best fit so that σ varied from $-2.5 \times 10^{-2} \text{Cm}^{-2}$ (at infinite distance between membranes) to $-1.3 \times 10^{-2} \text{Cm}^{-2}$ at $\ell = 1$ nm, which can be explained if lateral displacement of charge occurred away from the region of appression (see below). The intermembrane distances of 1 or 2 nm are unrealistic but can be justified, for example, by taking into account the existence of structured water on both surfaces which gives rise to a repulsive force corresponding to a total distance of 2 to 3 nm (19).

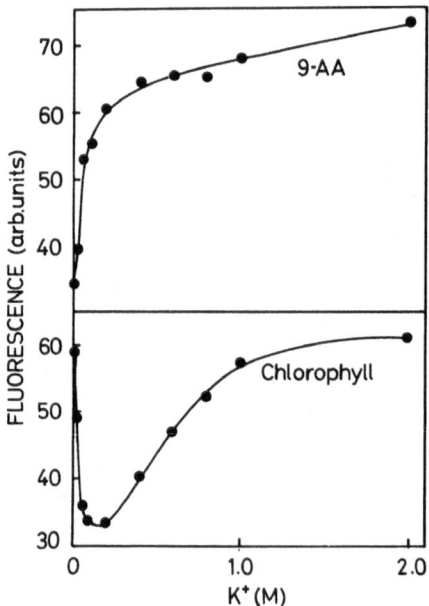

Fig. 5 Simultaneous measurement of chlorophyll and 9-aminoacridine fluorescence yields using isolated pea thylakoid membranes as a function of external levels of KCl.

To directly show that the salt induced chlorophyll fluorescence, and therefore thylakoid stacking, changes are not controlled by ψ_O, measurements were made in the presence of 9-aminacridine. It has previously been shown that this monovalent organic cation becomes non-fluorescent when it is distributed in the diffuse layer generated by a negatively charged surface (20). Reduction of ψ_O by addition of cations leads to the displacement of dye molecules from the diffuse layer and a concomitant increase in its fluorescence yield. Such an increase can be seen in Fig. 5 when isolated thylakoids are subjected to an increasing level of K^+. On the other hand, simultaneous measurement of chlorophyll fluorescence from the same cuvette but with a different detector, clearly shows the "dip" phenomenon and a lack of correlation between the two fluorescence signals.

LATERAL CHARGE DISPLACEMENT

As hinted by the calculations and experiments of Fig. 4, it is possible that the stacking of thylakoids involves a reduction of the negative surface charge density. There are also two other main features of thylakoid organisation which suggests that the coming together of two adjacent membrane surfaces on the addition of electrolytes is not simply due to the screening of a fixed quantity of electrical charges as visualized in the classical Derjaguin-Landau-Verwey-Overbeek (DLVO) theory for colloid aggregation.

i) Not all the membranes become appressed (stacked and unstacked regions form) indicative of heterogeneity in the surface properties.

ii) Where appression does occur the distance between adjacent surfaces is very short, being less than 4 nm.

If the DLVO theory could simply be applied then, because of the relatively large surface charge density (3), the inter-membrane distance would probably be greater than 10 nm. In fact the calculations of Sculley et al. (19) and ourselves (11) indicate that a membrane distance of 4 nm requires a very strong van der Waals attractive interaction with an extremely low electrostatic repulsion. The attractive force is increased if the protein to lipid ratio is high and if a substantial amount of protein extends well beyond the bilayer (11). Such conditions are probably found in the appressed regions (21,22). The requirement for low electrostatic repulsion can be met if the surface charge density is low in the regions where membrane appression occurs. This could be accomplished two ways; either by neutralization of surface charges by cation binding or by lateral displacement of charges to a non-stacking region. The former possibility has been favoured by Duniec et al. (23) while I have favoured the lateral charge displacement concept (24). This latter concept is depicted in Fig. 6 and requires that those components carrying net negative charge migrate to regions where stacking does not occur.

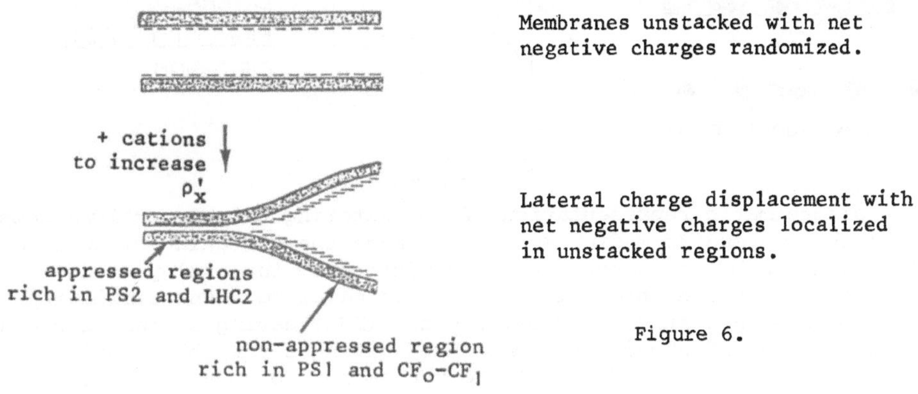

Membranes unstacked with net negative charges randomized.

+ cations
to increase
ρ'_x

Lateral charge displacement with net negative charges localized in unstacked regions.

appressed regions
rich in PS2 and LHC2

non-appressed region
rich in PS1 and CF_O-CF_1

Figure 6.

The charge displacement model has the following support:

i) The restacking of the thylakoids requires a fluid membrane and is very sensitive to temperature changes (25).

ii) In unstacked membranes photosystem two (PS2), photosystem one (PS1), the chlorophyll a/b light harvesting complex (LHC) and other components, notably cytochrome b-f (cyt b-f) and ATP synthetase (CF_O-CF_1) complexes are randomised in the plane of the membrane. Addition of cations and the resulting induction of stacking, causes the various complexes to partition into either the appressed or non-appressed regions (24). Those which become localized in the appressed regions (PS2, LHC and maybe some cyt b-f) taken together presumably have a low net surface charge while components having a significant level of surface charge are located in the non-appressed lamellae (PS1, CF_O-CF_1 and cyt b-f). This lateral distribution of the complexes has been shown by various membrane fragmentation studies, but particularly emphasised by the recent work of Albertsson, Andersson and colleagues (26) using a phase-separation technique.

iii) The stacking of thylakoids by increasing ρ_x' contrasts with the more trivial stacking induced by neutralizing the net surface charges on the membrane by strong ionic binding or by lowering the pH to the isoelectric point of 4.3. In this case more extensive stacking occurs which is independent of lipid fluidity and which maintains a randomisation of complexes (25,27).

iv) The separation of PS2 and LHC from PS1 during cation induced membrane stacking leads to a decrease in energy transfer to PS1 and thus explains the increase in the F_m level of fluorescence which comes from PS2 chlorophylls. Such an increase is not observed when the membranes are stacked by lowering the pH to 4.3 (24).

v) The postulated movements of complexes in response to changes in ρ_x' are observed in freeze-fracture (e.g. 28).

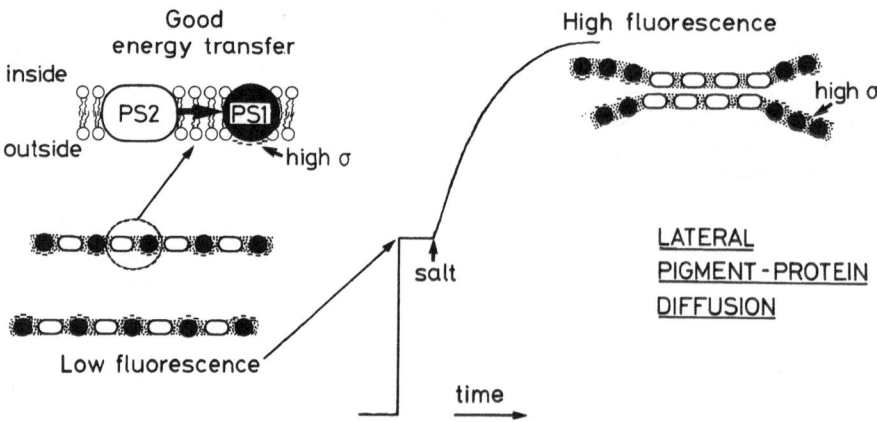

Fig. 7 A diagrammatic representation of how altering the electrolyte composition of the suspending medium so as to change the screening of surface electrical charges brings about lateral diffusion of integral pigment-protein complexes in the chloroplast thylakoid membrane. For simplicity only two complexes are shown, photosystem two (PS2) having a high chlorophyll fluorescence yield and photosystem one (PS1) having a low chlorophyll fluorescence yield and therefore able to act as a fluorescence quencher.

CONSEQUENCES

The above arguments indicate that the organisation of the thylakoid membrane and the spatial relationships between various intrinsic proteins are governed, at the first approximation, by long range forces of attraction and repulsion. Analyses of the cation induced conformation and fluorescence phenomena gives rise to two important conclusions:

i) That it is now possible to relate thylakoid membrane structure, as seen in the electron microscope (Fig. 8A) with the lateral distribution of protein complexes as shown in Fig. 8B.

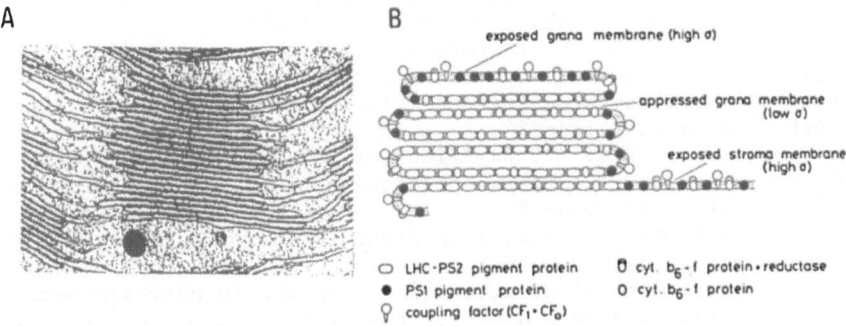

Fig. 8 A. Electron micrograph of spinach thylakoids x50,000

Fig. 8 B. Model of granal and stromal thylakoids showing protein complex distribution. σ = surface charge density.

ii) That the organisation shown in Fig. 8 can easily be disturbed by changing the electrostatic properties of the system. In this article we have only considered changes in ρ_x^t but an alternative way to perturb the electrostatics of the system would be to change the surface charge density on a particular component. In this context it has been possible to predict that phosphorylation of the LHC leads to its migration from the appressed to the non-appressed region due to increased electrostatic repulsion (3). This mechanism of controlling membrane organisation is the subject of the paper presented at this meeting by Alison Telfer.

ACKNOWLEDGMENTS

I wish to thank all those colleagues who have worked with me and helped me over the course of several years to develop and test the ideas presented above. I also acknowledge the financial support of the Science and Engineering Research Council, Agriculture and Food Research Council and the Royal Society.

REFERENCES

1. McLaughlin, S.G. (1977) Electrostatic potentials at membrane-solution interfaces. Curr. Top. Membr. Transp. 9, 71-155
2. Barber, J. (1980) Membrane surface charges and potentials in relation to photosynthesis. Biochim. Biophys. Acta 594, 253-308
3. Barber, J. (1982) Influence of surface charges on thylakoid structure and function. Ann. Rev. Plant Physiol. 33, 261-295
4. Barber, J. (1982) The control of membrane organisation by electrostatic forces. BioScience Rep. 2, 1-13
5. Mercer, F.V., Hodge, A.J., Hope, A.B. and McLean, J.D. (1955) The

structure and swelling properties of <u>Nitella</u> chloroplasts. Aust. J. Biol. Sci. 8, 1-18

6. Nakatani, H.Y., Barber, J. and Forrester, J.A. (1978) Surface charges on chloroplast membranes as studied by particle electrophoresis. Biochim. Biophys. Acta 504, 215-225

7. Nakatani, H.Y. and Barber, J. (1980) Further studies of the thylakoid membrane surface charges by particle electrophoresis. Biochim. Biophys. Acta 591, 82-91

8. Rubin, B.T. and Barber, J. (1980) The role of membrane surface charge in the control of photosynthetic processes and involvement of electrostatic screening. Biochim. Biophys. Acta 592, 87-102

9. Cowley, A.C., Fuller, N.L., Rand, R.P. and Parsegian, V.A. (1978) Measurement of repulsive forces between charged phospholipid bilayers. Biochem. 17, 3163-3168

10. Parsegian, V.A. (1973) Long-range physical forces in the biological milieu. Ann. Rev. Biophys. Bioeng. 2, 221-255

11. Rubin, B.T., Chow, W.S. and Barber, J. (1981) Experimental and theoretical considerations of mechanisms controlling cation effects on thylakoid membrane stacking and chlorophyll fluorescence. Biochim. Biophys. Acta 634, 174-190

12. Izawa, S. and Good, N.E. (1966) Effect of salt and electron transport on the conformation of isolated chloroplasts. II Electron microscopy. Plant Physiol. 41, 544-552

13. Homann, P. (1969) Cation effects on fluorescence of isolated chloroplasts. Plant Physiol. 44, 932-936

14. Murata, N. (1969) Control of excitation transfer in photosynthesis. II Magnesium ion dependent distribution of excitation energy between two pigment systems in spinach chloroplasts. Biochim. Biophys. Acta 189, 171-181

15. Barber, J. (1976) Ionic regulation in intact chloroplasts and its effect on primary photosynthetic processes. in: The Intact Chloroplast, Vol.1, Topics in Photosynthesis. J. Barber, ed., pp 88-134, Elsevier, Amsterdam

16. Gross, E.L. and Hess, S.C. (1974) Correlation between calcium ion binding to chloroplast membranes and divalent cation-induced structural changes in chlorophyll <u>a</u> fluorescence. Biochim. Biophys. Acta 339, 334-346

17. Gross, E.L. and Prasher, S.H. (1974) Correlation between monovalent cation induced decrease in chlorophyll <u>a</u> fluorescence and chloroplast structural changes. Arch. Biochem. Biophys. 164, 460-468

18. Nakatani, H.Y., Barber, J. and Minski, M. (1979) The influence of the thylakoid membrane surface properties on the distribution of ions in chloroplasts. Biochim. Biophys. Acta 545, 24-35

19. Sculley, M.J., Duniec, J.T., Thorne, S.W., Chow, W.S. and Boardman, N.K. (1980) The stacking of chloroplast thylakoids. Quantitative analysis of the balance of forces between thylakoid membranes of chloroplasts and the role of divalent cations. Arch. Biochem. Biophys. 201, 339-346

20. Chow, W.S. and Barber, J. (1980) Salt dependent changes of 9-aminoacridine fluorescence as a measure of charge densities of membrane surfaces. J. Biochem. Biophys. Methods 3, 173-185

21. Ford, R.C., Chapman, D.J., Barber, J., Pedersen, J.Z. and Cox, R.P. (1982) Fluorescence polarization and spin label studies of the fluidity of stromal and granal chloroplast membranes. Biochim. Biophys. Acta 681, 145-151

22. Kühlbrandt, W. (1984) Three dimensional structure of the light harvesting chlorophyll a/b protein complex. Nature 307, 478-480

23. Duniec, J.T., Israelachvili, J.N., Ninham, B.W., Pashley, R.M. and Thorne, S.W. (1981) An ion-exchange model for thylakoid stacking in chloroplasts. FEBS Lett. 129, 193-196

24. Barber, J. (1980) An explanation for the relationship between salt-

induced thylakoid stacking and the chlorophyll fluorescence changes associated with changes in spillover of energy from photosystem II to photosystem I. FEBS Lett. 118, 1-10

25. Barber, J., Chow, W.S., Scoufflaire, C. and Lannoye, C. (1980) The relationship between thylakoid stacking and salt induced chlorophyll fluorescence changes. Biochim. Biophys. Acta 591, 92-103

26. Albertsson, P-A., Andersson, B., Larsson, C. and Akerlund, H-E. (1982) Phase partition - A method for purification and analysis of cell organelles and membrane vesicles. Meth. Biochem. Anal. 28, 115-150

27. Chow, W.S. and Barber, J. (1980) Further studies of the relationship between cation-induced chlorophyll fluorescence and thylakoid membrane stacking changes. Biochim. Biophys. Acta 593, 149-157

28. Staehelin, L.A. (1976) Reversible particle movements associated with unstacking and restacking of chloroplast membranes in vitro. J. Cell Biol. 71, 136-158

SURFACE POTENTIAL IN ENERGIZED MITOCHONDRIA

Hagai Rottenberg

Hahnemann University, Department of Pathology
and Laboratory Medicine
Philadelphia, Pennsylvania

INTRODUCTION

Oxidative phosphorylation is widely believed to be mediated by proton transport (1). Substrate oxidation is coupled to proton extrusion from the mitochondrial matrix, while ATP synthesis is coupled to proton uptake into the matrix. According to the chemiosmotic hypothesis, the transmembranal proton electrochemical potential gradient is the direct and only driving force for ATP synthesis. The mitochondrial inner membrane contains several very active electroneutral proton-substrate co-carriers (2) which result in the reduction of the pH gradient. Therefore, in mitochondria, membrane potential is the major component of the proton electrochemical gradient, $\Delta \tilde{\mu}_H$ (4). In fully coupled mitochondria the magnitude of membrane potential is over 150 mV. On both sides of the mitochondrial inner membrane there is also a negative surface potential. However, under physiological conditions the magnitude of the surface potential is small (5-20mV). Moreover, if protons equilibrate between the bulk and the membrane surface, the magnitude of the proton electrochemical potential should be identical at the bulk and at the adjacent surface. In recent years it was suggested that coupling by proton transfer may occur by more direct pathways in addition or instead of the bulk solutions (5-8). One possibility is that surface protons do not equilibrate with the bulk so that the coupling proton current is carried on the surface (7,9). If this was the case, we would expect that energy conversion would be associated with more positive external surface potential since the mitochondrial surface charge strongly depend on surface pH (see below). Indeed, recently several laboratories reported changes in surface charge on energization of mitochondria and other energy converting membranes (10-13). However, the reported changes for mitochondria indicated that the surface on the cytosilic side become more negative, which is opposite to the predictions of the above mentioned model. Because of our interest in the possible existence of direct coupling modes (5,8), we decided to examine the reports on energy dependent surface potential changes in mitochondria.

The methods currently in use for the measurement of surface potential in biological membranes depend on the partition of charged probes (zeta potentials can be estimated from vesicle electrophoresis measurements. However, for membrane of complex shape and structure, such as the inner membrane of mitochondria, it is not possible to equate the zeta potential with the surface potential). The estimation of external surface potential from the apparent dissociation constants or partition coefficients of charged probes is valid only if the probe is not transported across the membrane, since in that case the apparent binding includes also accumulation of free probes in the vesicle and binding to the inner face of the membrane (14). If membrane potential is large and negative, as is the case in mitochondria, and the probe is permeable, most of the apparent binding would be due to the membrane potential and not due to the external surface potential (14). Previous investigators have used the fluorescent probe ANS and the spin probe Cat_{12} for the estimation of surface potential in mitochondria (10,11). We have found that both of these probes permeate the mitochondrial membrane and that the observed increase of the apparent binding is fully explained by a membrane potential induced uptake (18,19). Moreover, we have found that the lantanide ion terbium is not transported by the mitochondria and that its binding to the membrane surface is not affected by energization, indicating no change in surface potential (20).

RESULTS

ANS is a negatively charged fluorescent probe that has been used widely in mitochondria and other biological membrane (15-17). It is well known that energization of mitochondria induce quenching of the probe fluorescence which was shown to result form a decrease of the binding constant (15). This was recently interpreted (10) as due to an increase of the surface charge. In non-energized mitochondria the fluoresence and binding of ANS indeed appear to correlate with the surface charge. Figure 1A shows a titration of the fluorescence with salts. Monovalent cations enhance the fluorescence at high concentration while divalent cations enhance the fluorescence at the mM range; trivalent cations enhance the fluorescence in the µM range. In figure 1C these data are fitted to the Gouy–Chapman theory. The monovalent cations titration fit the prediction of the model assuming a surface charge density of 0.14 $\mu C/m^2$. However, the multivalent cation do not fit the equations with the same surface charge, indicating that their effect is not due to screening alone but are also due to binding. Indeed, calcium and lantanum are able to reverse the sign of surface charge due to their binding to the membrane surface. This fact can be utilized to construct a null point titration of the surface potential. This is shown in Figure 1B. The fluorescence is titrated with NaCl in the absence and presence of calcium. Thus, screening of the potential is approached from both positive and negative values. The extrapolated fluorescence intensity indicate the intensity at zero surface potential (F_o). Since, at low concentration of dye the fluorescence is proportional to the binding constant and the binding constant is a function of the surface potential (see below), than

$$\Psi_s = 59 \log (F/Fo) \ (mV).$$

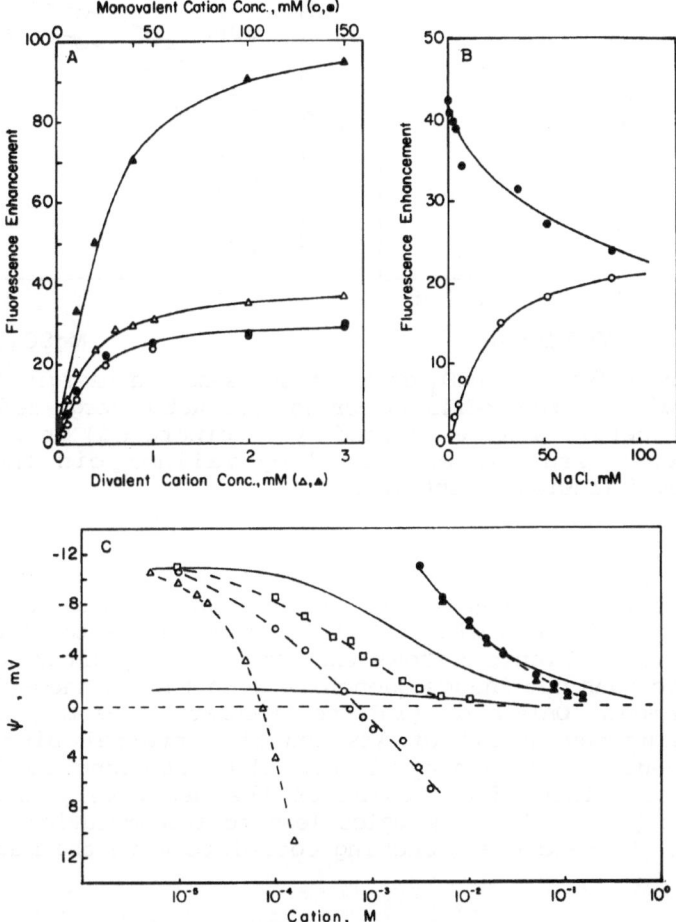

FIGURE 1: Salt dependence of ANS fluorescence and the calculated surface potential (Robertson and Rottenberg ref. 18). A: Fluorescence of ANS (8.3 μM) in mitochondrial suspension (1.9 mg P/ml) at pH 7.2 (10 mM Tris-HCl, 0.25 M sucrose) as function of NaCl (0), KCl(●) MgSO$_4$ (Δ) and CaPO$_4$ (▲). B: Null-point titration for Ψ_o determination. Conditions as in. NaCl was added in the absence (0) and presence (●) of 250 μM CaCl$_2$. C: Effect of Na$^+$(●), K$^+$ (▲) Mg^{2+}(□), Ca^{2+}(○) and La^{3+} (△) on the calculated Ψ_s. The solid lines are calculated from the Gouy-Chapman theory (see text).

If the fluorescence quenching observed on energization is due to increased surface charge, screening with salts should enhance the fluorescence to its original, non-energized value. Figure 2 shows that this is not the case. Whether energized by succinate oxidation, ATP hydrolysis or valinomycin-induced potassium-diffusion potential the difference in fluorescence between energized and non-energized mitochondria is independent of the salt concentration. This fact rules out the interpretation of the change as due to an increased surface charge.

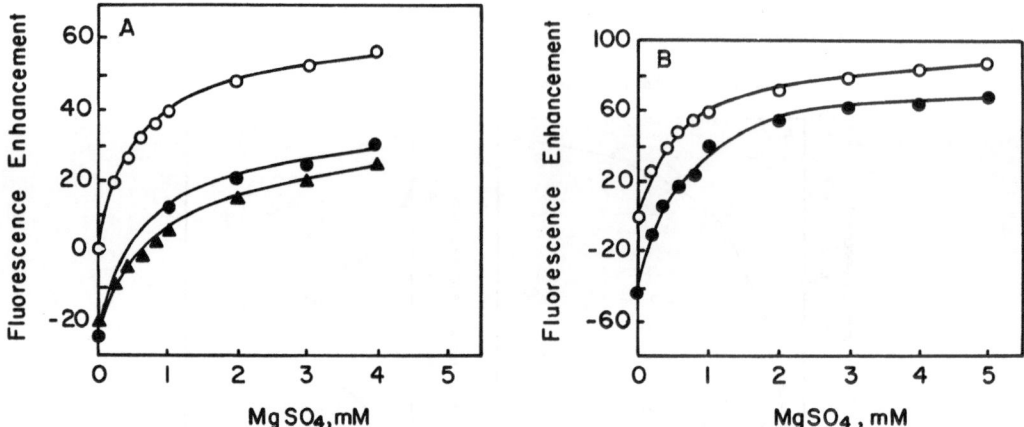

FIGURE 2: Effect of MgSO$_4$ on ANS fluoresence in energized and non-energized mitochondria (Robertson and Rottenberg, ref. 18). A: non-energized (o) energized by succinate (▲) or ATP (●) B: non-energized (o), energized by valinomycin induced potassium diffusion potential (●).

To show that the ANS response is due to transmembrane transport we incubated the mitochondria with ANS at low temperature to slow down the rate of transport. Figure 3 shows that immediately after addition of ANS, succinate does not induce quenching even though uncoupler (CCCP) induce enhancement. Only after prolonged incubation, as the fluorescence rises following the uptake of ANS and its internal binding, does succinate produce its quenching effect. Thus the quenching is clearly due to potential-induced extrusion of the negatively charged ANS. Collapsing the potential by uncoupler lead to reaccumulation of the ANS. The extent of fluorescence quenching correlate with the magnitude of

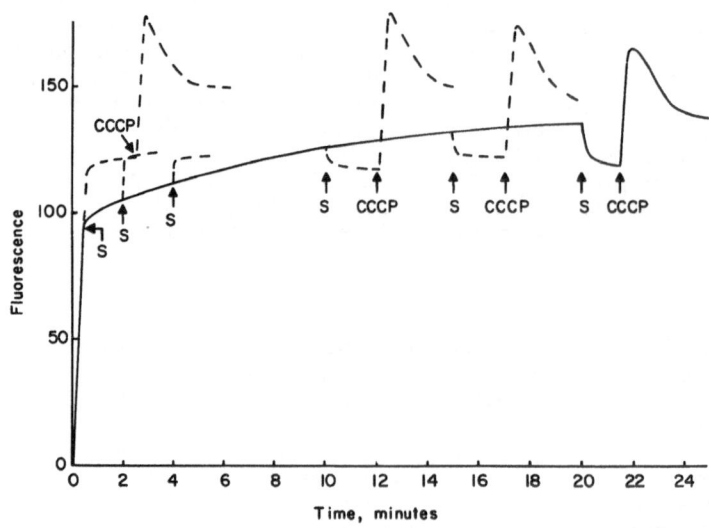

FIGURE 3: Effect of incubation time, at 10°c, on the succinate induced fluoresence quenching (Robertson and Rottenberg, ref. 18). Succinate, when added(s), was 10 mM and CCCP 1 uM.

membrane potential (not shown, ref. 18). Thus, ANS quenching is energized mitochondria is fully explained as a result of transmembrane potential induced transport and could not be be interpreted as due to increased surface charge.

Cat$_{12}$ is a positively charged spin probe which distribute between the bulk and the membrane surface. The apparent partition coefficient of charge probes depend on the surface potential according to the relation:

$$K = K^0 \exp\ (-ZF\Psi s/RT)$$

The partition can be calculated from the signals of the bound and free species. Figure 4A show the pH dependence of the partition. As the pH increases, the partition coefficient increases. This change is due to increased surface charge at high pH as indicated by the effect of salts. MgCl$_2$, as other salts abolish the pH–dependent increase in the apparent partition. Based on these data the pH dependence of the surface charge in non-energized mitochondria are shown in Figure 4B. These data are compatible with estimates by other probes (18).

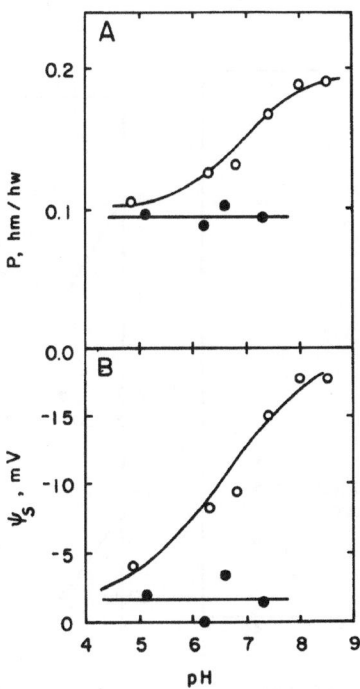

FIGURE 4: Effect of pH on the partition of Cat$_{12}$, 4-(N-N-dimethyl-N-dodecyl-) ammonium-2,2,6,6,-tetramethylpiperidine-1-oxyl) (Hashimoto, Angiolillo and Rottenberg, ref. 19). Mitochondria (8.1 mg/ml) was suspended in 0.25 Ml sucrose and 10 mM MES or HEPES buffer without Mg^{2+}(o) with 5 mM MgCl$_2$ (●). A: ratio of bound/free signal. B: The calculated surface potential.

When mitochondria are energized by ATP there is a considerable reduction in the amplitude of the free probe signal. This effect is inhibited and reversed by oligomycin and uncouplers and was interpreted as due to an increase in the negative surface charge (11). However, as Figure 5 shows, the total signal amplitude is reduced and not just the free probe. Moreover, as Figure 6 shows, the decrease in the free probe signal is associated with even a larger decrease in the bound probe signal. In fact, after a 10 min incubation period, the bound probe signal disappears almost completely. We interpreted these data as indicating transmembranal potential-induced uptake of the probe. The highly concentrated probe, which is bound internally, lead to line broadening due to spin-spin exchange. Indeed, in submitochondrial particles where the potential is positive, ATP has no effect on probe distribution (not shown, ref. 19). Thus, Cat_{12} apparently is transported by mitochondria, and therefore cannot be used to estimate energy dependent surface-potential modulation.

In order to estimate the surface potential in energized mitochondria, it was necessary to find a probe that is not transported by the mitochondria. We have found that the phosphorescent lantanide

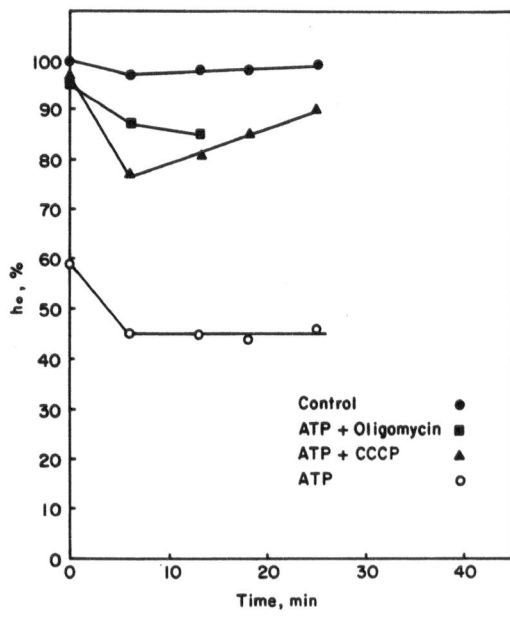

FIGURE 5: Time course of the effect of ATP on total signal intensity (Hashimoto, Angiolillo and Rottenberg, ref. 19). No addition (●), 10 mM ATP (o), ATP+oligomycin (■), ATP + CCCP (▲).

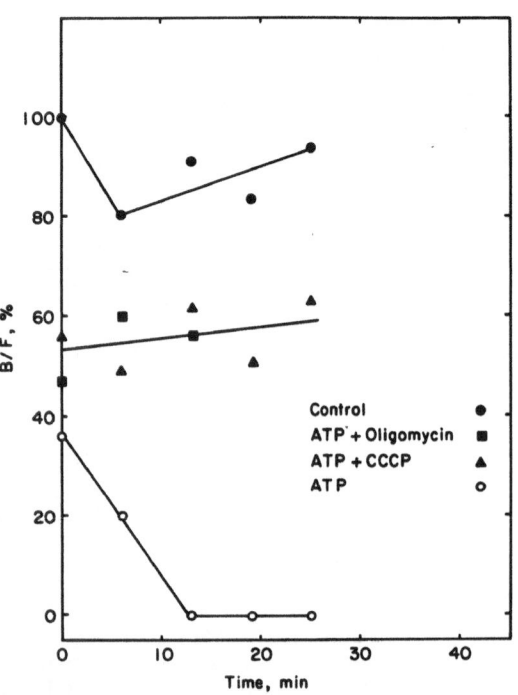

FIGURE 6: Time course of the effect of ATP on the bound/free ratio. Results are from the same experiment as Figure 5.

terbium bind to the mitochondrial membrane but is not transported (20). Figure 7 shows the concentration dependence of Tb^{3+} binding in energized and non-energized mitochondria. There was no significant effect of energization on binding indicating no potential-induced transport, nor an increase in surface charge.

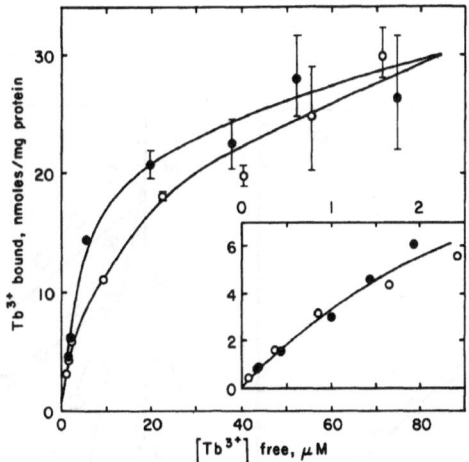

FIGURE 7: Binding of Tb^{3+} to mitochondria (Hashimoto and Rottenberg, ref. 20). Extent of binding was determined by assays of the free Tb^{3+} in the suspension supernatant. De-energized (●) mitochondria (antimycin, CCCP); energized (○) mitochondria (1 mM succinate).

That the binding itself is dependent on the surface potential can be seen from the effect of salts on the phosphorescence of the membrane bound ion (Figure 8). Similar to Figure 1, cations affect the probe binding, depending on their charge, due to both screening and binding to the membrane.

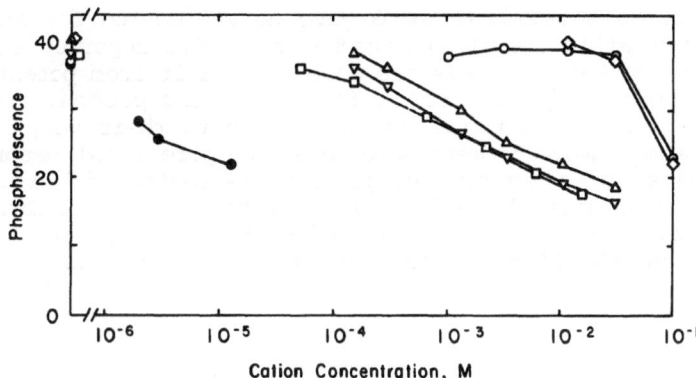

FIGURE 8: The effect of salt concentration of the phosphorescence of the slow phase of mitochondrial bound Tb^{3+} (Hashimoto and Rottenberg, ref. 20). NaCl(◊), KCl(○), $MgCl_2$ (▲), $Cacl_2$ (∇), $SrCl_2$ (▫) and $LaCl_3$ (●).

Figure 9 shows the phosphorescence decay of membrane bound Tb in energized and non-energized mitochondria. The slow phase, which arises from bound Tb^{3+} at low-affinity, high-capacity sites (not shown ref. 20), is equally affected by salts in energized and non-energized mitochondria, indicating similar surface potential. The fast decay phase arises from high-affinity, low-capacity sites, is less salt sensitive and disappear on energization, probably due to potential dependent internalization of these sites.

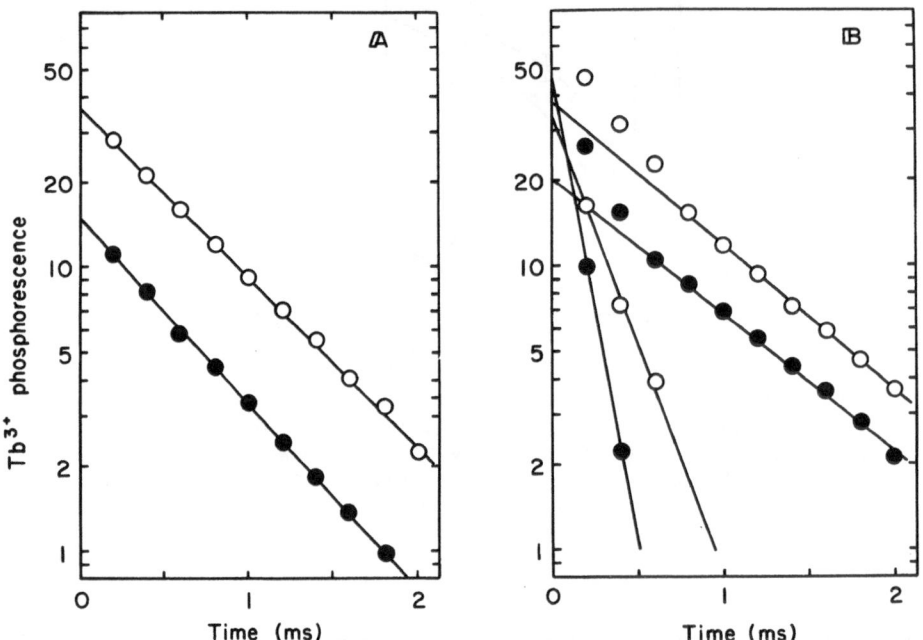

FIGURE 9: Salt effect on the phosphorescence decay of membrane bound Tb^{3+} in energized (A) and de-energized (b) mitochondria (Hashimoto and Rottenberg, ref. 20). No Salt (o) with 120 mM NaCl(●).

DISCUSSION

The studies summarized above gave no indication for modulation of surface potential in energized mitochondria. The reported effect on ANS fluorescence and Cat_{12} signals are shown to result from potential-induced extrusion and uptake and internalization of the probes. It should be emphasized that mitochondrial membrane, due to their very high protein content, are much more permeable to ions than pure lipid membrane. Often the suitability of a surface charge probe is tested with liposomes, and it is applied to mitochondria and other biological membranes on the assumption that the probe is impermeable. Such assumptions are generally unjustified and should be tested thoroughly in each case.

The fact that mitochondrial energization is not associated with a significant change in surface charge, does not rule out completely a surface proton current as a coupling pathway. However, it probably indicates that the surface membrane is equilibrated with the bulk and does not constitute a separate compartment for protons. Most likely, the direct proton coupling involves collision between redox and ATPase proton pumps which facilitate direct proton transfer and do not affect the delocalized surface charge of the membrane (8).

Acknowledgements: The work summarized in this paper was performed in collaboration with Dr. D.E. Robertson, Dr. K. Hashimoto and Mr. P. Angiolillo (references 18,19,20). It was supported by Grants GM 28173 from NIGMS and AA-3442, AA-5662 from NIAAA.

REFERENCES

1. Mitchell P. (1979) Science $\underline{206}$, 1148-1159.
2. Klingenberg M. (1970) in Essays in Biochemistry (P.N. Campbell and F. Dickens, eds.) Vol. $\underline{6}$, Academic Press, N.Y. pp. 117-159.
3. Rottenberg H. (1975) J. Bioenerg. $\underline{7}$, 63-76.
4. Nichols D.G. (1982) Bioenergetics, Academic Press, London.
5. Padan E., and Rottenberg H. (1973) J. Europ. Biochem. $\underline{40}$, 431-437.
6. Westerhoff H.V., Melandri B.A., Venturoli G., Azzone $\overline{G.F.}$, Kell D.B. (1984) FEBS Lett. $\underline{161}$, 1-5.
7. Kell D.B. (1979) Biochim. Biophys. Acta $\underline{544}$, 55-99.
8. Rottenberg, H. (1985) in Modern Cell Biol. Vol. $\underline{4}$ (B. Satir, ed.), Allen Liss, New York, in press.
9. Haraux F., De Kouchovsky Y. (1982) Biochim. Biophys. Acta $\underline{679}$, 235-247.
10. Aiuchi T., Kamo N., Kurihara K., Kobatake, Y. (1977) Biochemistry $\underline{16}$, 1626-1630.
11. Quintanilha A.T., Packer L. (1977) FEBS Lett $\underline{78}$, 161-165.
12. Quintanilha A.T., Packer L. (1978) Arch. Biochem. Biophys. $\underline{190}$, 206-209.
13. Archbold G.P.R., Farrington C.L., Lapin S.A., McKay M., Malpress F.H. (1980) Biochem. Int. $\underline{1}$, 422-427.
14. Rottenberg H. (1984) J. Memb. Biol. $\underline{81}$, 127-138.
15. Azzi A. (1975) Q. Rev. Biophys. $\underline{8}$, 2$\overline{36}$-316.
16. Haynes D.H. (1974) J. Memb. Biol. $\underline{17}$, 341-366.
17. Slavic J. (1982) Biochim. Biophys. Acta $\underline{694}$, 1-24.
18. Robertson D.E. and Rottenberg H. (1983) J. Biol. Chem. $\underline{258}$, 11039-11048
19. Hashimoto K., Angiolillo P. and Rottenberg H. (1984) Biochim. Biophys. Acta $\underline{764}$, 55-62.
20. Hashimoto K. and Rottenberg H. (1983) Biochem. $\underline{22}$, 5738-5745.

A REEVALUATION OF THE SURFACE POTENTIAL OF THE INNER MITOCHONDRIAL MEMBRANE USING ESR TECHNIQUES

Grant Hartzog, Rolf J. Mehlhorn and Lester Packer

Membrane Bioenergetics Group, Lawrence Berkeley Laboratory

and University of California, Berkeley, California 94720

ABSTRACT

The basis for the measurement of the surface potential of the inner mitochondrial membrane using electron paramagnetic probes is discussed. The effects of probe permeability in the mitochondrial membrane and the consequences of potential-driven uptake of probe are considered. The problem of probe reduction and destruction is discussed and corrective procedures for this phenomenon are presented. Data are presented on the permeability of the mitochondrial membrane to cationic, amphiphilic spin probes designated Cat_n. The uptake of these probes is compared to that of a freely permeable phosphonium ion probe. Using this data the repartitioning of the Cat_n probes upon energization of mitochondria with ATP is explained in terms of changes in surface potential of the inner mitochondrial membrane.

INTRODUCTION

A surface potential is the electrical potential extending into an aqueous phase due to immobilized charges on a membrane polar/apolar interface.[1] This is not to be confused with a transmembrane potential which is due to a charge separation across a membrane. While a surface potential may not directly contribute to the transmembrane potential in the mitochondrion, it is of interest for several reasons. Any change of surface potential at an energy-transducing membrane concurrent with a change in the energization of that membrane may reflect a structural change in the membrane due to its new energy state. Indeed, changes of surface potential have been implicated in control of energy-producing membranes.[2,3] Furthermore, since transmembrane potentials are generally determined by observing the transmembrane distribution of charged, hydrophobic molecules, which usually bind to membranes in a

surface potential-dependent manner, a knowledge of changes in surface potential is crucial to the correct interpretation of the changes in distribution
of the probe molecule and hence, the transmembrane potential itself.[4]

Surface potentials have been measured by two general methods. The first
uses the electrophoretic mobilities of membrane preparations to estimate
surface charge which is correlated with surface potential.[5] The second
relies upon the determination of the partitioning of charged, impermeable,
extrinsic probes which are specific to one monolayer of the membrane in
question and have environment-dependent spectra. The spectroscopic methods
using this type of probe include: fluorescence, resonance Raman, phosphorescence and electron spin resonance.[6-8]

These various methods are in good agreement for the case of the unenergized mitochondrion, giving a surface potential of -10 to -20 mv.[7,9] For
energized mitochondria however, there is disagreement. Previously, it was
reported that upon energization, the surface potential became more negative
by 10 to 20 mv.[7] However, data from ESR, electrophoretic and fluoresence
experiments have been recently reinterpreted with the conclusion that the
surface potential of the mitochondrial inner membrane does not change upon
energization.[9] The purpose of this report is to examine these criticisms and
reevaluate the ESR method of surface potential determination.

MATERIALS AND METHODS

A. Reagents

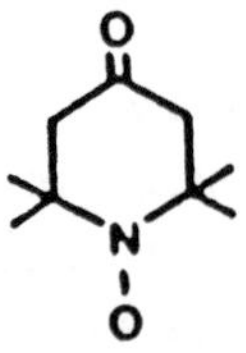

Tempone

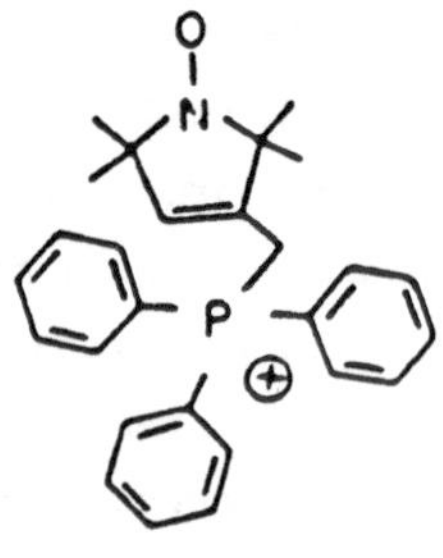

K Ø

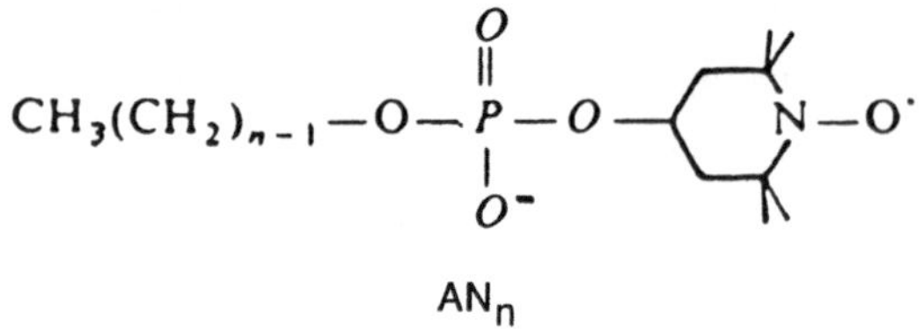

$CH_3(CH_2)_{n-1}-O-\overset{\overset{O}{\|}}{\underset{\underset{O^-}{|}}{P}}-O-$ 〔N-O⁺

AN_n

$CH_3(CH_2)_{n-1}-\overset{|}{\underset{|}{N^+}}-$ 〔N-O⁺

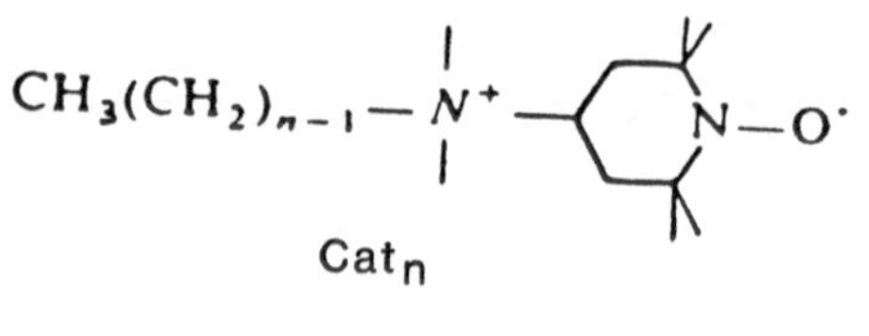

Cat_n

All reagents, with the exception of spin labels, were of analytical grade and were obtained from the Sigma Chemical Company. Spin labels were synthesized by published methods,[10,11] except for KØ, which was generously provided by K. Hideg.

B. Preparations

Male Fischer rats were obtained from Simonsen Laboratories, Gilroy, California. The rats were sacrificed, the liver was quickly removed and put in an ice-cold beaker containing 225 mM mannitol, 70 mM sucrose, 8mM Tris at pH 7.4 (buffer 1). The liver was minced, rinsed in buffer 1 and homogenized in a glass homogenizer with a tight-fitting pestle. The homogenate was then diluted to 200 ml and placed in four centrifuge tubes and spun for ten minutes at 370 g. The supernatant was retained and spun for ten minutes at 7500 g. The pellet from this wash was then resuspended in 100 mM KCl, 50 mM HEPES, 10 mM KH_2PO_4, 5mM $MgCl_2$ pH 7.4 (buffer 2) and washed twice more before being resuspended in buffer 1 for a final wash. All mitochondria were kept at 0-4°C throughout the procedure and were finally suspended in buffer 1.

ATP and ITP stock solutions were prepared at 0.25 M by dissolving the nucleotides in a pH 9.0 solution of .2 M sodium phosphate and adjusting to pH 6.80 $\pm$ .05 with 0.1 N NaOH. These solutions each contributed about 40 mM of cations to the final mitochondrial suspension.

C. Assays and Analytical Procedures

Mitochondrial volumes were determined as published.[4] Briefly, 1mM of Tempone, a spin label freely permeable to the mitochondrial membrane, was added to 50 microliters of mitochondria. The high field peak of this spectrum was compared to that of a sample which contained, in addition to the mito-chondria and spin label, 20 mM $Fe(CN)_6^{-3}$ and 80 mM K_2Mn-EDTA to quench the signal of probe outside of the mitochondria. The fractional volume of the aqueous volume within the mitochondria is the ratio of the high field peak heights of the quenched and unquenched samples, corrected for the osmotic effect of the quenching agent.

Mitochondrial respiratory control was determined by following oxygen consumption in a Rank oxygen electrode. An aliquot of mitochondria (20-50 microliters) was added to 3 ml of buffer which was continuously stirred and maintained at 30°C. After a baseline had been established, 10 microliters of 1M Na-K succinate at pH 7.4 were added. Once a constant rate of respiration was established, a 30 μl aliquot of 0.25 M ADP was added to the suspension.

The respiratory control ratio was calculated as the ratio of the ADP stimulated respiration to the normal respiration. Mitochondria were considered to
be coupled when the respiratory control ratio was above 1.8. This parameter
was checked periodically during the course of an experiment; all experiments
reported here refer to mitochondria with respiratory control ratios larger
than 1.8.

Determination of Surface Potentials

The problem of finding a mathematical description of the electrical
potential associated with a set of fixed charges bound to a planar surface
was solved by Guoy and Chapman in 1910. The Guoy-Chapman equation describes
the electrical potential, $\Psi(x)$ in the aqueous phase adjoining a planar colloidal particle or membrane surface as a function of distance, x, from that
surface, and describes the behavior of the potential in the presence of
simple electrolyte solutions where the valency of the anions and cations are
the same. It reduces to a much simpler expression at the surface (equation
1); this simpler expression is useful for assessing the influence of the
ionic composition of the aqueous phase on the magnitude of the surface potential.

$$\sinh\left[\frac{zF\,\Psi(0)}{2RT}\right] = \frac{A\sigma}{\sqrt{C}} \qquad (1)$$

where F/2RT is the usual Faraday constant, Ψ is the surface potential, σ is
the surface charge density, A is a constant, C refers to the ionic strength
of the medium, and z is the valency of the electrolyte in the aqueous phase.
At high ionic strength, the surface potential will be effectively screened by
ions to give a potential of about zero.

The estimation of surface potentials with ESR methods is based upon
measuring the partitioning of a spin-labeled amphiphilic and charged molecule
between the membrane and aqueous phases.[10,12] This partitioning is pictured
schematically in figure 1. Defining the partitioning coefficient P to be the
ratio of the amount of probe bound to membrane to the amount of probe in
solution, and P_0 to be this parameter at some reference potential $\Psi_0(x)$, we
can use the Boltzman relationship to write,

$$P/P_0 = \exp - \{zF\,[\Psi(0) - \Psi_0(0)]\,/RT\,\} \qquad (2)$$

This equation assumes that the only change between the two systems which will
affect the partitioning of probe is the surface potential difference between

the two states of the membrane. Equation (2) may be rearranged to yield, taking the reference potential as zero,

$$\Psi(0) = -(RT/F)(\ln P/P_0) \qquad (3)$$

Determination of Probe Partitioning

The measurement of the partitioning coefficient of an ESR probe is based on the fact that the lineshape of an ESR spectrum reflects the environment of the ESR probe. In an aqueous environment, the ESR probe is tumbling rapidly and gives a sharp, three line spectrum similar to that pictured at the top of figure 1.

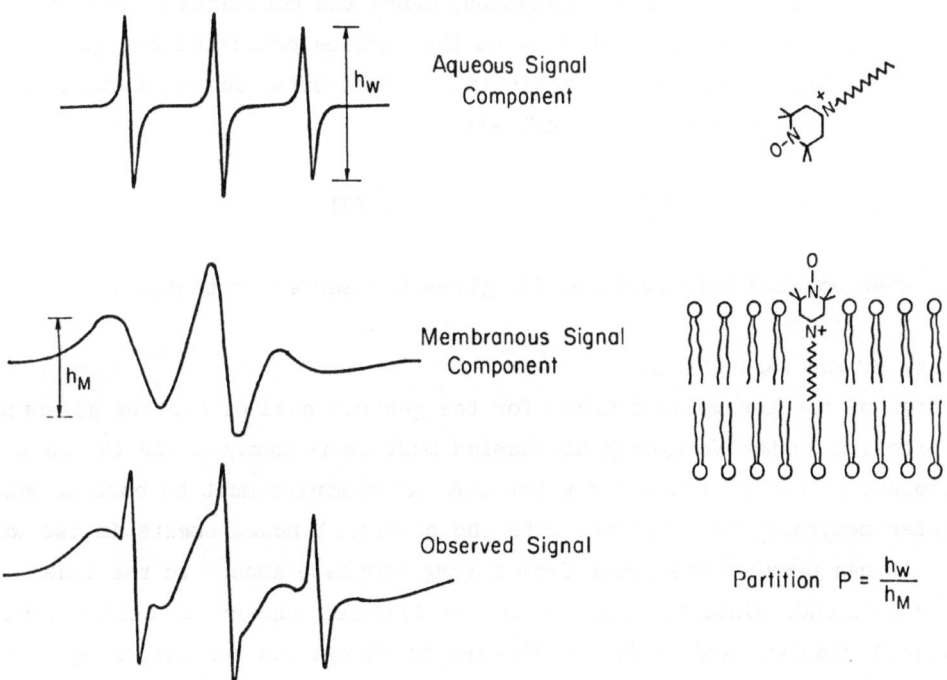

Fig 1. Probe phase equilibria and observed ESR spectra. The top spectrum is that observed for a CAT_n probe in an aqueous environment. In the middle spectrum, the signal of membrane-bound CAT_n is shown. The bottom spectrum shows a composite of both signals arising from CAT_n in an aqueous suspension of membranes.

As the motion of the probe becomes restricted and slower, the observed spectrum is broadened as shown in the middle of figure 1. When a system contains both an aqueous and membrane environment, the situation becomes more complicated. If the probe is not exchanging between the two environments on a fast time scale, then the observed spectrum will be a superposition of the bound and free signals. However, if the exchange rate is fast, the observed spectrum will reflect a weighted average of the different environments which the probe sees. Fortunately, the exchange of probe between the aqueous and membrane phases in biological systems is slow enough to resolve clear aqueous and bound signals in a single spectrum. Figure 1 shows an example of a superimposed spectrum with clearly resolved bound and aqueous components.

To a good approximation, h_m represents the line height of a peak due to membrane bound probe which is underlying a narrower peak due to aqueous probe. h_w represents the height of a peak due to aqueous probe. Both h_w and h_m are proportional to the amount of probe in their respective states. However, the constants of proportionality are not equal. Therefore, the relationship of partitioning to line heights is $h_m/h_w = P(k_m/k_w)$ where the quantity in parentheses is the ratio of the constants of proportionality. We assume that no property of the membrane other than its surface charge is altered under conditions of energization; hence the constants of proportionality are taken to be constant when the surface potential changes. By taking the ratio of two surface potential measurements, these constants drop out of the equations and we are left with,

$$P/P_o = (h_m/h_w)/(h_m/h_w)_o \qquad (4)$$

which, when coupled with equation (3) gives the surface potential.

Validity of the Assumptions

Some of the assumptions made for the general case of surface potential determination in the foregoing discussion must be reexamined for the case of mitochondria. First, the outer mitochondrial membrane must be considered. The outer membrane can interfere with the potential measurements in two ways. First, it can prevent the probe from having complete access to the inner membrane. Second, since the outer membrane does not support an energy-linked electrical gradient and it is not thought to change its surface charge, the amount of probe bound to that membrane should be constant. Probe binding to the outer membrane will therefore result in an underestimation of any fractional change in bound probe on the inner membrane and, hence, an underestimation of the change in potential. Previous studies have shown that for the Cat_n probes, these problems can be neglected, i.e., mitoplasts yielded results indistinguishable from those seen with mitochondria.[7]

The assumption that the partitioning coefficients used in equations 2-4 are invariable as a function of membrane energization state rests on the belief that energization does not change the fluidity of the inner membrane. If such changes did occur, the basic lineshape of the signal from bound probe should change to reflect that change. A more viscous membrane would yield a broader bound component in the composite spectrum while a more fluid membrane would give a narrower bound component. In the present studies, no changes in the intrinsic linewidth of the bound component were detected.

The assumption that the probe used is impermeable to the inner membrane must be verified. In fact many charged, lipophilic spin probes are permeable to the inner mitochondrial membrane. In this study, direct evidence for permeability of the Cat_n probes will be presented. The permeability of probe to the inner membrane impedes quantitative analysis of the surface potential since the probe may respond to surface potentials at both membrane interfaces as well as transmembrane potentials. However, rapid mix experiments provide the possibility of evaluating the surface potential on a short time scale before equilibration across the membrane is complete.

Finally the processes of reversible probe reduction and irreversible probe destruction must be considered. It is well known that mitochondria supplied with substrate cause high rates of probe reduction.[13] Because of this, surface potential studies have generally been done with ATP as the energizing substrate. But even ATP may increase rates of probe reduction, if only because of its effect in accelerating the rate of superoxide radical formation.[14] To bypass this problem, experiments are done in the presence of low concentrations of ferricyanide ($Fe(CN)_6^{3-}$) which seems to be effective at preventing probe reduction, although only to a partial extent (see below).

RESULTS AND DISCUSSION

Permeability of Mitochondria to Cat_n

The strategy used to determine if the spin probe is permeable to the inner membrane was essentially the same as that used for volume measurements.[4] Probe and impermeable quenching agents (Mn-EDTA and ferricyanide) were added to mitochondria. Any observed signal under these conditions is due to probe within the mitochondrial matrix. The spectra of these samples were collected as a function of time to reveal the amount of internalized probe. By taking a number of data points over time, it is possible to estimate the half time of probe flip-flop across the membrane. A complication of this method is that the longer chain probes such as Cat_9 are essentially completely bound so that the remaining internalized aqueous signal is too small to accurately measure. Therefore, uptake of only the shorter chain probes was studied.

The half-time for internalization of Cat_1 was estimated to be about 3.5 minutes (figure 2). The final amplitude of the internalized CAT_1 in energized mitochondria was 0.66% of the total signal. The volume of these particular

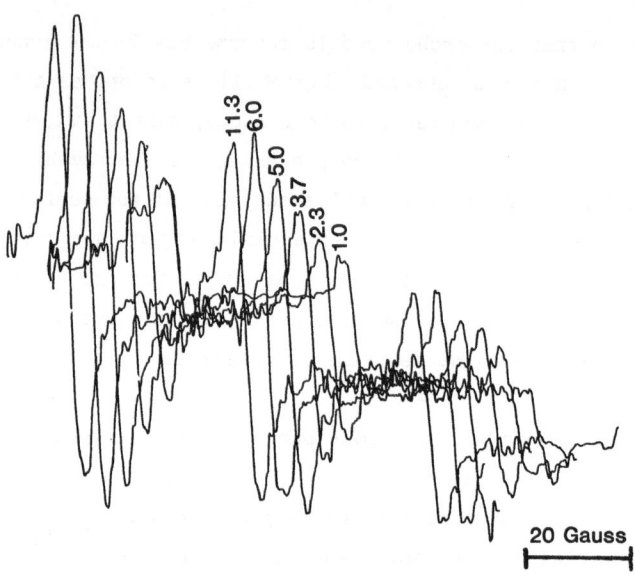

Fig. 2. Intramitochondrial ESR signal of the cationic spin probe CAT_1 in the presence of 10 mM ATP. The time (min) corresponding to the appearance of the maximum position of the midfield line is shown above the spectra.

mitochondria as measured by Tempone was 0.22% of total sample volume. The estimation of mitochondrial volume with Tempone is likely to be an underestimate because viscosity effects were neglected. These viscosity effects are significant as reflected in the line broadening of the aqueous ESR spectrum of Tempone in the mitochondria, but are far less important for the broader intrinsic line widths of CAT_1. Furthermore, it is possible that the amplitude of the aqueous Cat_1 low field peak is overestimated because of a bound signal, an effect that can be ruled out for the clearly resolved free and bound spectral features of Tempone. Such a bound signal could arise from inefficiently quenched probe bound to the outside of the membrane. In view of the above considerations the internalized Cat_1 was probably less than 0.66%/0.22% = 3 fold concentrated. Therefore, if energized mitochondria concentrate Cat_1 by some process linked to the transmembrane electrical potential, they do so by a factor of at most 3.

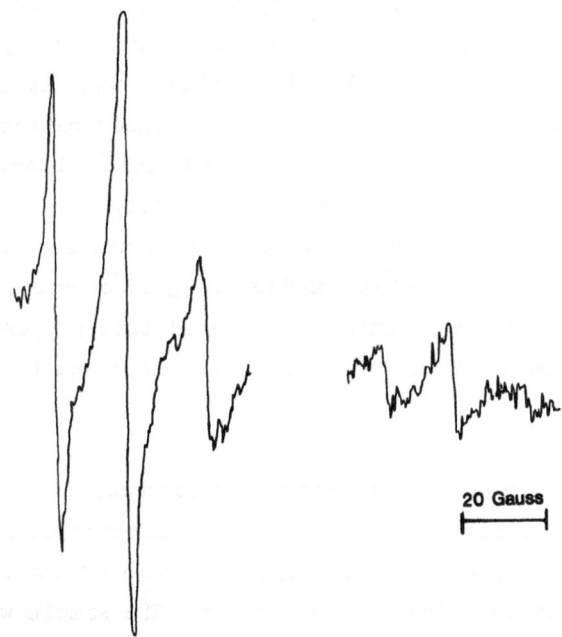

Fig. 3. ESR spectra of the spin-labeled phosphonium ion KØ in 10 mM ATP-energized mitochondria (left hand spectrum) and subsequent to treatment with 5 µM FCCP. The probe concentration was 60 µM.

To confirm that the mitochondria were supporting a transmembrane potential under the conditions used to measure Cat_1 permeability, the permeability of the charged hydrophobic probe KØ was checked. The use of KØ has been described previously. When comparing the internal spectra of the energized mitochondria with those of mitochondria pretreated with FCCP, a large decrease in signal amplitude in the presence of FCCP was observed (figure 3). It had been suggested that Cat_1 was being taken up and concentrated within the matrix of energized mitochondria to such an extent that it was self quenching its signal. To test this possibility, we performed the Cat_n permeability experiments in the presence of 1mM Tempone. If Cat_n uptake by a potential-driven process were substantial, then a high matrix concentration of Cat_1 would be expected to cause a quenching of the Tempone signal. However, no broadening of the Tempone signal was observed when this experiment was conducted with Cat_1, Cat_6 or Cat_9. Therefore, we conclude that there is a transmembrane potential present but that the observed Cat_1 uptake is not consistent with potential-driven uptake of the probe.

47

Probe Reduction

Another consideration crucial to the analysis of probe partitioning is that of probe reduction and destruction. Nitroxide spin probes may be reversibly reduced to a non-paramagnetic state. They may also participate in other reactions leading to irreversible loss of probe signal. To prevent probe reduction, ferricyanide is often added to mitochondria. However, at best, this is only an effective strategy in the short term. Ferricyanide quenches probe signal at concentrations in the millimolar range.[4] It also seems to contribute to probe destruction when present in concentrations of the order of ten millimolar (G.Hartzog, R. Mehlhorn unpublished observation). The basis for the addition of ferricyanide is its ability to oxidize reduced probe. While charge-charge effects prevent close contact of ferricyanide with the mitochondrial membrane, the Cat_n probes are good carriers of electrons between the membrane and ferricyanide. Because of this property, ferricyanide is used up as probe moves electrons from the membrane and to the ferricyanide in the bulk phase.

The following experiment illustrates the concepts discussed above. A preparation of mitochondria was incubated in the spectrometer with 1 mM Cat_9. When all signal had disappeared, the sample was removed and ferricyanide was added to give 2 millimolar final concentration. The sample was then put back in the spectrometer revealing a transient return of probe signal. Although the initial signal amplitude was not restored, the decay kinetics of this second signal resembled those of the first. Thus, mitochondria may reversibly reduce Cat_9 which in turn may shuttle electrons between the membrane and ferricyanide.

Given the above observations, it is possible that a previous report of signal loss upon energization of mitochondria in the presence of Cat_{12} was due to probe reduction rather than probe uptake and self quenching as was suggested. We postulated that the rate of probe reduction in the presence of mitochondria and ATP is higher than that in the presence of mitochondria alone. To test this hypothesis, spectra collected at various times after energization of mitochondria were double integrated. The height of the second integral of an ESR (first derivative) spectrum is directly proportional to the concentration of probe in the sample cavity. If the signal amplitude is reduced due to quenching, the height of the second integral will be unaffected as long as the spectrum was acquired over a wide range of magnetic field strength (100 gauss). This was confirmed by using stock solutions of Cat_9 in asolectin vesicles and showing the relationship of second integral height and probe concentration to be linear up to 100 millimolar, a concen-

tration at which probe signal is drastically broadened. If the signal height
of a spectrum is reduced due to probe reduction, however, the height of the
second integral will also be reduced as reduced probe is not paramagnetic and
therefore undetectable.

In two experiments, we treated mitochondria with 1 mM Cat_9 and either
10 mM ATP or 10 mM ITP. In the ITP treated samples, by five minutes after
probe addition, total probe signal had declined by 6.8%($\pm$3.2%). For the ATP
treated samples, this value was 36.6%($\pm$15.8%). This confirms previous reports
of increased rates of probe reduction in the presence of ATP. Interestingly,
in the ATP treated samples, the signal amplitude began to increase again
fifteen to twenty minutes after probe addition. Although the signal never
returned to its original amplitude, it is possible that after the ATP had
been exhausted the rate of ferricyanide reoxidation exceeded the rate of
reduction by the respiratory chain, thereby increasing the concentration of
oxidized detectable probe. After an initial decrease, the surface potential
of the mitochondria also became less negative with time, and, surprisingly,
eventually reached a magnitude 11 millivolts more positive than the potential
prior to energization. An interesting observation made during the course of
these experiments was that the probe AN_{12}, a negatively charged lipophilic
spin probe, was quite resistant to probe reduction. Because of its structural similarity to the Cat_n probes, the resistance of AN_{12} to reduction may
be related to its negative charge. Unfortunately, it was found to be difficult to use for surface potential determinations due to high permeability and
response to the transmembrane potential of mitochondria.

Estimation of the Mitochondrial Inner Membrane Surface Potential

Given our data regarding Cat_n permeability and reduction, we attempted
to estimate both the absolute mitochondrial surface potential and the change
in the surface potential upon energization. We reasoned that by measuring
the surface potential as soon as possible after probe addition, problems of
probe uptake and probe reduction would be minimized. The effects of probe
reduction and probe uptake will result in opposite effects upon the partitioning P, i.e., probe reduction is likely to occur preferentially at the
membrane surface and result in a decreased partitioning ratio while probe
uptake will result in a larger value of P.

Surface potentials were measured in two ways: absolute potentials were
measured by comparing samples at low and high salt concentrations and relative potential differences were determined by direct comparison of spectra
from energized and unenergized mitochondria. Furthermore, measurements were

made relatively quickly after probe addition (within 30 to 50 seconds).

In the course of gathering this data, it was observed that there were essentially no differences between surface potentials measured in ATP-treated and ITP-treated mitochondria. In order to de-energize the mitochondria, it was necessary to treat them with FCCP. The transmembrane potential as measured with $K\emptyset$ is greatest in the presence of ATP, less in ITP treated mitochondria, and least in mitochondria treated with both ITP and FCCP. It is possible that FCCP influences the surface potential of the inner membrane since, as a weak acid, it exists in a neutral and a negatively charged form. The negatively charged species would be expected to increase probe binding in the ITP-treated sample and therefore, decrease the difference between the estimated surface potentials of the ITP- and ATP-treated samples. ITP was used in the control experiments so that the salt concentration of the samples would be identical to that in the ATP experiments.

Table 1

Absolute Surface Potentials

Sample	P low salt	P high salt	Ψ
Mitochondria	0.81 ± 0.04 $N=3$	0.22 ± 0.04 $N=3$	-34mV
Mitochondria, +FCCP	0.46 ± 0.05 $N=4$	0.18 ± 0.05 $N=4$	-25mV

$$\Delta\Psi = \text{(energized - deenergized)} = -9\text{mV}$$

Relative Surface Potential Changes

Sample	h_m/h_w	$\Delta\Psi$
Mitochondria+ATP	1.01 ± 0.13 $N=5$	-14mV
Mitochondria+ITP+FCCP	0.59 ± 0.06 $N=5$	

Notes: FCCP = 5 μM, ATP = ITP = 5 μM. Cat_9 =50μM
N = number of measurements.

Table 1 presents the results of two experiments. In the first, the absolute surface potentials of untreated and FCCP treated mitochondria were compared. Note that, although the difference in absolute potentials is -9 millivolts, the relative potential difference under low salt conditions is -15 millivolts. This discrepancy is due to the difference in the high salt values of P. In the second experiment, the relative surface potential difference between ATP- and (ITP+FCCP)-treated mitochondria was found to be -14 millivolts. The small difference in the relative potential differences between these two experiments is not surprising given our earlier results with KØ.

Three sources of possible error must be considered: FCCP perturbations of probe binding, probe reduction effects and probe uptake. The effect of FCCP on P should be minimal for two reasons: at 5 μM, there is not enough FCCP present to bind more than 10% of the available probe. At the high values of bound to free probe in this system, this effect would be small. Secondly, under high salt conditions no significant difference was seen in probe binding between FCCP-treated and untreated mitochondria. Therefore, probe binding due to FCCP seems to be minimal. Error due to probe reduction is also negligible since, although probe was being reduced, the ratio h_m/h_w was essentially unchanged in the first few minutes after probe addition. This indicates that, even if probe is being reduced at different rates in the membrane and aqueous environments, the probe is in rapid equilibrium between these two phases so that the partitioning of the remaining probe is unchanged. There is no good way to quantitatively estimate the error due to probe internalization. However, as was argued earlier, this perturbation should be small on the time scale of the experiment. Furthermore, since any probe internalization would be expected to increase the bound signal at the expense of the aqueous signal, the observed partitioning can be used to set an upper limit on the change in probe binding due to a change in the surface potential following energization of the inner mitochondrial membrane. We therefore conclude that the surface potential of the inner mitochondrial membrane becomes more negative by no more than 14 millivolts upon energization.

CONCLUSIONS

Recently, Hashimoto and co-workers reviewed previous studies of the surface potential of the inner mitochondrial membrane and its response to energization with ATP.[9] Reviewing the ESR technique of surface potential determination using the Cat$_n$ probes, they concluded that the method was flawed. Hashimoto et. al. observed a decrease in probe signal after energization and attributed that decrease to potential-driven uptake of probe with subsequent quenching of internalized probe. We examined this suggestion in three different ways: Cat_n uptake in the presence of 1mM Tempone did not quench the signal of the internalized Tempone. When observed directly, Cat_n uptake was found to occur but not in a manner directly responsive to the transmembrane potential or likely to result in self quenching of probe. Finally, double integrations of spectra showed a signal decrease in energized mitochondria to be due to probe reduction rather than self quenching. Based on our findings, we conclude that the surface potential of the mitochondrial membrane becomes more negative upon energization. In our particular preparation, this value is about 14 millivolts. However, the magnitude of this change is likely to vary with method of preparation.

Acknowledgments: This work was supported by the National Institutes of Health (AG-04818) through the U.S. Department of Energy under contract No. DE-AC03-76SF00098.

REFERENCES

(1) S. McLaughlin, Electrostatic Potentials at Membrane-Solution Interfaces, Current Topics Membrane Trans. 9: 71-126 (1977)

(2) G. Schaefer and G. Rowohl-Quisthoudt, Influence of Surface Potentials on the Mitochondrial H^+ Pump and on Lipid Phase Transitions, J. Bioenergetics 8: 73-81 (1976).

(3) R. J. Mehlhorn and L. Packer, Inactivation and Reactivation of Mitochondrial Respiration by Charged Detergents, Biochim. Biophys Acta 423: 382-397, (1976).

(4) R. J. Mehlhorn, P. Candau and L. Packer, Measurements of Volumes and Electrochemical Gradients with Spin Probes in Membrane Vesicles, Methods Enzymol. 88: 751-762 (1982).

(5) N. Kamo, M. Morotsugu, K. Korihara and Y. Kabatake, Significance of Surface Potential in Interaction of 8-Anilo-1-naphthalenesulfonate with Mitochondria: Fluorescence intensity and Zeta Potential, Biochemistry 16: 1626-1630 (1977).

(6) B. Ehrenberg and Z. Meiri, The Bleaching of Purple Membranes Does not Change their Surface Potential, FEBS Letters 164: 63-66 (1983).

(7) A. T. Quintanilha and L. Packer, Surface Potential Changes on Energization of the Mitochondrial Inner Membrane, FEBS Letters 78: 161-165 (1983).

(8) D. E. Robertson and H. Rottenberg, Membrane Potential and Surface Potential in Mitochondria, J. Biol. Chem. 258: 11039-11048 (1983).

(9) K. Hashimoto, P. Angiotillo and H. Rottenberg, Membrane Potential and Surface Potential in Mitochondria: Binding of a Cationic Probe, Biochem. Biophys. Acta 764: 55-62 (1984).

(10) R. J. Mehlhorn and L. Packer, Membrane Surface Potential Measurements with Amphiphilic Spin Labels. Methods Enzymol. 56: 515-526 (1979).

(11) G. S. B. Lin, R. I. Macey, and R. J. Mehlhorn, Determination of the Electrical Potential at the External and Internal Bilayer-Aqueous Interface of the Human Erythrocyte Membrane Using Spin Probes, Biochim. Biophys. Acta 732: 683-690, (1983).

(12) J. D. Castle and W. L. Hubbell, Estimation of Membrane Surface Potential and Charge Density from the Phase Equilibrium of a Paramagnetic Amphiphile, Biochemistry 15: 4818-4831 (1976).

(13) A. T. Quintanilha and L. Packer, Surface Localization of Sites of Reduction of Nitroxide Spin-labeled Molecules in Mitochondria, Proc. Natl. Acad. Sci. USA 74: 570-574 (1977).

(14) H. Nohl and D. Hegner, Do mitochondria produce oxygen radicals in vivo? Eur. J. Biochem. 82: 563-567 (1978).

FUNCTIONAL INTERACTION OF ANIONS AND CATIONS WITH THE RECONSTITUTED

ADENINE NUCLEOTIDE CARRIER FROM MITOCHONDRIA

Reinhard Krämer

Institut für Physikalische Biochemie
Universität München
München, FRG

INTRODUCTION

The adenine nucleotide carrier from the inner mitochondrial membrane
is influenced by various factors pertaining to the surrounding hydrophobic
and hydrophilic phases as well as by transmembrane parameters. The composition of the surrounding water phases with respect to monovalent and polyvalent anions and cations[1,2] and the composition of the surrounding lipid
phase with respect to phospholipid headgroups[3,4], cholesterol content[5] and
fluidity of the membrane represent the main factors modulating the transport activities of the ADP/ATP-carrier protein. The transmembrane parameters, i.e. substrate gradients and membrane potential, are the most important parameters which regulate the transport function in vivo[6,7].

A reconstituted system, consisting of the purified adenine nucleotide carrier embedded in phospholipid vesicles, is very suitable for dis-
criminating between these different parameters and for studying their
influence on the carrier function. In the following, an overview shall
be given with respect to only one particular aspect of this field, namely
the various types of interaction of the reconstituted adenine nucleotide
carrier with charged molecules from the water phases and from the lipid
surfaces. The ADP/ATP carrier is an interesting object for these studies,
because it bears a relatively high amount of charged amino acids at its
hydrophilic surface[8] and because it accepts substrates with an unusually
high amount of 3-4 negative charges. This transport system is furthermore
a suitable object for studying these phenomena per se, since it is one of
the most abundantly occurring proteins and has been very extensively investigated in the past twenty years[9].

MATERIALS AND METHODS

The preparation of lipids and liposomes and the isolation, reconstitution and assay of ADP/ATP-carrier protein were performed as described
previously[1]. All cationic activators as well as the polylysine agarose
were purchased from Sigma. Extrapolation of true exchange velocities was
carried out according to ref.[7]. The discrimination between right-side-out
and inside-out oriented carrier proteins has already been described in
detail[10].

RESULTS AND DISCUSSION

Interaction with Anions

Anions interact with the adenine nucleotide carrier in three ways : (i) as substrates, (ii) as competitive inhibitors, and (iii) as activators.

Anions as substrates. Binding of the anionic substrates ADP and ATP represents the first step in the series of translocation events and shall not be discussed here.

Anions as competitive inhibitors. Binding of adenine nucleotides to the active site of the translocator is competitively inhibited by various anions[1]. These anions can be arranged in a sequence of increasing competitive effect due to their position in a lyotropic order and also due to their increasing charge. The K_i values for the anions are given in Table 1.

Anions as activators. Apart from this pure K_m effect of the competing anions on the kinetics of the ADP/ATP carrier, an unexpected influence of anions on the V_{max} of the adenine nucleotide transport was discovered in the reconstituted system[1]. A sufficiently high concentration of anions turned out to be essential for functional ADP/ATP exchange. The nucleotide transport shows sigmoidal dependence on the stimulating anions (Fig. 1). Under the assumption that the adenine nucleotide carrier binds a certain number of anions, thereby changing its functional state from low activity (without anions) to high activity (with bound anions), the activation process can be analyzed according to classical enzyme kinetics[1]. The graphical evaluation formally resembles a Hill diagram.

Table 1. Comparison of inhibition (K_i) and activation ($K_{0.5}$) of ATP/ATP exchange by anions (taken from [1]). K_i = inhibition constant for competitive inhibition by the respective anions, $K_{0.5}$ = concentration of activator which causes half maximum activation.

Anion	K_i (mM)	$K_{0.5}$ (mM)	$K_{0.5}/K_i$
F^-	250	82	0.27
Cl^-	85	39	0.46
HCO_3^-	21	31	1.48
HPO_4^{--}	14	15	1.07
SO_4^{--}	12.5	5.5	0.44
ClO_4^-	11	6.5	0.59
$Aconitate^{3-}$	5.2	16.5	3.2
$P_2O_7^{4-}$	0.9	7	7.8
GTP^{4-}	0.25	6	24
ATP^{4-}	0.015	4	265

As shown in Fig. 2, activation by anions can be described by a Hill coefficient $n = 2$.

Again the activating anions can be arranged in a sequence of increasing stimulation effect (Table 1).

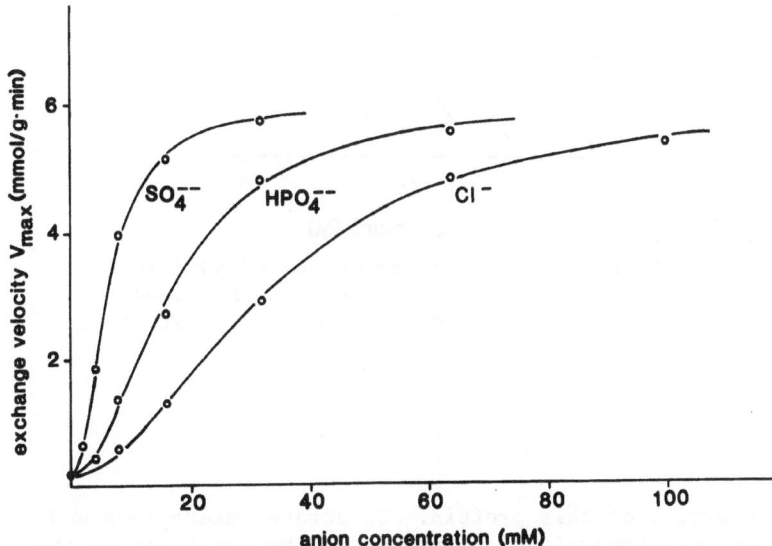

Fig. 1. Influence of anions on the exchange velocity of ATP. External ATP (200 uM) was exchanged against internal ATP (15 mM) in the presence of 0.25 uM bongkrekic acid. The external ions were substituted by sucrose of the same osmolarity. The respective ions were then added prior to the exchange assay under constant osmotic pressure (from [1]).

On the basis of the following results, inhibition and activation by anions are shown to be due to different mechanisms : (i) The order of competitive effect and the order of strength of activation are different, as shown by the large variation in the ratio of $K_{0.5}/K_i$ in Table 1. (ii) Inhibition and activation were shown to take place at different sites, since anions compete with substrates both at the outer (cytosolic) and at the inner (matrix) site, whereas activation is observed solely by interaction with the cytosolic side of the translocator. (iii) Stimulation is not mediated by the lipids in general, since it has been proven to be side-specific, i.e. the adenine nucleotide carrier is activated by anions only at its c-side and not at its m-side.

Stimulation by anions is therefore based on direct interaction of the activators with presumably two binding sites at the c-side of the protein. This recently discovered phenomenon thus represents a new aspect in the

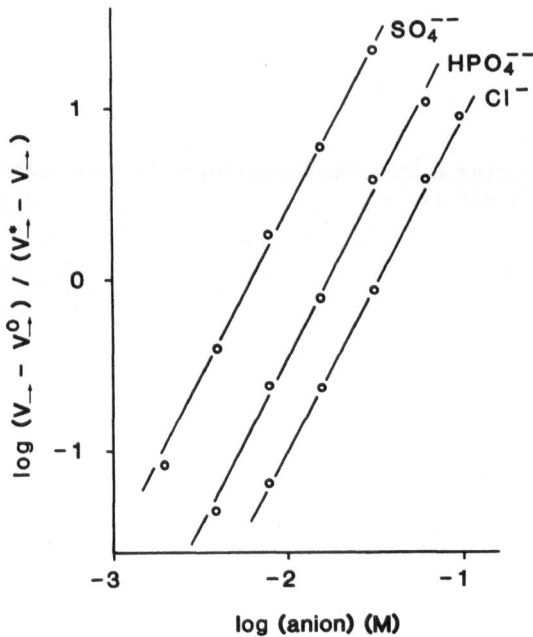

Fig. 2. Activation of the reconstituted ADP/ATP
exchange by anions. The data are taken
from Fig. 1, calculation as described in [1].

asymmetric properties of this protein. Substrate interaction and competitive inhibition are symmetric with respect to the two faces of the protein, whereas anion stimulation is strictly asymmetric and restricted to the c-side of the adenine nucleotide carrier.

Interaction with Cations

Cations may influence the carrier activity by modulating the actual surface potential in the neighborhood of the translocator protein (see next chapter). Besides this completely unspecific effect the transport activity is modulated by the presence of divalent cations[2]. This is, however, due to complexation of the substrates and shall not be further discussed here.

Quite recently, a completely different class of effectors of the reconstituted adenine nucleotide carrier has been discovered. When trying to show stimulation of intact mitochondria by anions, we were not successful, i.e. the carrier was nearly fully active also without anions. However, when we used mitoplasts, i.e. mitochondria in which the outer membrane has been stripped off, we suddenly found the same stimulation by anions as that observed in the reconstituted system. Thus some substance in the intermembrane space must permanently activate the adenine nucleotide carrier. This cannot be done by simple anions, since the outer membrane is freely permeable to these molecules. We thus had to look for a substance large enough to be restricted by the outer membrane and present in sufficiently high amounts. When testing cytochrome c, we found a drastic stimulation of the reconstituted ADP/ATP exchange by this compound (Fig. 3). Comparison with bovine serum albumin proved that this stimulation is not unspecific. The adenine nucleotide carrier activity is very low without a sufficiently high amount of activators; in Fig. 3 the basic exchange rate was measured in the presence of 3 mM Pipes, pH 6.5, 1 mM Na$_2$SO$_4$, 100 mM sucrose; full stimulation ($\cong$ 100%) was reached on addition of 30 mM Na$_2$SO$_4$.

A list of compounds tested for stimulation of the reconstituted ADP/ATP exchange is given in Table 2. It becomes obvious that the stimulation which is caused by cytochrome c, lysozyme, histone and protamin, but not by myoglobin, lactalbumin and serum albumin, is not dependent on the size of the molecule, but requires a positively charged, i.e. cationic polypeptide. Then we tried to simplify the activating substances to cationic amino acids and polyamines. Lysine did not activate, lysyl-lysine only very poorly, lysine-tripeptide was already a good activator and the polylysines were the best stimulators of all. The polyamines spermine and spermidine also activate. Thus it becomes obvious that activation is possible only with cationic substances having a minimum of two positive charges. We finally tried to simplify the system even further by using diaminoalkanes which should represent the minimum structure necessary for activation. The data of Table 2 show that, in fact, the diaminoalkanes, at least the shorter ones, stimulate the ADP/ATP exchange, although with relatively low affinity.

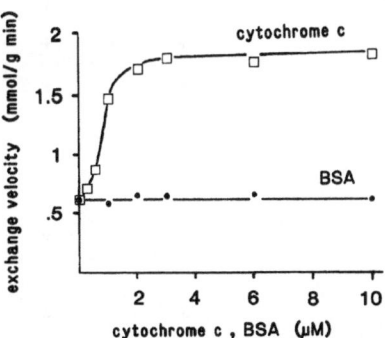

Fig. 3. Influence of cytochrome c and bovine serum albumin on the reconstituted ADP/ATP exchange. 100% means full activation by 30 mM Na$_2$SO$_4$. Activity without added protein was measured in the presence of 3 mM Pipes.

Activation ($K_{0.5}$) of the Reconstituted Adenine Nucleotide Exchange by several organic molecules. The $K_{0.5}$ values represent the concentration of activator that is necessary for half-maximum stimulation of the ATP/ATP exchange. Nucleotide transport was measured at pH 6.5.

Substances	Molecular weight	Charge at pH 6.5	Activating effect	$K_{0.5}(M)$ (approx. value)
(a) Proteins				
Cytochrome c	13400	+	+	10^{-6}
Lactalbumin	17400	0	0	−
Myoglobin	16900	0	0	−
Lysozyme	13900	+	+	10^{-6}
Serum albumin	66000	−	0	−
Histone	37000	+	+	10^{-7}
Protamin	5000	+	+	10^{-6}
Polylysine	1000–4000	+	+	2×10^{-6}
Polylysine	4000–15000	+	+	2×10^{-8}
Polylysine	15000–30000	+	+	5×10^{-9}
Polylysine	30000–70000	+	+	5×10^{-9}
(b) Amino acids, polyamines				
Lysine	146	+	0	−
Lys–lys	289	+	+	10^{-4}
Lys–lys–lys	433	+	+	5×10^{-6}
Putrescine	86	+	+	5×10^{-4}
Spermidine	142	+	+	5×10^{-6}
Spermine	198	+	+	10^{-6}
(c) Diaminoalkanes				
1,4 diaminobutane (putrescine)	86	+	+	3×10^{-4}
1,6 diaminohexane	110	+	+	5×10^{-4}
1,7 diaminoheptane	122	+	+	7×10^{-4}
1,8 diaminooctane	134	+	+	8×10^{-4}
1,10 diaminodecane	162	+	−	−

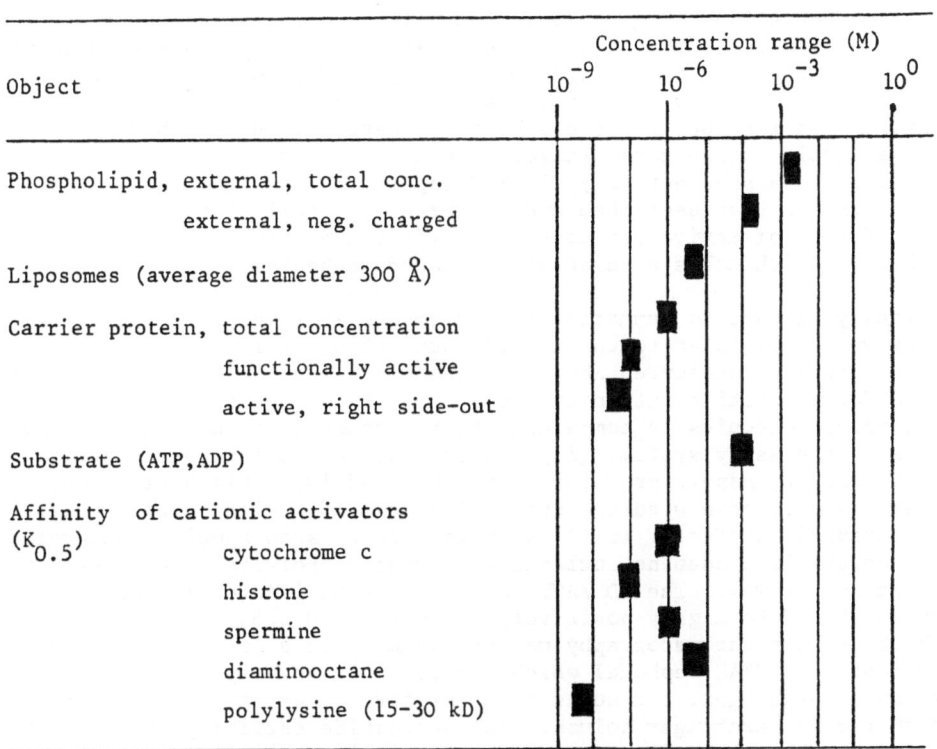

Fig. 4. Correlation of average concentrations of carrier protein, phospholipids and nucleotides in the transport assay with the observed activation constant $K_{0.5}$.

Although activation by these cations seems to be clear, an important question arises, particularly when considering the negatively charged membrane surrounding the adenine nucleotide carrier: Is the carrier protein really the direct target in the interaction with these polycations ? In principle, the activators can interact with all substances present in the assay system, i.e. phospholipids or liposomes, substrate and carrier protein. In Fig. 4 the average concentrations of all these substances in the reconstitution assay are compared with the $K_{0.5}$ values of some of the cationic activators. Using these data together with some additional experiments, we obtained a number of good arguments that in fact the protein is the actual target.

Phospholipids as the target of stimulation are excluded on the grounds of the following arguments: (i) Activation should take place by direct interaction of the activating molecule with its target, the concentration of activating molecules should be at least identical to or presumably higher than the concentration of its target. The concentration of cationic activators, however, is below the concentration of phospholipids as well as below that of negatively charged phospholipids, even below that of liposomes. (ii) If we use membranes with a surface made cationic by the addition of cetyltrimethylammonium-bromide, the same dependence on cationic activators can be demonstrated, which rules out a direct influence of these cations by surface attachment to negatively charged phospholipids.

Substrates can also be ruled out as mediators for the activation by cations. Although the argument concerning the large discrepancy between actual concentrations (Fig. 4) does not hold true in this case, since possibly a very small fraction of the negatively charged substrate in the close neighborhood of the carrier may be repeatedly bound and "activated" by the cations, another experiment carried out by us excludes such a mechanism. If the substrates are bound by cations, thereby becoming at least partially charge-compensated, this substrate-activator complex should lead to a substantially altered apparent K_m of the exchange process. This has been described and analyzed in detail elsewhere[4]. However, addition of activators does not shift the apparent K_m (experiment not shown), which makes a substrate-mediated mechanism very unlikely.

Finally, I want to summarize the arguments in favor of correlating the activator molecules to the adenine nucleotide carrier as their target. (i) The argument presented above for the exclusion of lipids can now be applied for a positive interpretation. Only the concentration of nucleotide carrier molecules is comparable to the concentration of activating cations in the assay system. (ii) In our lab we routinely reconstitute several other transport proteins from mitochondria. This activation, however, is only seen with the nucleotide carrier, which further rules out an unspecific effect. (iii) We were able to show specific adsorption of the solubilized adenine nucleotide carrier to polylysine agarose (experiment not shown). The ADP/ATP carrier with an isoelectric point of about 10 should be highly positively charged at pH 6.5, the pH value at which the column chromatography was performed. In a corresponding control experiment with DEAE-Sephacel which – comparable to polylysine agarose – is an anion exchanger, the nucleotide carrier, as expected, passes freely through the ion exchanger column. The nucleotide carrier protein can be eluted from the polylysine agarose by high ionic strength in the presence of detergent in a functionally active form. Therefore this column can be used for an affinity chromatography of the carrier protein based on the specific interaction of the translocator with polycations.

When looking at the list of activators given in Table 2, a question arises which should be taken seriously. Several of the newly discovered activators are well-known 'fusogens' and thus we have to ask ourselves whether perhaps they exert their influence by some kind of fusion effect. There are, however, a number of reasons which make a fusion-mediated process very unlikely: (i) Purified egg yolk phospholipids are not membranes suitable for inducing fusion events since they carry only a very small amount of negatively charged phospholipids. (ii) Activation can still be observed in membranes positively charged by insertion of cetyltrimethyl-ammonium-bromide, as described in the previous section. (iii) An unspecific mechanism mediated by fusion should also be unspecific with respect to various carrier proteins. However, taking the aspartate/glutamate carrier as example, this activation could not be demonstrated. (iv) The most striking piece of evidence is the fact that this activation could be shown to be reversible. When the reconstituted proteoliposomes, activated by the addition of spermine, are passed through cation exchange columns under appropriate conditions, the eluted liposomes are back in the non-activated state. They can be activated again by adding spermine once more.

Interaction with Charged Hydrophobic Molecules

When testing the influence of charged lipid molecules on the adenine nucleotide carrier activity, one has to consider the strong effect on the apparent transport affinity K_m due to interaction of the highly negatively charged substrates ADP and ATP with the surface charges. This has been quantitatively evaluated in the reconstituted system[4]. Besides these unspecific effect on the K_m, there are pronounced specific interactions of surface charges with the ADP/ATP carrier embedded in the phospholipid membrane leading to modulation of the exchange velocity V_{max}.

Negatively charged lipids. The stimulation of the adenine nucleotide transport by anionic phospholipids, such as phosphatidylserine, cardiolipin and even phosphatidic acid, was detected several years ago. Negatively charged phospholipids in general and cardiolipin in particular drastically enhance the rate of adenine nucleotide transport in the reconstituted system[3]. This observation has recently been elucidated as a side-specific interaction of negative surface charges (anionic phospholipids) with the carrier protein (Table 3). Negatively charged phospholipids strongly activate the adenine nucleotide carrier, particularly when present at the internal (matrix) side of the carrier protein; this is made obvious by the significantly increased rates of the inside-out oriented carrier proteins under the influence of externally added negatively charged lipids (Table 3). The right-side-out oriented proteins are only slightly activated by negative surface charges.

Table 3. Maximum Exchange Velocity of the Reconstituted ADP/ATP Exchange in Liposomes with Different Surface Charges.
The liposomes consisted of egg-yolk phospholipids/cholesterol = 85/15 (M/M). The surface charges are introduced by addition of cetyltrimethylammonium-bromide (positive) or dicetylphosphate (negative) to the preformed liposomes from the outside. The surface potential was measured by fluorescence of 2-p-toluidinyl-naphthalene-6-sulfonate in the presence of 20 mM Na^+. The exchange velocity (V_{max}) was determined for right-side-out oriented carrier proteins (RSO, cytosolic face to the outside) and for inside-out oriented carrier molecules (ISO, matrix face to the outside). Thereby the influence of the external surface charges on the different sides of the carrier protein can be discriminated[4].

Surface potential (mV) (external side)	V_{max} of ATP/ATP exchange in % (RSO)	(ISO)
− 10 (egg yolk lipids)	100	20
− 22	121	41
− 32	120	50
− 3	98	7
+ 6	92	8

Positively charged lipids. In order to elucidate further this specific effect, the reconstituted system was used as it offers the possibility to test the sidedness of the phospholipid-protein interaction by introduction of positive charges into the membrane. In contrast to anionic phospholipids, cationic lipids decrease the carrier activity - but again the influence was predominantly on the matrix side of the protein (ISO carrier). This is also shown in Table 3.

These results can be explained by a definite structural asymmetry of the carrier embedded in the phospholipid membrane. The active site, which is physiologically exposed to the cytosol, is located at a considerable distance from the plane of the membrane, whereas the opposite site seems to be in close proximity to the membrane surface[4]. It should be pointed out that the side-specific stimulation by negatively charged phospholipids predominantly on the matrix side of the protein closely resembles the situation in mitochondria, where the anionic phospholipids cardiolipin and phosphatidylinositol are mainly located at the inside of the mitochondrial inner membrane[11]. These conditions would activate the adenine nucleotide carrier, as has been shown here.

The influence of surface charges on the ADP/ATP transport represents an additional aspect in the asymmetric properties of this important carrier protein and is furthermore the first example of a side-specific modulation by phospholipids of the function of a transmembrane protein.

REFERENCES

1. R. Krämer and G. Kürzinger, Biochim.Biophys.Acta 765, 353:362 (1984).
2. R. Krämer, Biochim.Biophys.Acta 592, 615:620 (1980).
3. R. Krämer and M.Klingenberg, FEBS Lett. 119, 257:260 (1980).
4. R. Krämer, Biochim.Biophys.Acta 735, 145:159 (1983).
5. R. Krämer, Biochim.Biophys.Acta 693, 296:304 (1983).
6. R. Krämer and M. Klingenberg, Biochemistry 19, 556:560 (1980).
7. R. Krämer and M. Klingenberg, Biochemistry 21, 1082:1089 (1982).
8. W. Bogner, H. Aquila and M. Klingenberg, FEBS Lett. 146, 259:261 (1982).
9. M. Klingenberg, in: "The Enzymes of Biological Membranes", Vol.4, A.N. Martonosi, ed., Plenum Publ. Corp., pp.511:553 (1985).
10. R. Krämer and M. Klingenberg, Biochemistry 18, 4209:4215 (1979)
11. G. Daum, Biochim.Biophys.Acta 822, 1:42 (1985).

SURFACE ELECTRIC PROPERTIES OF CYANOBACTERIAL THYLAKOIDS

[1] K. Kalosaka, [2] B. R. Velthuys, [2] J. M. Briantais
and [1] G. C. Papageorgiou

[1] National Research Center Demokritos
Department of Biology, Athens, 153 10 Greece, and
[2] Laboratoire de Photosynthese, C. N. R. S.
91190 Gif-sur-Yvette, France

INTRODUCTION

All biological membranes, including the thylakoid membrane of photosynthetic organisms, carry fixed electric charges on their surfaces, which are due to exposed ionized groups of molecules that are integrated in the membrane.[1-6] The reported surface charge density is in the region of 1-3 $\mu C/cm^2$, in the particular case of thylakoids, the light transducing membrane vesicles of oxygenic photosynthetic organisms. When the electrically charged surface is in contact with a solution of electrolytes, two layers of electric charge are formed, with one layer localized on the surface plane of the membrane and the other, the diffuse layer, localized in the solution.

The Gouy-Chapman theory correlates the surface charge density σ and the electrostatic potential at the surface ψ_o with the bulk phase concentration Cb of a symmetric (z-z) electrolyte by means of the following equation,

$$\sigma = - (8RT\varepsilon_r\varepsilon_o)^{\frac{1}{2}} Cb^{\frac{1}{2}} \sinh(zF\psi_o/2RT) \qquad \text{Equation 1}$$

where the physical magnitudes ε_r, ε_o R, T, F have their usual meaning.

The Gouy-Chapman equation correlates the microscopic quantities σ, ψ_Θ with an easily determined macroscopic quantity, the bulk phase concentration of a symmetric electrolyte in the bathing solution. Its theoretical limitations have been discussed in detail by McLaughlin[1] and Barber,[2] but its validity in actual suspensions of membrane vesicles has been demonstrated in several systems and by several methods, including optical spectroscopy (absorption and fluorescence) of extrinsic and intrinsic chromophores, rates of interaction of ionic solutes with membrane surface constituents, and particle electrophoresis. In the case of the thylakoid membrane, experiments have been carried out, almost exclusively, with broken membrane fragments isolated from higher green plants. In our laboratory, we have studied the electrostatic properties of crude preparations of cyanobacterial thylakoids, applying the 9-aminoacridine fluorescence method of Barber and Chow.[7] Published results,[8] as well as unpublished experiments based on the same method, performed in our laboratory, with several species, indicate that cyanobacterial thylakoids have more negative surfaces (approx. 9-11 $\mu C/cm^2$) compared to higher plant thylakoids.

In the present work, we describe experiments performed with contamination-free thylakoid fragments isolated from the unicellular cyanobacterium Anacystis nidulans. The electrostatic properties of these membranes were studied in terms of the electron donation by ferrocyanide to photooxidized P700, and by particle electrophoresis.

MATERIALS AND METHODS

Isolation and purification of thylakoid membrane fragments

The unicellular cyanobacterium Anacystis nidulans was cultured photoautotrophically in the medium C of Kratz and Myers.[9] The cultures were aerated with a mixture of 2.5 % CO_2 in air, and illuminated with 3400 lux of light obtained from incandescent lamps. Cultures, 3-4 days old, were harvested by centrifugation at 3500 x g for 6 min, and purified thylakoid fragments were prepared according to Omata and Murata,[10] applied with some modifications. Bacteria were collected at a concentration of 100 μg Chl a /ml, were washed with 60 ml buffered medium (Hepes.NaOH 5 mM, pH 7.0) and were resuspended in a medium consisting of 600 mM sucrose, 2 mM EDTA and 5 mM Hepes.NaOH, pH 7.0. This suspension was incubated with 0.1 mg/ml egg-white lysozyme for 2 h at 34 °C. At the end, the cells were collected by centrifugation (3500 x g; 6 min), were washed with 60 ml of a buffered medium consisting of 600 mM sucrose, 20 mM Hepes NaOH, pH 7.0, and were finally resuspended in 22 ml of the same medium.

The cells were disrupted by means of a single passage through a French pressure cell at 6000 psi, and the resulting homogenate was incubated with 5 μl of 0.1 % w/v DNAase I (Sigma), in the presence of 10 mM Na-acetate, pH 6.0 and 1 mM $MgCl_2$. The homogenate was subsequently fractionated by centrifugation. Unbroken cells and large fragments were sedimented at 4500 x g for 10 min and the pellet was resuspended in 600 mM sucrose, 5 mM Hepes. NaOH, pH 7.0. The sucrose concentration was then raised to 48 % w/v, a 5 ml aliquot was placed in the bottom of a 12 ml centrifuge tube, and it was overlayered with 2.1 ml of each of the following sucrose solutions, 45 %, 30 %, 20 % w/v. Centrifugation was carried out at 130 000 x g for 18 hours at 4 °C, in a swinging bucket rotor. All sucrose solutions used for density gradient centrifugation were made in 10 mM NaCl, 5 mM EDTA and 10 mM Hepes. NaOH, pH 7.0.

The separated thylakoid membranes were withdrawn from the centrifuge tube with a Pasteur capillary pipette. The thylakoid fragments were diluted 2-fold with 100 mM sorbitol, 1mM Hepes, 0.8 mM KOH, pH 7.5 (Hepes-sorbitol buffer). The thylakoids were precipitated at 100 000 x g for 30 min, and they were washed and resuspended in the same medium at a final Chl a concentration of 0.5-0.6 mg/ml.

Spectrophotometric measurements

The kinetics of P700 oxidation and reduction were followed by means of absorption spectrophotometry, using the split beam spectrophotometer described by Velthuys.[11] The measuring wavelength was 702 nm, with a half-band width of 1 nm. Actinic illumination was supplied by an EG&G FX101 μs Xe flash. Two Corning glass filters (C.S. 4-96) were positioned between the flash and the reaction cuvette, which provided a pathlength of 2 mm. Samples were made at 42 μg Chl a/ml, in Hepes-sorbitol buffer.

Electrophoretic mobility measurements

Electrophoretic mobility of thylakoid fragments (11 µg Chl a/ml) was determined in the Hepes-sorbitol medium with a Rank Brothers Mark II apparatus, at 25 °C. The fragments moved under the force of an externally applied field of 680 volts.m -1, in a cylindrical cell 7.9 cm long. The image of moving particles was picked up by a microscope and was projected on a monitor screen, which allowed the observation of individual particles as they moved across the microscope field. The motion of at least 30 individual particles was monitored in order to calculate the electrophoretic mobility u according to the equation,

$$u = \frac{dl}{Vt} \qquad\qquad \text{Equation 2}$$

where d is the distance travelled by the particle at time t, under the force of the applied in voltage, and l is the length of the electrophoresis chamber.

RESULTS AND DISCUSSION

Typical time courses of P700 reduction by 0.8 mM K-ferrocyanide, during the dark time that ensues saturating µs flash illumination of Anacystis thylakoid fragments, are shown in Fig. 1. DCMU and methyl viologen were present in the reaction mixture to block replenishment of the intersystem pool of intermediates with pre-photosystem II and post-photosystem I electrons. In this way, virtually all electrons reaching P700 were provided by the tetravalent ferrocyanide anion.

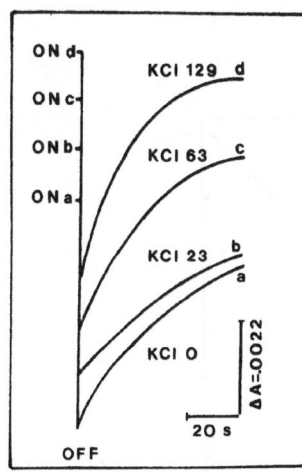 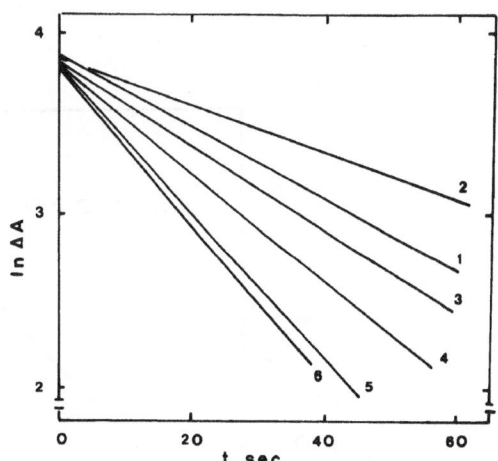

Fig. 1. Effect of KCl on the dark reduction rate of P700 in Anacystis nidulans thylakoid fragments by 0.8 mM K-ferrocyanide. The reaction mixture, buffered at pH 7.5, contained per ml the following: sorbitol 100 µmoles, Hepes 1 µmoles, KOH 0.8 µmoles, DCMU 16 nmoles, methyl viologen 60 nmoles and Chl a 42 µg.

Fig. 2. Semilogarithmic plots of P700 reduction kinetics. Numbers identifying curves stand for the following parametric KCl concentrations: 1, no KCl added; 2, 23 mM; 3, 63 mM; 4, 80 mM; 5, 129 mM; and 6, 156 mM.

According to Fig. 1, bulk phase electrolytes (here KCl) exert a pronounced effect on the rate of P700 reduction by ferrocyanide. The same is evident from Fig. 2, which displays semilogarithmic graphs of the reduction kinetics. The linearity of these graphs clearly shows that P700 reduction proceeds by means of a pseudo first order process, at all parametric KCl concentrations.

Apparent pseudo first order rate constants (in 1/s), obtained from measurements either in low, or in high electrolyte media, are plotted against the K-ferrocyanide concentration in Figs. 3 and 4, respectively. In low electrolyte medium, no reduction is observed up to 4–6 mM K-ferrocyanide.

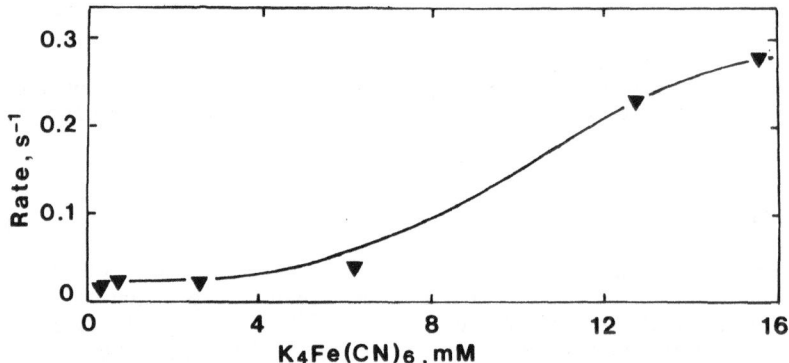

Fig. 3. Dependence of the P700 reduction rate by Anacystis thylakoid fragments on the concentration of K-ferrocyanide in the bulk aqueous phase. Assays were performed in the low electrolyte medium given in the legend to Fig. 1.

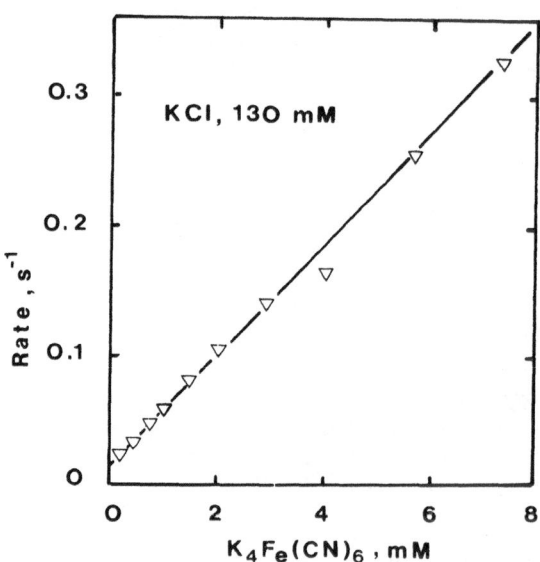

Fig. 4. Dependence of the rate of P700 reduction in Anacystis thylakoid fragments on the bulk phase concentration of K-ferrocyanide, in the presence of a 130 mM background concentration of KCl.

This reflects, most likely, the inaccessibility of the P700 sites on the thylakoid membrane surface to the reducing anions, as a result of coulombic repulsion. At higher K-ferrocyanide concentrations, screening of the membrane surface charge by the K^+ counterions becomes appreciable. This, in conjunction with the higher bulk phase concentration of ferrocyanide, makes possible the reduction of P700. When, on the other hand, KCl is present at a concentration that is adequate for membrane charge screening (Fig. 4), the rate constant of P700 reduction becomes a linear fucntion of the bulk phase concentration of ferrocyanide.

These results implicate electrostatic control of P700 reduction by ferrocyanide. Electrostatic theory provides that the collision frequency between membrane surface sites and small ions of like charge sign should depend on the bulk phase concentrations and the valencies of ionic species. In case of anions and negatively charges membrane surfaces, cations exert the most determining influence (cf. equation 1 and refs. 1,2). In Fig. 5, we show an experiment, in which we examine the dependence of the P700 reduction rate on the bulk phase concentrations of mono- and divalent metal chlorides. Theory would predict a monotonic rise of the reduction rate with increasing metal chloride concentration, because the increasingly more effective charge screening would bring more of the reducing ferrocyanide anion to the surface region. However, the concentration curves displayed in Fig. 5 pass through a minimum which is particularly prominent in the case of the monovalent Na^+ and K^+ cations. It should be noted, nevertheless, that the cation valence hierarchy, predicted by the electrostatic theory (divalents more effective than monovalents) is obeyed both in the ascending and in the descending branches of these concentration curves. We may infer, therefore, from this experiment that although electrostatic forces appear to govern electron donation to P700 by ferrocyanide, membrane charge screening

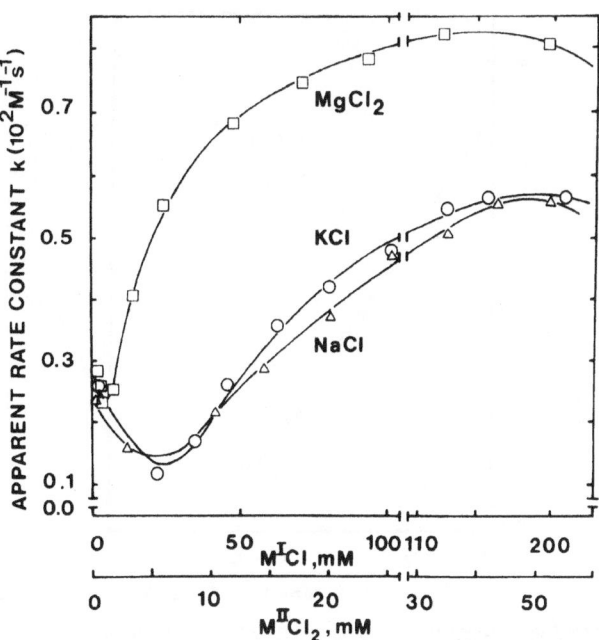

Fig. 5. The dependence of the apparent pseudo first order rate constant of P700 reduction by K-ferrocyanide on the concentration of mono- and divalent metal chlorides in the aqueous bulk phase.

becomes the dominant factor only above a certain threshold concentration of the counterion. Below that threshold, other electrostatic and secondary effects, such as charge neutralization and attendant conformation changes of membrane subunits may come into play. For example, it is possible that ferrocyanide donates electrons to P700 through a macromolecular intermediate, such as subunit III of Bengis and Nelson.[12] In such a case, the P700 reduction rate will depend not only on the surface area concentration of the reducing anion, but also on the interactions between the subunits of photosytem I reaction center. According to Barber[3] interactions between macroscopic electrically charged surfaces are controlled by the space charge density in the diffuse layer rather than by the surface potential.

The electrostatic potential generated by the thylakoid membrane forces a redistribution of ions along the normal to the surface, pulling in cations and pushing away anions, according to Boltzmann's distribution law. As a result, apparent rate constants determined on the basis of the bulk phase concentrations of ionic reactants should differ from the true rate constants which depend on the surface area concentrations. Employing the Gouy-Chapman theory and assuming only surface charge screening by counterions, Itoh[13] derived the following expression, which links the surface charge density (σ) the apparent rate constant (k) and the true rate constant (k^0), to the activity coefficient (γ_b) the valence (z) and the bulk phase concentration (Cb) of the reacting anionic solute (here ferrocyanide):

$$\log k/\gamma_b = \log k^0 - 0.074 z Cb^{-\frac{1}{2}} \qquad \text{Equation 3}$$

Fig. 6 displays a plot of the apparent rate constants for P700 reduction by ferrocyanide against the inverse square root of the ionic strength in the bulk phase. The data used correspond to the ascending portions of the curves shown in Fig. 5, obtained in the presence of KCl and

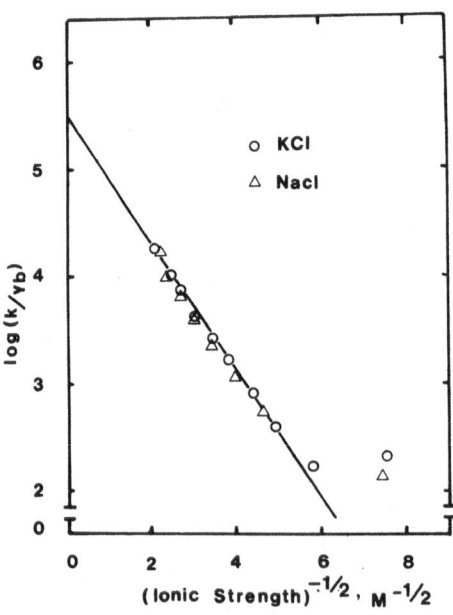

Fig. 6. Plot of $\log k/\gamma_b$ against the inverse square root of the ionic strength for the reaction between ferrocyanide and photooxidized P700 in Anacystis thylakoid fragments. The apparent rate constant values (k) were obtained from data shown in Fig. 5.

NaCl. The straight line obtained indicates that charge screening is indeed the predominant controlling factor for P700 reduction above a threshold monovalent counterion concentration. The absence of charge neutralization contributions is further supported by the fact that points obtained from the NaCl and KCl labeled curves of Fig. 5 fall on the same straight line.

From the slope of the straight line in Fig. 6 we calculated a surface charge density of $\sigma = -2.0 \ \mu C/cm^2$. The true rate constant ($k^0 = 3.16 \times 10^5$ $M^{-1}s^{-1}$) was obtained from the ordinate intercept at infinite ionic strength. To calculate the surface electric potential of the thylakoid membrane near the P700 site, we employed the following equation of Itoh:

$$\psi_0 = -15(\log k^0 - \log k/\nu_b) \qquad \text{Equation 4}$$

The results are displayed in Fig. 7.

It should be mentioned, that the values we obtained for σ and ψ_0 for the P700 site of Anacystis thylakoids are higher than the corresponding values obtained by Itoh[13] for sonicated spinach thylakoids ($\sigma = -0.86$ $\mu C/cm^2$; $\psi_0 = -38$ mV, at 0.01 M ionic strength). Indeed, the surface electricity of Anacystis thylakoid fragments may be higher since in our

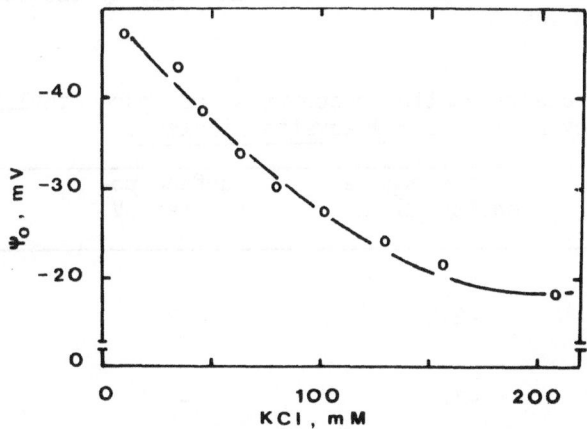

Fig. 7. Dependence of the estimated surface potential ψ_0 of Anacystis thylakoid fragments, calculated according to Equation 4, on the KCl concentration of the suspension medium.

calculations we ignored monovalent cation concentrations below 40 mM.

The ψ_0 potential, which is defined as the electrostatic potential at the hydrodynamic plane of shear (estimated to lie about 2 A^0 from the particle surface)[14] was calculated from electrophoretic mobility (u) measurements by means of the following equation,

$$\zeta = 3\eta \ u/2 \ \varepsilon_r \varepsilon_0 \ H \qquad \text{Equation 5}$$

where η is the viscosity coefficient of the suspension medium (assumed to be equal to that of H$_2$O at 25^{0}C) and H the Henry function of dimension.

In Fig. 8, we show the effect of the KCl concentration in the suspension

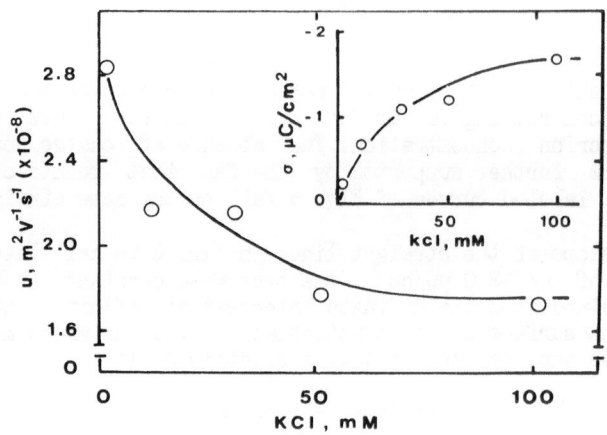

Fig. 8. Effect of KCl concentration in the suspension medium on electrophoretic mobility of Anacystis thylakoid fragments. Surface charge density values in the insert were calculated according to Equation 1. Chl a = 11 µg/ml; other conditions, as in Fig. 1.

medium on the electrophoretic mobility of thylakoid fragments. The electrophoretic mobility falls off with increasing KCl concentration, evidencing a decrease in the electrostatic potential at the hydrodynamic plane of shear. Finally, in Table 1 we present comparative data for the surface charge density and electrostatic potential of pure thylakoid fragments, obtained in the presence of 30 mM KCl, by the reaction rate and the electrophoresis methods.

Table 1. Surface electrostatic properties of pure thylakoid fragments isolated from the cyanobacterium Anacystis nidulans.

Method of measurement	Surface charge density $\mu C/cm^2$	Surface potential mV	potential, mV
P700 reduction by K-ferrocyanide	–2.0	–43	–
Electrophoretic mobility	–1.1	–	–28

Acknowledgements: We acknowledge, with gratitude, technical assistance by Dr. S. Phung-Nhu-Hung in the electrophoresis experiments.

REFERENCES

1. S. McLaughlin, Electrostatic potentials at membrane-solution interfaces, in: Current Topics in Membranes and Transport, F. Bronner and A. Kleinzeller, eds., vol. 9, Academic Press, New York (1977).
2. J. Barber, Membrane surface charges and potentials in relation to photosynthesis, Biochim. Biophys. Acta 594:253 (1980).
3. J. Barber, Influence of surface charges on thylakoid structure and function, Ann. Rev. Plant Physiol. 33:261 (1982).

4. P. S. Nobel and H. C. Mel, Electrophoretic studies of light induced charge in spinach chloroplasts, Arch. Biochem. Biophys. 113:695 (1966).
5. F. V. Mercer, A. J. Hodge, A. Hope and J. D. McLean, The structure and swelling properties of Nitella chloroplasts, Austr. J. Biol. Sci. 8:1 (1955).
6. H. Y. Nakatani, J. Barber and J. A. Forrester, Surface charges on chloplast membranes as studied by particle electrophoresis, Biochim. Biophys. Acta 504:215 (1978).
7. W. S. Chow and J. Barber, Salt dependent changes of 9-aminoacridine fluorescence as a measure of charge densities of membrane surfaces, J. Biochem. Biophys. Methods 3:173 (1980).
8. K. Kalosaka and G. C. Papageorgiou, Surface electric properties of thylakoid fragments isolated from vegetative and heterocystous cyanobacteria, in: Advances in Photosynthesis Research, C. Sybesma, ed., vol. 2 M. Nijhoff/Dr. W. Junk Publishers, The Hague (1984).
9. W. A. Kratz and J. Myers, Nutrition and growth of several blue-green algae, Am. J. Bot. 42:282 (1955).
10. T. Omata and N. Murata, Isolation and characterization of the cytoplasmic membranes from the blue-green alga (cyanobacterium) Anacystis nidulans, Plant & Cell Physiol. 24:1101 (1983).
11. B. R. Velthuys, A third site of proton translcation in green plant photosynthetic electron transport, Proc. Natl. Acad. Sci. US 75:6031 (1978)
12. C. Bengis and N. Nelson, Subunit structure of chloroplast photosystem I reaction center, J. Biol. Chem. 252:1564 (1977).
13. S. Itoh, Surface potential and reaction of membrane bound electron transfer components. I. Reaction of P700 in sonicated chloroplasts with redox reagents, Biochim. Biophys. Acta 548:579 (1979).
14. M. Eisenberg, Gresalfi T., Riccio T and S. McLaughlin, Adsorption of monovalent cations to bilayer membrane containing negative phospholipids Biochemistry 18:5213 (1979).

CHANGE OF SURFACE POTENTIAL AND INTRAMEMBRANE ELECTRICAL FIELD INDUCED BY
THE MOVEMENTS OF HYDROPHOBIC IONS INSIDE CHROMATOPHORE MEMBRANES OF
Rhodopseudomonas sphaeroides STUDIED BY RESPONSES OF MEROCYANINE DYE AND
INTRINSIC CAROTENOIDS

Shigeru Itoh

National Institute for Basic Biology
Nishigonaka, Okazaki 444, Japan

INTRODUCTION

Change of the ionic conditions or pH of the outer medium changes
surface potential of energy transducing membranes as expected from the
Gouy–Chapman theory and affect electrical field within the membrane under
some conditions (1,2). This induces a change of the redox states of
electron carriers inside membrane depending on their intramembrane
localization (3,4). Thus, change of surface potential is expected to
affect the movement of electrons and ions within the membrane as well as
the reaction rates at the membrane surface (3). Vice versa, energization
of the membranes are expected to induce changes of surface potential value
as experimentally suggested in some membrane systems (3,5).
    We have been using a series of merocyanine dyes which respond to the
change of surface potential as well as membrane potential (5,6). The dyes
change their distributions between the membrane surface and the outer bulk
phase by apparently responding to the change of surface potential. They
seem to respond to the difference of electrical potential in the site in
some depth inside the surface boundary region of the membrane with respect
to that in the outer aqueous phase (5). Movements of the dyes across the
membrane were essentially negligible (6). Therefore, the dyes can be good
probes of the electrostatic state of the membrane surface. In this study
changes of surface potential and intramembrane eleectrical field were
analyzed under various conditions in chromatophore membranes of photosynthetic bacteria through the measurements of absorption changes of merocyanine dye and the intrisic carotenoid pigment. In the presence of hydrophobic anions such as tetraphenylboron (TPB⁻), intramembrane electrical
field was found to be modified in a different manner from that induced by
ordinay ionophores. Changes of electrical field in the surface and central
regions of the membrane are found to be different each other. Profiles of
electrical potential inside chromatophore membranes under these condtions
are estimated. The responses of the potential probes were well interpreted
by the three capacitor model of the membrane which was proposed in model
bilayer membranes (8).

ABBREVIATIONS: CCCP, carbonylcyanide-m-chlorophenylhydrazone. TPB⁻,
tetraphenylboron⁻. TFPB⁻, tetrakis(4-fluorophenyl)boron⁻.

MATERIALS AND METHODS

Chromatophore vesicles were prepared from cells of <u>Rhodopseudomonas sphaeroides</u> green mutant (2). Light harvesting protein complexes I and II were prepared from chromatophores after the treatment with triton-X100 and the purification by polyacrylamide gel electrophoresis. Absorption changes of carotenoids and merocyanine dye (NK2274) were measured with Hitachi 557 dual wavelength spectrophotometer or with a split beam flash spectrophotometer constructed in this laboratory. The merocyanine dye was purchased from Nippon Kanko Shikiso Lab., Okayama. Tetraphenylboron and its fluoroderivative tetrakis(4-fluorophenyl)boron (TFPB⁻) were purchased from Dojindo Lab., Kumamoto.

RESULTS AND DISCUSSION

Surface Charge Density of Various Energy Transducing Membranes

Table 1 summarizes the values of net surface charge density estimated in various membrane systems at neutral pH region. In chloroplast thylakoid membrane, analyses of the effects of monovalent salts on the reaction rates of various membrane electron transfer components gave larger values of net surface charge density for the outer surface $(-1.1 \sim -1.4 \ \mu C/cm^2)$ than the inner surface $(-0.8 \sim -1.1 \ \mu C/cm^2)$, suggesting different distribution of charges. On the other hand, larger values of net surface charge density were estimated from the comparison of the effects of mono- and divalent cations on the fluorescence yield of intrinsic chlorophyll or 9-aminoacridine. The values obtained by the reaction method seem to be somewhat overestimated since they do not fit the case with divalent cation salts well, although the effects of monovalent salt can be most correctly predicted with these values. Our experiences tell that the larger surface charge density value is obtained when we use the difference of the effectiveness of mono- and divalent salts in the estimation of net surface charge density. The value which is the best fit for the cases with mono- and divalent salts, in turn, fit poorly to the case with trivalent cations. On the other hand, the values obtained by fluorescence probes may be a slight overestimation since the nonspecific binding of divalent cations and fluorescence porobes to the surface may also contribute to the surface charge density. The surface domains measured by different methods also seem to vary depending on the methods, e. g., the reaction method measures the local charge density in the vicinity of the reaction site while the fluorescence probe method usually measures an average of entire membrane surface. However, the comparison of the values of net surface charge density seems to be meaningful between the values obtained by similar estimation methods in spite of these difficulties. Different surface charge density values were also estimated for the inner and outer surfaces, in chromatophore membranes of photosynthetic bacteria, <u>R. sphaeroides</u>, from the comparison of the effects of mono- and divalent cation salts on the intramembrane electrical field monitored by the absorbance change of carotenoid. These results indicate that the surfaces of these membranes are heterogeneous with respect to charge distribution.

The values of net charge density estimated for various biological membranes (Table 1) are not very large $(-0.4 \sim -4.6 \ \mu C/cm^2$, which values correspond to $-0.25 \sim -2.9$ electronic charges per 1,000 square angstroms) except for the case of membranes of <u>H. halobium</u> which organism is known to grow in the high salt medium. Surface potential values under physiological conditions, if it may be represented by 0.1 M monovalent salt, are estimated to be $-27 \sim -48$ mV with $-2 \sim -4 \ \mu C/cm^2$ net surface charge density. The value is not very large but is not negligible, especially when the correct value of electrochemical free energy difference involved in the functioning of the membrane molecules are required.

Table 1. Values of net surface charge density of various membranes at neutral pH (6.5–8.5).

Membranes	q ($\mu C/cm^2$)	Methods (ref.)
inner membrane of chloroplast (thylakoid membrane) (spinach and pea)		
outer surface	$-1.1 \sim -1.4$	reactions of $\underline{Q}$ (13), cyt $\underline{f}$,cyt b_{559} (15), $\underline{P700}$ (14)
	-0.95	change of pH optimum of Fd–NADP reductase (16)
inner surface	$-0.8 \sim -1.1$	reaction of $\underline{P700}$ in thylakoid fragments (14–16)
total membrane	-3.0	merocyanine dye partition (5)
	-2.5	fluorescence of 9–aminoacridine (18,19) and intrinsic chlorophyll (18,20)
	-1.6	ion–exchange ability (21)
chromatophrore membrane of photosynthetic bacteria		
outer (periplasmic) surface	-3.0	carotenoid absorbance change (1) ($\underline{R.\ spaeroides}$)
	-2.2	reaction of cyt c_2 (4) ($\underline{R.\ sphaeroides}$)
inner (cytoplasmic) surface	-4.6	carotenoid absorbance change (22) ($\underline{R.\ sphaeroides}$)
intramembrane components	-1.6	E_m of cyt c_2, bacterio-chlorophyll dimer (4) ($\underline{R.\ sphaeroides}$)
	-1.2	E_m of cyt c_{555} (3) ($\underline{C.\ vinosum}$)
purple membrane of $\underline{H.\ halobium}$ surface of bacteriorhodopsin	-13.9	absorption change of bacterio-rhodopsin chromophore(12)
mitochondrial inner membrane		
outer surface	-0.38	phosphorescence of Tb^{3+} (23)(rat liver)
outer surface	-3.3	fluorescence of 9–aminoacridine (27) (Jerusalem artichoke)
smooth microsome membrane (rat liver)		
outer surface	-0.62	reaction of arylsulfatase (24)
total membrane	-1.73	ion–exchange ability (25)
plasmalemma membrane (wheat root)		
outer surface	-1.6	fluorescence of 9–aminoacridine (26)
inner surface	-3.4	fluorescence of 9–aminoacridine (26)

Merocyanine and Carotenoid as Probes of Membrane Potential

Fig. 1 shows the light-induced difference absorption spectrum of chromatophore membrane in the presence of merocyanine. Absorption change between 410 and 520 nm is mainly induced by the absorbance change of intrinsic carotenoids responding to the change of intramembrane electrical field (inside positive direction), and the one centering around 550 nm, by the absorbance change of merocyanine dye (increase of the membrane bound form). Both types of absorbance changes are shown to be proportional to the membrane potential created by illumination or by diffusion potential and are shown to be reliable probes of membrane potential change (5,6). Fig. 2A shows the effect of an ionopohore valinomycin, which is known to dissipate membrane potential by increasing the membrane conductivity of pottasium ion, on the light-induced responses of carotenoid and merocyanine. As the increase of valinomycin concentration absorbance changes of carotenoid and merocyanine were decreased in a parallel manner as expected. Fig. 2A shows the effects of illumination light intensity on

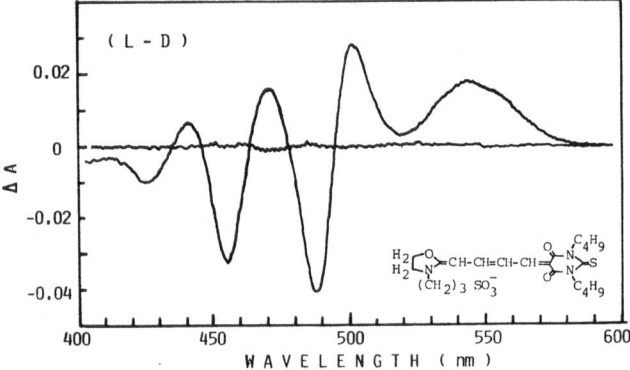

Fig.1. Difference absorption spectra of intrinsic carotenoid and merocyanine dye induced by illumination in chromatophore membranes of R. sphaeroides green mutant. The reaction mixture contained 6 μM merocyanine, chromatophores (10 μM bacteriochlorophyll), 5 mM Tricine-NaOH buffer (pH 7.8) and 50 mM NaCl.

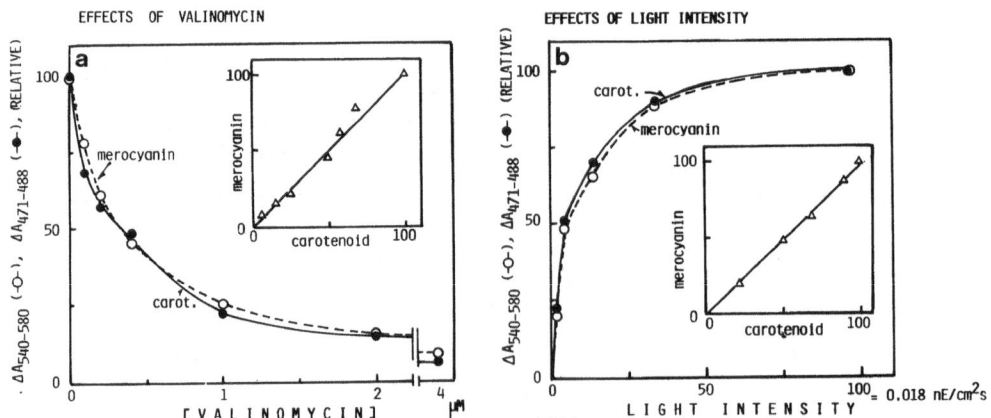

Fig. 2A. Effects of valinomycin on the extents of light-induced absorbance changes of carotenoid and merocyanine. B. Effects of light intensity on the extents of light-induced absorbance changes of carotenoid and merocyanine. Experimental conditions were similar to those in Fig. 1.

the responses of merocyanine and carotenoid. It is clear that both responses correlate very well under various illumination intensities. The results in Fig. 2 suggest that the electrical field sensed by each type of potential probe was proportional to the change of membrane potential under different illumination intensity or in the presence of valinomycin. Similar results were obtained with gramicidin S and CCCP (not shown). Intrinsic carotenoid can be assumed to respond to the change of electrical field in the hydrophobic central region of the membrane. On the other hand, the merocyanine molecule which has a localized negative charge on one end is expected to respond to the field change in the surface region of membrane.

Change of Surface Potential (Potential Change in the Surface Boundary Region) of the Membrane

Fig. 3 shows the response of a merocyanine dye to the change of surface potential induced by salt addition in the suspension of chromatophores of R. sphaeroides. The scale for the surface potential on the right-hand ordinate was calculated from the known surface charge density of -3.0 $\mu C/cm^2$ for this membrane (1) and concentration of KCl in the outer medium according to the Gouy-Chapman theory. At each concentration of KCl, illumination of chromatophores induced the additional absorbance change in parallel with the membrane potential change. Broken line indicates the sum of these light and salt-induced absorbance changes. We can estimate the light-induced change of surface potential (actually it seems to be the change of electrical field within the boundary region of the membrane on the outer surface) from the right-hand scale. The difference between the broken and solid lines in the figure indicates the light-induced surface potential changes of $+11 \sim +3.5$ mV at KCl concentrations of 10-100 mM. The dependency on the salt concentration changes depending on the molecular formulae of merocyanine dyes (5). Under these conditions light-induced membrane potential of 260 mV (inside positive) was estimated from the absorbance change of the intrinsic carotenoid. These results suggest

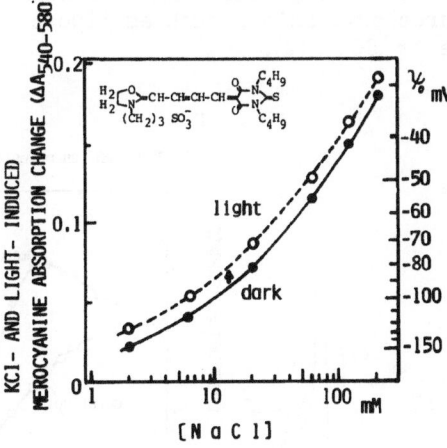

Fig. 3. Absorbance change of merocyanine dye induced by NaCl addition in the dark (solid line), and sum of light-induced and the NaCl-induced absorbance changes (broken line). Scale for the surface potential on the right-side ordinate was calculated from the net surface charge density value of -3.0 $\mu C/cm^2$ using the Gouy-Chapman eqation (3). Reaction mixture contained 0.01 mM $MgSO_4$, 1 mM tricin-NaOH buffer (pH 7.8), chromatophores equivalent to 20 μM bacteriochlorophyll and varied concentrations of NaCl.

that this merocyanine dye senses 2-8 % of the field change imposed on the
total low-dielectric part of the membrane, probably sensing the field
change of the binding site just beneath the surface inside the outer boundary region of the membrane. Similar conclusions were also obtained with
other merocyanine dyes (5). These results in Fig. 1 indicate that the
extent of surface potential change depends on the salt condition of the
medium, although the accurate estimation of the surface potential change
requires some more compex analysis of the results (5). It may be more
adequate to represent the change of electrostatic characteristics of the
membrane surface in terms of the change of net surface charge density than
to represent it in terms of surface potential value itself. The date in
Fig. 1 can roughly be explained by assuming a change of surface charge
density from $-3.0 \ \mu C/cm^2$ to $-2.7 \ \mu C/cm^2$.

Change of Intramembrane Electrical Field by the Movement of Hydrophobic Ion inside the Membrane

Effects of hydrophobic ions on the responses of carotenoid and merocyanine were studied. Fig. 4 shows the effects of a hydrophobic anion,
tetraphenylboron (TPB^-) on the responses of merocyanine and the intrinsic
carotenoids during illumination. In the presence of low concentration of
TPB^-, the response of the carotenoid decreased, while that of merocyanine
increased. This result is very different from the effect of other type of
ionophores. It is suggested that at low concentrations of TPB^-, movements
of TPB^- within the membrane increased the field change on the outer surface
region of the membrane, in which merocyanine senses the field change, but
decreased it in the central region of the membrane, in which carotenoid
senses the field change. Under these conditions, the time response of the
merocyanine (half time of about 5 ms) was fast enough to follow the potential change induced by continuous illumination.

In the measurement of the flash-induced absorbance change of carotenoids, the idea above was further confirmed. TPB^- induced a very fast
decay phase within a few milliseconds after the flash excitation (i. e., in
a time range in which merocyanine can not respond) (Fig. 5). However, the
following decay rate was only slightly affected. Higher concentrations of
TPB^- gave a larger rapid decay phase with almost the same decay rate
constants of the rapid and slow phases. Similar biphasic decays can be
induced with other hydrophobic anions such as fluoro derivative of TPB^-
($TFPB^-$), dipiclylamine or SCN^- (2).

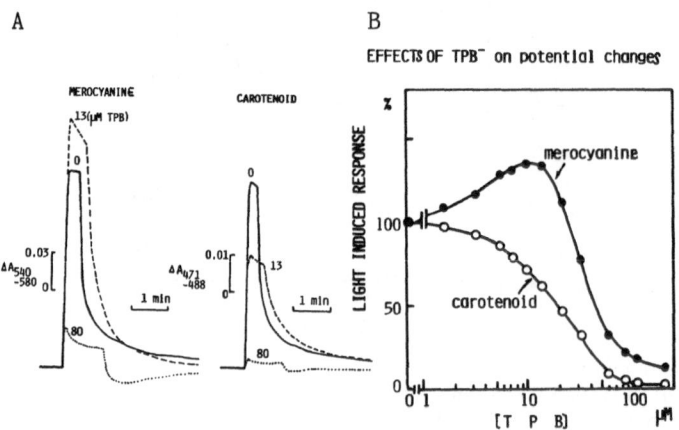

Fig. 4A. Effects of TPB^- on the light-induced responses of merocyanine and
carotenoid in chromatophores. B. Dependences of the extents of absorbance
changes of merocyanine and carotenoid on the TPB^- concentration. Reactions
were performed at 200 mM NaCl in a medium similar to those in Fig. 3.

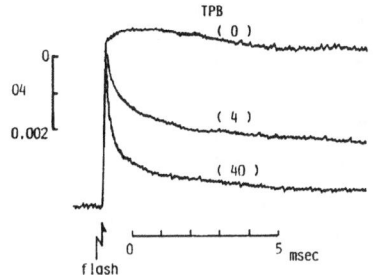

TPB
(0)

(4)

(40)

flash   0   5 msec

Fig. 5. Effect of TPB⁻ on the flash-induced absorbance change of carotenoid. Reaction mixture contained 50 mM Tris buffer (pH 7.8), 50 mM NaCl, 0.2 μM antimycin A and 15 μM bacteriochlorophyll.

Table 2.

KINETIC PARAMETERS OF TPB TRANSPORT THROUGH MEMBRANES

k_i (s^{-1})	N_t (pmole/cm^2)	ß (10^{-3}cm)	TPB conc. (μM)
Chromatophores			
224	0.84 (=2.3/RC)	0.174	2.4
206	3.81 (=20/RC)	0.016	120
PC bilayer membrane (ref. 9)			
130	0.18	0.90	0.1

k_i, rate const. of TPB transfer inside membrane
N_t, total amount of TPB inside membrane
ß, partition coeff. of TPB

These effects of TPB⁻ seem to be explained by postulating the rapid movement of TPB⁻ within the membrane according to the three capacitor model of the membrane proposed in the model lipid bilayer membranes by Anderson et al.(8). According to the model, TPB⁻ exists in potential energy minimum (potential well) which exists just beneath each surface. Upon the application of inside positive membrane potential, TPB⁻ moves rapidly from one potential well just beneath the outer surface, to the other one close to the inner surface (Fig. 6) and rapidly dissipates the electrical field in the central but not in the surface region of the membrane. Equilibration of TPB⁻ concentrations between each well and the outer aqueous phase on each side of the membrane is expected to be rather slow due to the cooperative interaction between TPB⁻ molecules (8). This can be seen by the rate of slow phase in the decay kinetics of carotenoid. Biphasic decay kinetics of carotenoid can be explained by postulating the response of carotenoid to the field change in the central region of the membrane. On the other hand, change of electrical field in the boundary region, to which merocyanine responses, is expected to become slightly less by TPB⁻ movements after a single flash excitation and larger after the movements of large amount of TPB⁻ molecules between the wells, as long as the membrane potential is maintained at a high level under continuous illumination (Fig. 6). Kinetic parameters of TPB⁻ movement in the membrane calculated according to Benz and Nonner (9) were shown in Table 2. The rate constant of TPB⁻ movement was similar to that observed in lipid bilayer or nerve membranes (8,9). The lower partition coefficient value estimated in chromatophore membranes may be due to the use of high concentration of TPB⁻ or due to the decrease of actual concentration of TPB⁻ in the bulk medium after the binding of TPB⁻ to the membranes. These characteristics of the TPB⁻ movement are assumed to be very different from those of merocyanine which prbably moves between the bulk medium and the binding site in the outer surface region of the membrane, by dissipating a portion of intramembrane electrical field. This idea is consistent with the observation that the movement of merocyanine, at concentrations used in this study, induced almost no effect on the intramembrane electrical field sensed by carotenoid.

Similar biphasic decay kinetics of the carotenoid have been reported under phosphorylating conditions in chromatophores (10) or in chloroplasts (11). In chloroplasts extent of the faster decay phase was shown to be dependent on the number of activated ATP-synthetase under phosphorylating conditions, but not under nonphophorylating conditions (11). Potential wells for protons are postulated inside the CFo moiety of the enzyme in analogy with those proposed in the study of hydrophobic anions (11).

Existence of potential wells for TPB⁻ inside membrane can be tested

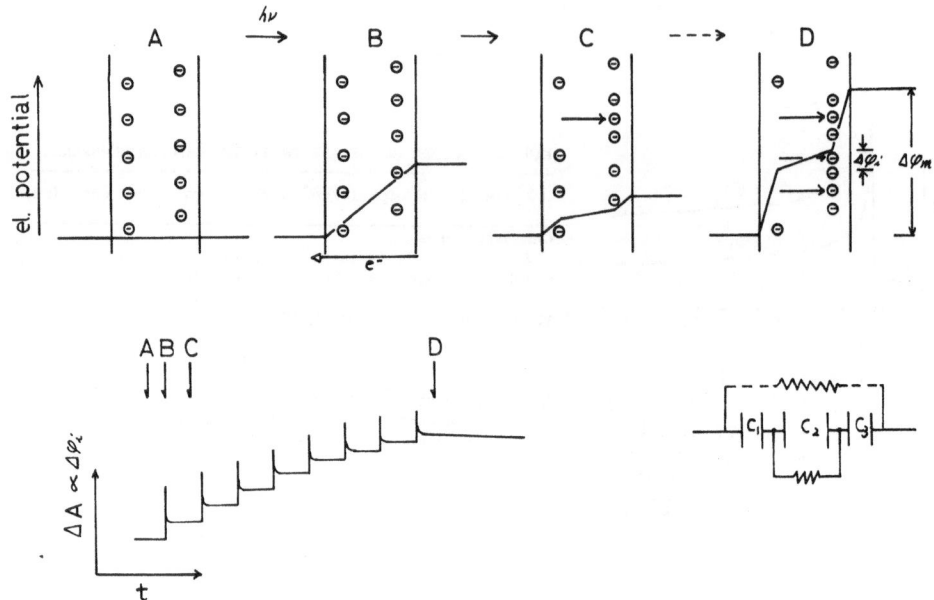

Fig. 6. Schematic model for the movement of TPB$^-$ and change of electrical potential profile across the membrane under repetitive flash excitation. Lower left: schematic time course of absorption change of carotenoid. Lower right: Three capacitor model of the membrane. C_1 and C_3 are composed of the ions in the diffuse layer in the outer medium and in the potential well on each side of the membrane, and c_2 between the ionic layers in the potential wells. A, schematic representation of the electrical potential across the membane in the dark nonenergized state. B, at the time just after the flash excitation before the intramembrane movement of TPB$^-$. C, after the intramembrane movement of TPB$^-$. D, after the multiple flash excitations or under continous illumination. In cases C and D, intramembrane potential difference between the wells ($\Delta\varphi_i$) is expected to be decreased to the level at which $\Delta\varphi_i$ equilibrates with the difference of TPB$^-$ concentrations within these wells. Slower equilibration of TPB$^-$ concentration in the well to that in the outer bulk medium on each side of the membrane (accompanying the movements of other ions and the change of potential in the edge part) is assumed to return C and D states slowly to nonenergized state A in the dark. TPB$^-$ molecules were assumed to be situated in the potential wells just beneath the surfaces as postulated by Anderson et al. (8) and to rapidly move between the wells down the electrical potential created by the flash induced electron flow. Note that potential changes are confined to the edge parts of the membrane in cases C and D. Carotenoid responds to $\Delta\varphi_i$, whlie merocyanine to the potential change in the boundary region in the outer surface. $\Delta\varphi_m$ represents membrane potential (potential difference between the bulk aqueous phases).

from another aspect by studying the conductivity of the membrane (decay rate of the carotenoid absorbance change) supported by the membrane permeability of CCCP (Fig. 7). It is known that membrane conductance supported by CCCP decreases on addition of TPB$^-$ in lipid bilayer membranes since TPB$^-$ localization in the potential wells makes interior of the membrane electrostatically more negative and will decrase the number of anionic CCCP inside membrane (8). In the absence of TPB$^-$, the decay rate of carotenoid absorbance change was accelerated by CCCP after the flash excitation (Fig. 7), indicating the increase of membrane conductivity supported by the rapid permeation of CCCP anion across the membrane. Addition of TPB$^-$ induced the biphasic decay even in the presence of CCCP. The rapid phase showed

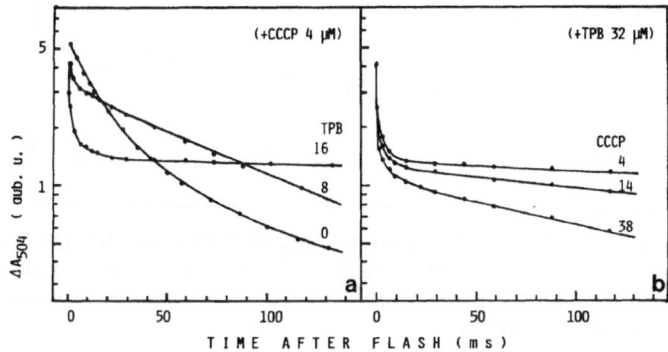

Fig. 7A. Effect of TPB⁻ on the decay kinetics of flash-induced absorbance change of carotenoid in the presence of 4 µM CCCP. B. Effect of CCCP in the presence of 32 µM TPB⁻. Experimental conditions were similar to those in Fig. 5. Concentrations of reagents added were shown in µM.

similar decay rate as that observed in the absence of CCCP. On the other hand, the decay rate of the slow phase decreased as the TPB⁻ concentration increased, although the decay rate was still sensitive to CCCP in the presence of higher concentration of TPB⁻. It, therefore, supports the idea that both the rapid and slow phases of the carotenoid absorbance change reflect the change of electrical field in the central part of the membrane between the potential wells for TPB⁻. The rapid decay represents the change of field by the rapid electrophoretic movements of TPB⁻ between the potential wells. It continues untill the potential difference between the wells eauilibrates with the concentration difference of TPB⁻ between the wells. The slow phase represents the field change which occurs during the equilibration of TPB⁻ concentration in each well with those in the bulk medium in either side of membrane. During the slow phase, potential difference between the wells will be in equilibration with the difference of TPB⁻ concentrations between the wells, and therfore, will decrease in parallel with the decay of electrical field in the boundary region of the membrane or with the decay of membrane potential. That the decay rate supported by the permeability of CCCP was decreased by TPB⁻ addition (slow decay phase) gives another support for the potential well model since the result suggests that TPB⁻ permeation into the potential wells creates high negative boundary potential which decreases the concentration of CCCP inside membrane. Rapid dissipation of the electrical field in the central part of the membrane by the TPB⁻ movement, which is faster than that of CCCP, is also assumed to decrease the electrophoretic movement of CCCP inside membrane.

Effect of Tetrakis(4-fluorophenyl)boron (TFPB⁻) on the Carotenoid Band Shift and Proton Uptake

Change of intramembrane electrical field and its effects on the reactions performed inside membranes were studied in the presence of fluoro-derivative of TPB⁻ (TFPB⁻), which shows higher rate of intramembrane movement than TPB⁻. Excitation of chromatophores with repetitive single turn-over flashes induced a large carotenoid absorbance change (Fig. 8) which corresponds to about 250 mV inside positive membrane potential change. On addition of 1.5 µM TFPB⁻, the absorbance change induced by the first flash was depressed to be 26 % of the control, while the pseudo-steady state level attained after 16 flashes was not affected significantly (73 % of the control). At a very high concentration, the absorbance change appeared to be almost completely suppressed in this time range. Also in the case of TFPB, very rapid decay of carotenoid band shift (with a half decay time of 7.5 µs) was observed within a few tens of microseconds just after each

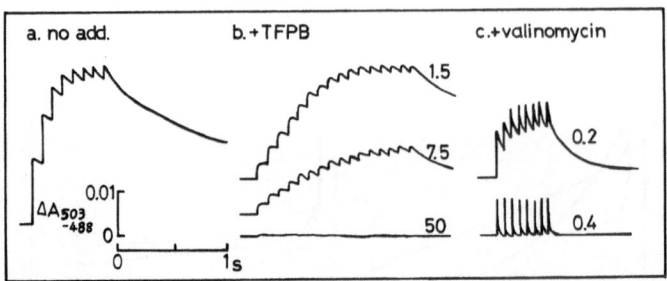

Fig. 8. Effects of TFPB⁻ on the carotenoid absorbance change induced by
successive flash excitations. In b and c TFPB⁻ and valinomycin,
respectively, were added to give final concentrations as indicated by
numbers (in μM) in the figure.

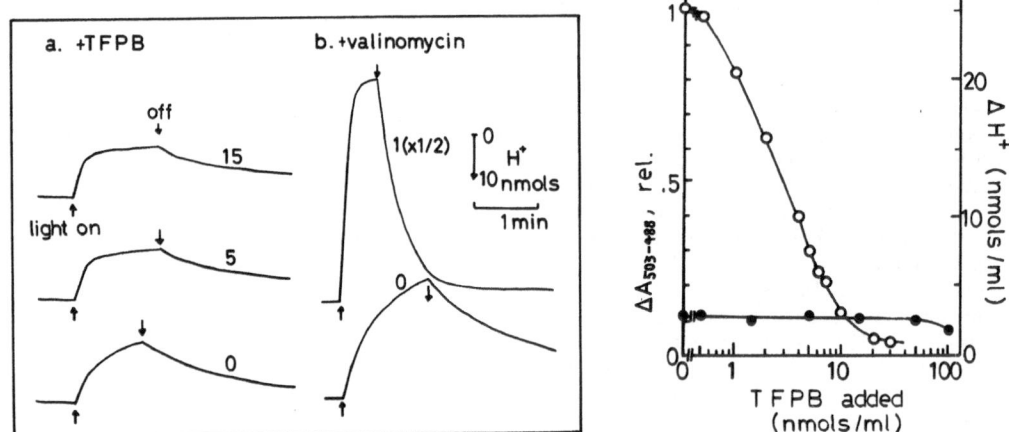

Fig. 9. Effects of TFPB⁻ (a) and valinomycin (b) on the light-induced
proton uptake of chromatophore vesicles. Acidification of the medium
(pH 6.5) containing 0.1 M NaCl (a) or KCl (b) and chromatophores were
measured. Concentrations of TFPB⁻ and valinomycin are shown in the figure
in μM.

Fig. 10. Effects of TFPB⁻ on the extent of light-induced proton uptake
(closed circles) and the carotenoid absorbance change (open circles)
induced by continuous illumination.

flash excitation showing biphasic decay as observed with TPB⁻ (not shown).
The rate was much faster than that observed with TPB⁻. These effects were
quite different from those of valinomycin, which severely depressed the
pseudo-steady state level by accelerating the decay rate with a small
effect on the initial extent induced by each flash.

Fig. 9 shows the effect of TFPB⁻ on the proton uptake into chromatophores induced by a continuous illumination, monitored by the acidification
of the unbuffered suspending medium. Addition of TFPB⁻ did not enhance the
extent of proton uptake even at the concentrations at which carotenoid band
shift was significantly depressed (Figs. 9 and 10). Only the rate of
proton uptake during the first ten seconds became faster by TFPB⁻ (Fig. 9).
This result was a little different from that of TPB⁻, which enhanced proton
uptake at the higher concentrations. The steady-state carotenoid absorb-

ance change observed in the presence of TFPB⁻ does not seem to be directly related to the membrane potential as estimated in the case with TPB⁻, since the depression of the membrane potential by membrane permeable ions usually induces the higher rate of electron transport and proton uptake as seen with valinomycin. The result suggests that TFPB⁻ is a different type of membrane permeable ion from ordinal ionophores. Change of local electrical field inside membrane induced by TFPB⁻ movement may also work to decrease the electron transport rate.

Effect of TFPB⁻ on the Local Intramembrane Field

Binding of TPB⁻ or TFPB⁻ to the membrane itself seems to change the local electrical field inside membrane (11). TFPB⁻ addition induced the blue shift of carotenoid molecules in the isolated light harvesting protein (LHII) complex, which is known to carry the field-responding carotenoid, as well as in chromatophores and intact cells (11). The carotenoid molecule seems to belongs to the same group of molecules which responds to the intramembrane electrical field. Similar shifts of carotenoids induced by the change of local field were also observed in other membranes having field sensitive carotenoid molecules (membranes of chromatophore of R. capsulata and spinach chloroplast) but not in chromatophores of C. vinosum which lack the LHII complex or in isolated light-harvesting protein complex I which lacks field-indicating carotenoids. TPB⁻ and TFPB⁻ seem to induce change of local field nearby field-indicating carotenoid molecules through bindings to the sites in the vicinity of these molecules.

CONCLUSIONS

Conclusions as follows can be made from the results in the present study.

1)   In the energy transducing membranes, extents of surface potential change induced by the energization apparently depend on the ionic condition of the medium. Therefore, the change of electrostatic characteristics of the surface may be compared more adequately when represented by the change of surface charge density value, which does not depend on the ionic conditions.

2)   Surface potential of the outer surface of chromatophore membrane becomes 3.5–11 mV more positive under energized state. This corresponds to the change of net surface charge density from -3.0 to -2.7 $\mu C/cm^2$.

3)   Carotenoid band shift senses the local intramembrane electrical field in the central part of the membrane and cannot directly be related to the membrane potential or to the average field change within the membrane in the presence of hydrophobic membrane permeable ions.

4)   Merocyanine dye senses the field change inside the membrane just beneath the outer surface.

5)   In the chromatophore membranes, no rapid conformational change of proteins or other molecules, which dissipates local field in the central part of the membrane and stores energy, seems to exist under nonphosphorylating conditions. ATP-synthetase may induce local charge movements in the central part of the membrane under phosphorylating conditions.

6)   By the movement of hydrophobic anions inside membrane, field change in the central region of the membrane is rapidly dissipated. This results in the decrease of membrane potential. On the other hand, higher local field can be induced in the boundary region of the membrane when the field strength in the central region is depressed and membrane potential is maintained at a similar extent, in the presence of hydrophobic ions. This type of field change may affect movements of electrons and charges inside the membrane.

7)   The movement of hydrophobic ion inside membrane can be well explained by the three capacitor model which assumes potential wells for hydrophobic

anions inside membrane close to the surfaces.
8)   Binding of the hydrophobic ions into the specific binding site, itself, will change the intramembrane local field and may affect energy coupling processes.

The modulation of the intramembrane field by hydrophobic ions seems to provide a new method for the study of energy coupling between electron transport and ATP-synthesis by providing means of selective modification of the field strength at different depth inside the membrane.

ACKNOWLEGEMENTS

The author thanks Dr. K. Shimada in Department of Biology, Faculty of Science, Tokyo Metropolitan Univ. for providing me LHI and LHII complexes, Drs K. Matsuura and Y. Fujita in the same laboratory, and Dr. K. Masamoto in Faculty of Education, Kumamoto Univ. for their kind discussions. Travel Grant from the Yoshida Science Foundation is also acknowleged.

REFERENCES

1.   K. Matsuura, K. Masamoto, S. Itoh and M. Nishimura, Biochim. Biophys. Acta 547:91 (1980).
2.   S. Itoh, Plant Cell Physiol. 23:595 (1982).
3   S. Itoh, Biochim. Biophys. Acta 591:346 (1980).
4.   K. Matsuura, K-I. Takamiya, S. Itoh and M. Nishimura, J. Biochem. 87:1431 (1980).
5.   K. Masamoto, K. Matsuura, S. Itoh and M. Nishimura, Biochim. Biopys. Acta 638:108 (1981).
6.   S. Itoh, Biochim. Biophys. Acta 593:212 (1980).
7.   S. Itoh and S. Morita, Biochim. Biophys. Acta 682:413 (1982).
8.   O. S. Anderson, S. Feldberg, H. Nakadomari, S. Levy and S. McLaughlin, Biophys. J. 21:35 (1978).
9.   R. Benz and W. Nonner, J. Memb. Biol. 59:127 (1981).
10.   S. Saphon, J. B. Jackson, V. Lerbs and H. T. Witt, Biochim. Biophys. Acta 408:58 (1975).
11.   S. Itoh, Biochim. Biophys. Acta 766:464 (1984).
12.   Y. Kimura, A. Ikegami and W. Stockenius, Photochem. Photobiol. 40:641 (1984).
13.   S. Itoh, Biochim. Biophys. Acta 504:324 (1978).
14.   S. Itoh, Biochim. Biophys. Acta 548:579 (1979).
15.   S. Itoh, N. Tamura, K. Hashimoto and M. Nishimura, in "The Oxygen Evolving System of Photosynthesis" p. 421 (1983).
16.   N. Tamura, S. Itoh and M. Nishimura, Plant Cell Physiol. 25:589 (1984).
17.   K. Masamoto, K. Matsuura, S. Itoh and M. Nishimura, J. Biochem. 89:397 (1981).
18.   J. Barber, Biochim Biophys. Acta 594:253 (1981).
19.   G. F. W. Searle and J. Barber, Biochim. Biophys. Acta 502:309 (1978).
20.   W. S. Chow and J. Barber, Biochim. Biophys. Acta 591:82 (1980).
21.   K. Masamoto, S. Itoh and M. Nishimura, Biochim. Biophys. Acta 591:142 (1980).
22.   K. Matsuura, K. Masamoto, S. Itoh and M. Nishimura, Biochim. Biophys. Acta 592:121 (1980).
23.   K. Hashimoto and H. Rottenberg, Biochemistry 22:5738 (1983).
24.   K. Masamoto, J. Biochem. 95:715 (1984).
25.   K. Masamoto, J. Biochem. 92:365 (1982).
26.   I. M. Møller, T. Lundborg and A.Bérczi, FEBS Lett. 167:282 (1984).
27.   I. M. Møller, W. S. Chow, J. M. Palmer and J. Barber, Biochem, J. 193:37 (1981).

INTERACTIONS OF DIVALENT CATIONS WITH PLANAR LIPID MEMBRANES

CONTAINING PHOSPHATIDYLSERINE

Dido Yova,
Vasso Karnavezou, and
George Boudouris

Technical University, Section of Physics, Zografou Campus
Gr 157-73, Athens

INTRODUCTION

Planar lipid membranes supported by millipore filters, are a model system particularly well suited for the investigation of transport phenomena across biological membranes. This model membrane system, that separates two essentially infinite compartments and allows control of the solution composition, it can be used to assay directly electrical potentials and currents associated with ion transport, since these processes are conveniently monitored electrically with high sensitivity [1,2].

There is evidence that membrane structure and composition influence ion transport through the membrane. The composition of the membrane will be reflected in the adsorption and translocation of an ion and in the membrane conductance [2]. On the other hand Ca^{2+} ions play an essential role in many membrane - related biological functions. Some of the observed modifications of ion permeabilities by Ca^{2+} have been attributed to a change in the surface potential of the membranes, while for others a direct action of the divalent ion on the transport system has been shown. Various hypotheses have been proposed to account for the critical role of calcium and it seems possible that some of the effects arise because calcium can bind to the negatively charged phosphatidylserine [3].

In order to contribute to the understanding of interactions of divalent cations with membranes containing phosphatidylserine we studied the effects of Ca^{2+} and Mg^{2+} on the electrical parameters of planar rat - brain lipids membranes, supported by a millipore filter. It is well known that these membranes contain mainly phosphatidylcholine phosphatidylethanolamine and phosphatidylserine [4]. Varying the ionic concentration of aqueous solutions bathing planar rat - brain lipid membrane, we measured the conductance and electrical potential across the membrane. Conductance measurements on model membranes can determine changes in surface potential and as a consequence changes in electrostatic potentials produced by Ca^{2+} or Mg^{2+}. From our results we can conclude that divalents cations specifically interact with rat - brain membranes, and their affinity follows the series $Ca^{2+} > Mg^{2+}$.

MATETIALS AND METHODS

For the membrane formation Millipore filters Type HAWP - 02500 (porous disks made of cellulose nitrate and acelate, pore size 0,45 nm, 79% porosity) were used. All organic solvents were reagent or analytical grade. Inorganic salts ($CaCl_2$ and $MgCl_2$) were obtained from Merck.

Brains without the cerebellum from 3 - 4 rats were homogenized in 3 volumes of Tris-HCl buffer, PH 7.4, by using an Omnimixer homogenizer. The homogenate was centrifuged at $1,000 \times g$ for 10 min and the supernatant fraction was centrifuged at $10,000 \times g$ for 10 min. The extraction of lipids from the pellets was performed according to well known methods[4,5]. The various lipid classes were separated by two-dimensional thin layer chromatography on silica gel H plates and they were identified using specific sprays. Chromatography of the neutral lipids was carried out by using silica gel G plates[4]. Inorganic phosphate determination was done according to a classical method, for microphosphorus determination[6]. It was found that the main phospholipids of rat-brain, expressed as percentage of whole brain, were: phosphatidylcholine 30.0%, phosphatidylethanolamine 27.5%, and phosphatidylserine 9.0%.

For the formation of the membranes, millipore filters Type HAWP-02500 were used. A cyclic section of the filter, 1cm diameter, was immersed into a solution of rat-brain lipids in decane (40 mg of lipids per milliliter of decane solvent) and was stored in the dark for 24 h, and then the solvent was removed by drying. Upon rehydration a planar membrane of relatively high resistance was formed. The amount of lipid incorporated into the filter was measured by comparing its dry weight before and after immersion in the lipids solution and it was found to be about 1.2 mg/cm^2.

Filter supported membrane was placed at a hole, 1 cm diameter, that separated two compartments of a glass chamber. By changing the solution composition of the two compartment, ion-transport studies were performed. The above mentioned membrane system allows accurate electrical measurements and control of the potential difference across it. One pair of calomel electrodes were used to apply the voltage and monitor the current across the membrane. The voltage source was a 1.35 V mercury battery divided through a 1 Kohm ten-turn potentiometer. Voltage step was applied between one of the electrodes and ground, while the other electrode was connected to the input of a Keithley 616, electrometer. All experiments were performed at room temperature and PH 7.4. Values of the membrane conductance were calculated from the slope of the steady state current-voltage relationship at zero current. To verify that electrode polarization, electrode and electrolyte resistance were negligible, control measurements were carried out in the absence of a membrane.

RESULTS

Rat-brain lipids filter supported membrane was found to have a resistance in the range $2.5 \times 10^7 - 4.5 \times 10^7$ $\Omega \cdot cm^2$. In the absence of membrane, electrode and electrolyte resistance was found to be of the order of $10^3 \Omega \cdot cm^2$.

We studied the time-course of the change in electrical conductivity of the membranes in solutions of CaCl$_2$ or MgCl$_2$ and tris-buffer. Figure 1 shows that the electrical conductivity or rat-brain model membranes increased from 1.5×10^{-8} mho·cm^{-2} to 4.5×10^{-8} mho·cm^{-2} on adding CaCl$_2$ and this rise was observed about 60 min after addition of the cation. In case where MgCl$_2$ was added conductivity increased from 2.10×10^{-8} mho.cm^{-2} to 4.2×10^{-8} mho.cm^{-2}, in the same time range. This rise in electrical conductivity of rat-brain membranes, by adding divalent cations is an evidence for their affinity to the membranes. Apparently the differences between the values of conductivity for the two ions must be attributed to their different affinities, to bind to the membranes.

The dependence of membrane conductance on the concentration of calcium and magnesium is illustrated in Fig.2. We can observe that from 1 mM to 10 mM concentration of divalent cations, the conductance varied little, from 10 mM to 0.5 M, the conductance presented a steep rise and for greater concentrations increased slowly, until it reached a plateau.

An increase in the concentration of divalent cations from 5 mM to 0.2 M, induced a rise in membrane's conductivity from 1.7×10^{-8} mho.cm^{-2} to 4.0×10^{-8} mho.cm^{-2} for calcium and from 1.0×10^{-8} mho.cm^{-2} to 3.7×10^{-8} mho.cm^{-2} for magnesium. The absolute value of the observed increase in

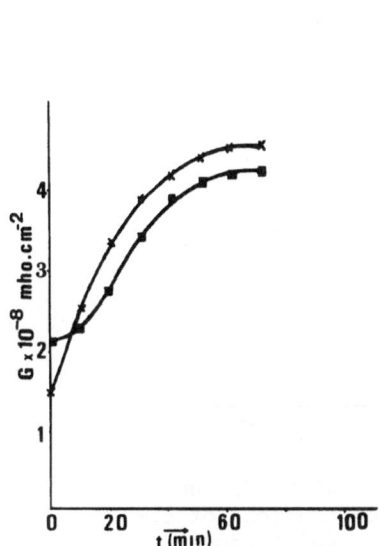

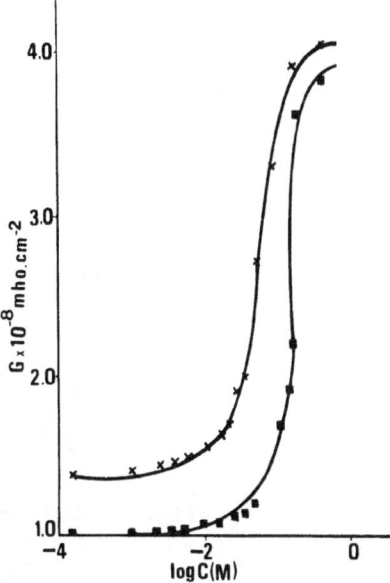

Fig.1. Time-course of change in the electrical conductivity of rat-brain model membranes. Aqueous medium 50 mM Tris-HCl, PH 7.4. Addition: 2.5mM CaCl$_2$ (crosses) or MgCl$_2$ (squares) at zero time.

Fig.2. Conductivity of rat-brain model membranes as a function of divalent cations concentration: Ca^{2+} (crosses) or Mg^{2+} (squares).

membranes electrical conductivity for Mg^{2+} was less than that for Ca^{2+}, so the sequence of the action of divalent cations on membrane conductance was Ca^{2+} > Mg^{2+}.

To evaluate further the selectivity of the ion transport system we measured the transmembrane potential difference appearing in the membrane for a concentration gradient of divalent cations, according to the well known Nerst equation:

$$V = \frac{RT}{ZF} \ln \frac{\alpha_1}{\alpha_2} = \frac{RT}{ZF} \ln \frac{C_1}{C_2}$$

in case of dilute solutions and the same ionic strength on both sides of the membrane. Fig.3 shows the potential difference across rat-brain model membrane, for a gradient concentration of divalent cations in 1mM - 1M range (1mM solution was used as reference). Solutions were continuously stirred by teflon coated bar magnets. Stirring of the aqueous solutions didn't seem to affect the value of membrane conductance. From the slope of the two lines, 19mV/decade in case of CaCl$_2$ and 16mV/decade in case of MgCl$_2$ we can see that we have subnerstian cationic activity following the series Ca^{2+} > Mg^{2+}.

The behavior of rat-brain lipids model membranes by adding divalent cations, as a function of the applied potential was studied by performing current-voltage experiments. As reported in Fig.4 the curve is about symmetric for Mg^{2+} ions, having a conductivity of 2.5×10^{-8} mho.cm^{-2}. For calcium ions it presents a different conductivity for positive and negative voltage values i.e. 4.7 10^{-8} mho-cm^{-2} and 5.4 10^{-8} mho-cm^{-2} correspondingly and a slight deviation from linearity at higher potentials.

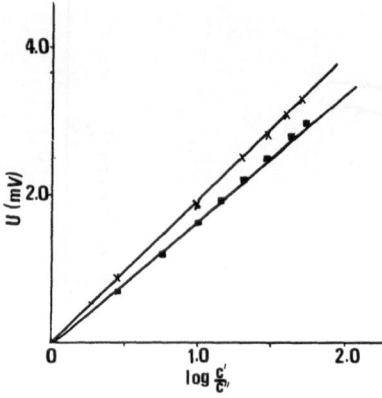

Fig.3.   Membrane potential V(mV)   under
         a concentration gradient of di-
         valent cations: Ca^{2+}(crosses)
         or Mg^{2+} (squares).

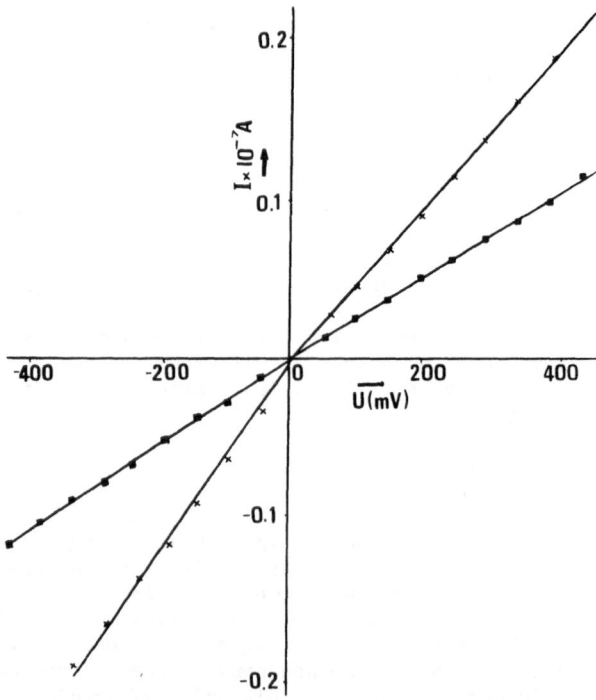

Fig.4.   Current-voltage relationship for rat-brain model mem-
         branes in 50mM Tris-HCl, PH 7.4, with 2.5mM of either
         salt: $CaCl_2$ (crosses) or $MgCl_2$ (squares).

The interaction of divalent cations Ca^{2+} and Mg^{2+} with membranes is of interest for several reasons. Ca^{2+} and other related divalent cations play an essential role in many membrane-related biological functions like nerve excitation or cell-cell communication. They alter surface potentials near voltage gated channels and it seems possible that some of their effects arise because they can bind to negative phosphatidylserine [3,7].

In our work we studied the effects of divalent cations on the electrical parameters of rat-brain model membranes, which they contain the negatively charged phosphatidylserine. The fixed negative charges on these membranes produce a negative electrostatic potential in the aqueous diffuse double layer adjacent to the surface and according to the Boltzmann relation, the concentration of a divalent cation in the aqueous phase at the surface of the membrane will be larger that the concentration in the bulk aqueous phase. Cations and membrane association was shown by the dependence of conductance on cationic concentration. Semilogarithmic coordinates were used to trace the theoretical ionic association curve $(G = f(\log C))$ and it was found that this curve corresponds to the experimental curves which have been obtained (Fig.2). An increase in the concentration of divalent cations should reduce the surface potential and should therefore increase the membrane conductance.

The rat-brain membrane potential U(mV), of model membranes formed in $CaCl_2$ or $MgCl_2$ in relation with a concentration gradient of either salt showed a typical sub-nerstian cationic activity following the series $Ca^{2+} > Mg^{2+}$, and this effect was a further evidence of specific interactions of divalent cations with the membranes [9].

As far as the current-voltage characteristics were concerned, in case of Ca^{2+} ion the curve presented an asymmetry to the positive and negative voltage values, and this result may present evidence for a membrane asymmetry, but this result must be further investigated by using alamethicin as a probe [7].

Our results may be summarized as follows:
- The electrical conductivity of rat-brain model membranes, increases with the addition of Mg^{2+} and to a greater extend of Ca^{2+}, due to the affinity of the cations to the membranes.
- The conductance depends on the cations concentration.
- Rat-brain membranes show a subnerstian cationic response, according to the series $Ca^{2+} > Mg^{2+}$.
- The I-V curve presents a symmetric behavior for Mg^{2+}, while in the case of Ca^{2+} it presents different slopes for negative and positive voltage values.

From our results we can conclude that cations interact specifically with rat-brain membranes.

REFERENCES

1. Shieh,P.K., Lanyi, J., and Packer, L. (1979), Methods Enzymol. 55, 604-613.
2. Benz, R. and Gisin, B.F. (1978), J.Membrane Biol. 40, 293-314.
3. McLaughlin, S., Mulrine, N., Gresalfi, T., Vaio G., and McLaughlin, A. (1981), J. Gen. Physiol. 77, 445-473.
4. Cotman, C., Blank, M.L., Moelh, A., and Snyder, F. (1969), Biochemistry 11, 4606-4612.
5. Breckenridge, W.C., Combos, G., and Morgan, I.G. (1971), Biochim. Biophys. Acta 266, 695-707.
6. Bartlett, G. (1975), J. Biol. Chem. 235, 466.
7. Hall, J.E., and Cahalan, M.D. (1982), J. Gen. Physiol. 79, 387-409.
8. McLaughlin, S. (1977), Curr. Top. Membr. Transp. 9, 71-144.
9. Hauser, H., Phillips, M.C., Levine, A., and Williams, R.J.P. (1975), Eur. J. Biochem., 133.

KINETIC ANALYSIS OF TIME RESOLVED PROTON

DIFFUSION ON PHOSPHOLIPID MEMBRANE

Menachem Gutman, Ester Nachliel and Michael Fishman

Biochemistry Department
Tel Aviv University
Ramat Aviv, Tel Aviv, Israel

INTRODUCTION

The transfer of protons between protogenic sites on a biomembrane is, biophysically, an extremely complex reaction. The complete event is a result of many intermediate steps where proton, discharged by one enzyme, collides with the membrane, binds to mobile buffer or phospholipid head group, diffuses as a free proton hydrate or "piggyback" carried on a protonated buffer molecule.

This complex biochemical event is still a combination of discrete chemical reactions in a defined space. The understanding of these elementary steps can lead to reconstruction of the whole process. In this manuscript we shall adopt this approach for studying and anlysing the characterisation of proton transfer at the interface of phospholipid membranes.

The dynamics of proton flux following a protogenic event on a bioenergetic membrane are influenced by the composition and structure of the membrane and the buffering molecules in the bathing solution (substrates and products present in the organelle).

Because of this complexity notions of local confinement of protons, in space and time, have been proposed. The investigation of all "special cases" which may lead to such confinement is a futile effort. Instead we shall analyse what is the most probable trajectory of a proton discharged by a protogenic event using the diffusion controlled rate constants of proton transfer at interfaces which we were measuring in the past few years (1,2,3,4). With these building blocks we shall reconstruct the time course of a proton diffusion in a system consisting of phospholipid membrane and demonstrate the role of the mobile buffer in the bathing solution.

The methodology of studying diffusion controlled proton transfer is the Laser Induced Proton Pulse supplemented by numerical solution of the differential equations pertinent for the chemical reactions under study. This subject has been recently reviewed (1,2) and will not be discussed here. We shall use the rates and equations applicable for the case and limit ourself in presenting the outcome of these computations. This approach facilitates the elucidation of the subject pertinent to bioenergetics - how fast will a proton generated on a surface equilibrate its electrochemical potential with that of the bulk.

Three subjects will be presented:

1.   The effect of buffer on proton transfer.

2.   The effect of surface bound buffer moities.

3.   The effect of the impenetrable surface of the membrane on diffusion at its immediate vicinity.

:   The combination of these three elements will be summarized to draw general conclusions on the characteristics of proton transfer in bioenergetic systems.

I.   The effect of mobile buffer molecules on the dynamics of acid base equilibrium in perturbed systems

All protogenic events consists of a perturbation of acid-base equilibrium. Some of the energy in the system is shifted from one form (redox, chemical potential, solute gradient) to produce (consume) protons in excess of those determined by the existing equilibrium. The relaxation of the system after the perturbation proceeds by proton transfer events which follow two parallel pathways; proton dissociation followed by diffusion and collisional proton transfer between reactants. The most simple system (scheme I) consists of only two reactants (besides the proton); a donor DH and acceptor A.

Scheme I

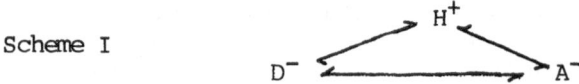

The collisional proton transfer is given by the base of the triangle. This scheme is too simple for biochemical applications, it ignores the multitude of mobile buffer moities present in any enzymic reaction. To include the buffering capacity of substrates, products and buffers we expand our model to scheme II. The collisional proton transfer reactions are given by the three sides of the external triangle.

Scheme II

Systems pertinent to this model have been studied experimentally by the Laser Induced Proton Pulse. The rate constants have been measured for many donors, acceptors and buffering molecules. The measurements were carried out either in ideal, dilute solutions (3) or where one or two of the reactants are adsorbed on micelles or phosphatidyl choline vesicles (4,5). These rate constants were employed to generate the tracings in Figure 1.

The system described in this figure consists of the proton donor (2 naphtol 3,6 disulfonate pK = 9.2, 2mM) and the proton detector (bromocresol green, pK = 5.0, 2mM) at pH=7. At time zero 8µM of the donor molecules (ROH) were dissociated (in experimental systems it is achieved by a short (10 ns) laser pulse).

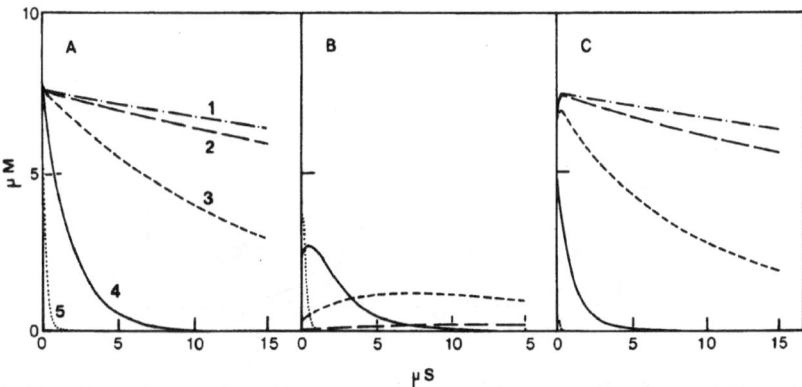

Figure 1:
        Computer simulation of real time dynamics of proton transfer in buffered solution. The calculations, based on experimentally measured rate constants, depict the event in solution containing 2mM of the proton emitter 2 naphthal 3,6 disulfonate; 2mM of the pH indicator, bromocresol green, and varying imidazol concentration, all at pH=7. At time zero 8uM of the proton emitter were dissociated, representing the effect of the laser pulse which discharges the proton from the emitter. The incremental concentration of the dissociated species is shown in Frame A. The discharged protons interact with the Imidazol (Frame B) or the indicator (Frame C). Both compounds demonstrate a phase of protonation, which can be very fast, followed by relaxation to the prepulse level. The lines represent the dynamics with increasing Imidazol concentration. 1(·—·—·) no imidazol; 2(——) 10 µM buffer; 3(---) 100µM buffer; 4(—) 1mM buffer; 5(·········) 10mM buffer. Note with 10mM buffer all transients shrink to submicrosecond event.

        The resulting perturbation of equilibrium consists of three elements; H^+ and RO^- concentration are above equilibrium, and within a few nanoseconds, the protonated indicator concentration increases too. In absence of buffer the relaxation of the system is rather slow (line 1). This is the relaxation of a system described in scheme I. Addition of a minute amount of imidazol (10 µM, 0.5% with respect to the indicator concentration) already change the regime of the system (scheme II) and accelerates the relaxations (line 2). At higher imidazol concentrations the relaxation becomes faster and faster. 10 mM buffers already make the event so fast that equilibrium is reached within 1us.

        Figure 2 represents another aspect of the buffered system - its initial state of protonation. At pH = 6, 90% of the imidazol buffer is H^+Imidazol, which has a profound effect on the whole system. The RO^- formed by the perturbation is rapidly protonated by H^+ Imidazol with subsequent rapid relaxation of RO^- and transient decrease of [H^+ Imidazol]. As the strongest base (RO^-) is rapidly protonated the free proton concentration has a longer lifetime, and more indicator is protonated.

        At pH = 8 where the imidazol is 90% deprotonated the life time of RO^- is prolonged while the indicator dynamics shrinks to a minor spike.

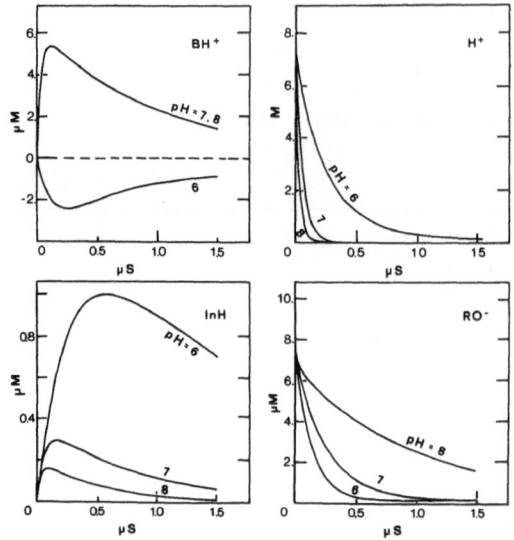

Figure 2

Simulation of the dynamics of all reactants participating in an acid-base equilibrium of a buffered solution. The dynamics were computed for solution containing 1mM of the proton emitter 2 naphthol 3,6 disulfonate, 20µM bromocresol green and 1mM Imidazol. The tracings represent the dynamics of the buffer (BH$^+$); Free protons (H$^+$); Indicator (HIn) and the proton emitter (RO$^-$). The pH of the system is indicated on the pertinent lines.

The dynamics studies as presented above were carried out with micellar bound indicator. As seen in Figure 3 imidazol accelerates the rate of deprotonation of the bound indicator and lowers the increment of protonated indicator. The line going through the experimental points is a theoretical curve calculated from the rate constants described in scheme II pertinent to the reactants present in the solution.

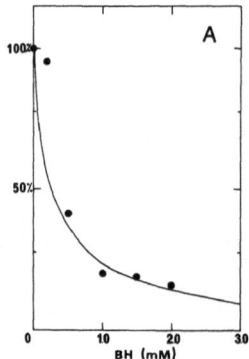 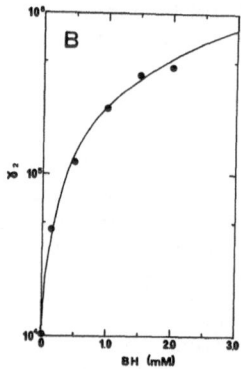

Figure 3

The effect of Imidazol on the kinetic parameters of protonation of a surface group on high molecular weight body. Bromocresol green was adsorbed on uncharged micelles (500 µM) of Brij 58, and the bulk was pulse protonated by discharging protons from unadsorbed proton emitter 2 naphthol disulfonate; (1mM, all at pH 7.5). The maximal amplitude of the indicator concentration decreases with increasing Imidazol (Frame A) while the rate of the relaxation to the equilibrium state is accelerated a hundred fold by 3mM of Imidazol (Frame B). The lines drawn through the experimental points are the theoretically predicted curves, based on experimentally measured rate constants.

II. The effect of immobile buffer moity

The effect of phospholipids head group on proton dynamics is shown in Figure 4. The upper trace documents the transient protonation of Bromocresol green adsorbed on neutral micelle. The lower trace, measured under identical conditions, depicts the dynamics when the micelle was spiked by six molecules of phosphatidylcholine. Extensive experimental measurements were executed to quantitate the effect of phosphatidyl choline, phosphatidyl serine and phosphatidic acid. The dynamics were analysed and the rate constants were calculated. Figure 5 documents the accuracy of this analysis, the extent of the measured signal (5A) and its relaxation rate (5B) are drawn as function of the phosphatidyl serine content of the mixed micelles. The line drawn through the experimental values is the theoretically predicted function.

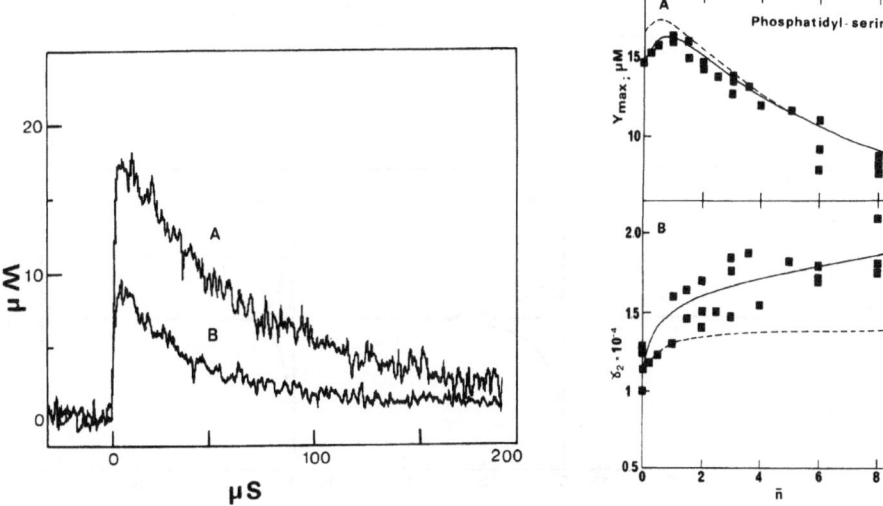

Figure 4 (Left)
The effect of immobile surface buffer on the dynamics of protonation of a macromolecular structure. The tracings represent time resolved dynamics of protonation of bromocresol green on Brij 58 micelles (A) or on a mixed micelle of Brij 58 plus 6 molecules of phosphatidyl choline (B). Note the amplitude in B is smaller and relaxes faster than in A.

Figure 5 (Right)
The effect of varying surface buffer capacity on the dynamics of protonation of macromolecular body. The experiment as described in Figure 4 was repeated with micelles containing varying phosphatidyl serine concentration. The maximal protonation of the indicator Ymax (upper frame) and rate of relaxation γ_2 (lower frame) are drawn as a function of the average number of phospholipids in the micelle. The continuous line drawn through the experimental points is the theoretically predicted curve, based on the kinetic parameters of the reactant with pk = 4.7 for the phosphatidyl serine carboxy group. The dashed line was computed for the same system but with pK = 4.85. Note how small a change affects the shape of the curve.

To ascertain that the rate constants are intrinsic properties of the reactants, not modulated by the micelle, we repeated the measurements using phosphatidyl-choline small unilamelar vesicles. The rate constants measured in the two systems were identical.

The effect of the immobile buffer - calculated for a surface made of phospholipid head groups is documented in Figures 6A and 7A. In these presentations we consider a rather simple case, the transient deprotonation of a donor attached to a surface made of phospholipids. The pK of the donor was set at pK = 8 and at zero time a small fraction (8 μM) was synchronously dissociated.

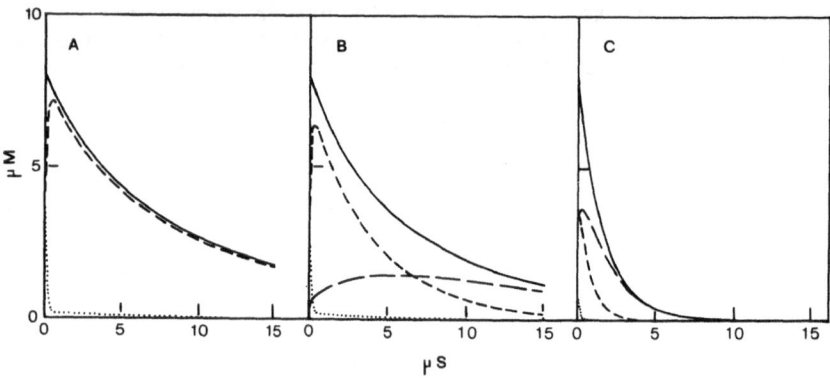

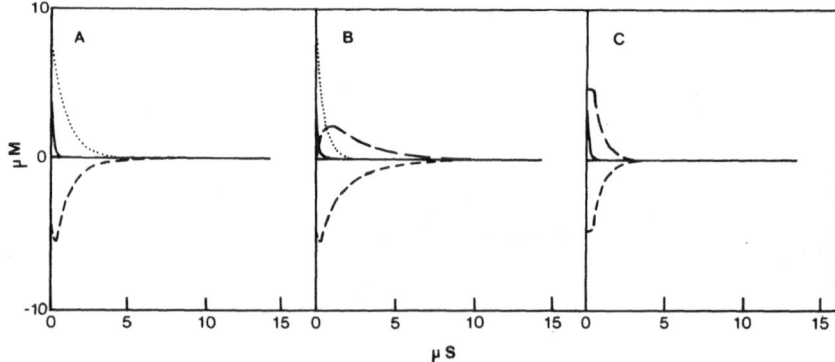

Figure 6

The effect of mobile buffer on the dynamics of surface-bulk proton transfer. The system described in this figure represents a phospholipid surface with the kinetic parameters of phosphatidyl serine (total buffer capacity 2mM). At time zero 8μM of a proton emitter was dissociated producing the RO⁻ species (——), free proton (······) and protonated phospholipids PSH (— —). In Frame A the mobile buffer (Imidazol) is zero. Frames B and C depict the events in the presence of 100 μM and 1mM Imidazol. The dynamics of protonation of Imidazol (Frames B and C) are given by (—— ——).

Figure 7

The effect of mobile buffer on the dynamics of surface-bulk proton transfer.

The figure represents the same events as in Figure 6 but the phospholipid surface has the kinetic parameters of phoaphatidic acid. A: No Imidazol; B: 100 μM Imidazol; C: 1 mM Imidazol. The marking of the tracings is as in Figure 6.

Figure 6 depicts the event taking place on a surface with rate constants and equilibrium parameters of phosphatidyl serine.At pH = 7 the carboxyl moities act as efficient proton acceptors thus only a few protons are effectively discharged, most of them are immediately (within less than 50 ns) trapped by the surface buffer. Figure 7A depicts the event under identical conditions, except that the surface buffer properties are those of phosphatidic acid. At pH 7 the phosphohead group is in its PAH^- state and acts as a proton donor with respect to RO^- generated by the perturbation. As a result RO^- is rapidly reprotonated and the lifetime of the discharged protons is 10 times longer.

The effect of imidazol added to the system described in Figure 6A is documented in frames B and C. Even small imidazol concentrations (10% of the phosphatidyl serine content) already affects the dynamics. In the absence of imidazol the acid base disequilibrium is practically limited to the surface moities - represented by the increments of RO^- and PSH. The amount of protons discharged to the bulk is small. 100 μM of imidazol changes the situation, and the disequilibrium is propagated to the bulk. After 5us nearly all RO^- moities are balanced by H^+ imidazol in the bulk. Higher imidazol concentrations accelerate the relaxations and the state of equilibrium is reestablished within 3 μs.

The effect of buffer on the dynamics of a surface with pK⟩pH is shown in Figure 7. 100 μM imidazol accelerate the relaxation of the free proton - mostly by trapping them forming $H^+·$ imidazol. The dynamics of RO^- and $PA^=$ are in essence unaffected. Higher concentrations shrink the perturbation to 1 us during which the main disequilibrium is the excessive protonation of the bulk ($H^+·$ imidazol) and deprotonation of the surface groups.

Comparison of Figures 6B and 7B demonstrate the other aspect of the calculations. Both figures represent dynamics identical in their initial state and reactants concentrations. Yet their time evolution is markedly different. That representing an acidic surface is characterized by a long surface (RO^-) bulk ($H^+·$ imidazol) disequilibrium. On a basic suface the relaxation of the same species is much faster.

Our studies indicate that the distribution of protons between bulk and surface are specific for the composition of the two phases. The rate constants of all relaxations are fast and we can definitely state that under physiological conditions equilibration will take less than microsecond, a time frame much shorter than the turnover of protogenic enzymes.

III. The effect of surface

The buffer concentration in most biochemical assays ranges between 10-100 mM. In the presence of such concentrations the state of disequilibrium resulting from a protogenic event has a time constant of a few tens of nanoseconds. Indeed, within such a time frame we have a space (reaction sphere) where the state of protonation differs from bulk phase, but its radius, is less than 200A. Considering the size of membranal proteins and the spacing between them it will be an extremely rare event that a proton discharged by one enzyme will react with another before it equilibrates with the bulk phase.

These calculations indicate that the first few nanoseconds will be crucial for the trajectory of a proton discharged on a surface - will it diffuse to the bulk or remain at the interface and spread by a two dimensional diffusion towards another enzyme (6). These are purely geometrical considerations and were accordingly investigated by another analytical approach.

On studying the effect of surface we were looking for a system where the probability of collision, between proton and surface target, will be determined only by geometrical constraint and not the chemical-physical properties of the specific surface.

For this purpose we envisioned a sphere with an impenetrable surface and generated a mobile point like particle at a certain place out of the space defined by the sphere. This mobile element moves a single step (l=1) in a random direction. This sequence is repeated as many times as we wish with only two exceptions. If the particle hits the target (placed at a defined coordinate) the run is terminated. If a given step carries the particle across the sphere surface, the step is aborted and another random direction is selected. The procedure is repeated for a population of many particle and the mean position after each step is calculated and recorded. The results of such computer experiments are given below.

1.   The mean position of a particle with respect to surface

A particle placed at a large distance from a surface will select at random any direction. On the other hand that which is in contact with the surface can take a step only away from the surface. Thus if we follow a particle for many steps we shall observe that its mean position with respect to the surface will be a function of time. If within a given period of time this particle had a single (or many) collisions with the surface, bias on its direction selection will shift its mean position away from the surface. The closer is the initial position to the surface the higher is the probability of reaching the surface and the stronger is the bias. This relationship is demonstrated in Figure 8.

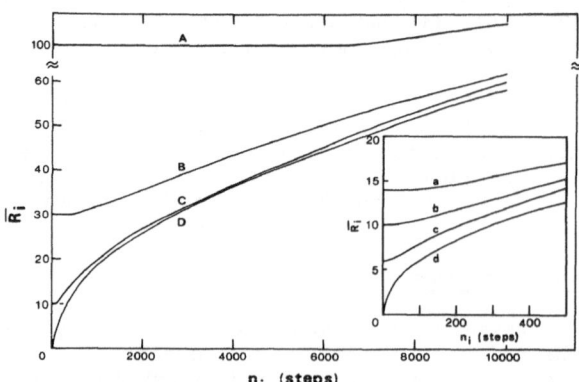

Figure 8
The motion of a particle near an impenetrable surface: The average distance to surface as function of time. A population of non-interacting particles was generated at a given distance from a flat surface (Ri at ni=o) and was allowed to diffuse at random. For each diffusion step the average distance was calculated and drawn as a function of the number of steps (ni). Line A initial distance to surface is 100 step units; B initial distance 30; C 10; D the particles were generated on surface. Insert:blow up of the motion during the first 500 steps the initial distance to surface is a-13; b-10, c-6, d-particles generated on the surface.

100

We depicted the distance to the surface of the average particle (calculated for a population of 1000) as it varies with time (number of steps). The various line represents particle produced at different initial distance from the surface. A common property of all these functions is the "floating" above the surface. Particles in contact with the surface will start floating on the first step and will slow this motion as the separation increases. Particles which are at a given distance will delay their floating for a certain time (number of steps). This delay is a direct consequence of the mechanism underlining the process. The particle must approach by random walk the surface before it feels its bias.

Based on these calculations we conclude that the most probable trajectory of a particle near a surface will be tangential to it.

The encounter of a particle with a given target

The probability of such collision is a function of time and relative position. At infinite time a collision with target will take place whatever is the distance between the two, but if the initial distance is small collision may occure at a shorter time. This can be expressed by the probability that encounter will happen within a given number of steps. Typical probability functions are shown in Figure 9. The value $P_{(H)}$ ni=1000 can be taken as a good measure for collision of particles with a target.

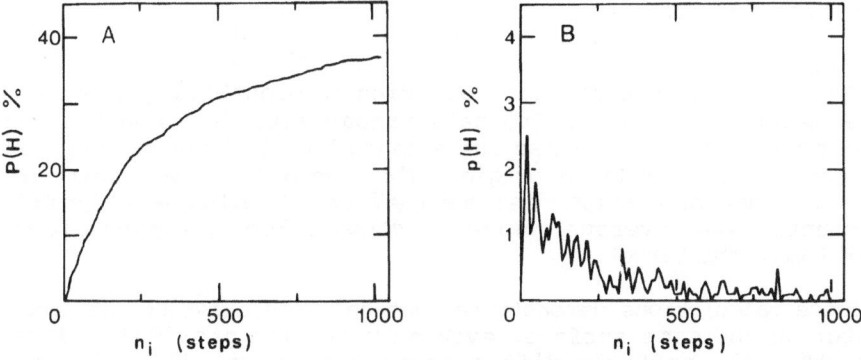

Figure 9
The probability function describing the collision of a mobile particle with a target on a surface. A) Ordinate: the probability of encounter within ni steps. Abcissa, number of steps which is equivalent to the time vector. B) Ordinate, the probability that the encounter will take place on the ni[th] step. The curves were computed for a particle generated on a flat surface, 5 step length from the target. Note at ni>1000 p(H)→0.

The effect of the initial position on the probability of collision is shown in Figure 10. As expected the probability decreases with increasing distance — but surprisingly the relative position has also a major effect. At identical separation a particle in bulk will have a higher chance of colliding with a target than a particle generated on the same surface. This conclusion is further supported by results presented in Figure 11, where the asimuthal angle is the independent variable. We drew in this figure two functions. The probability of hitting the target (solid line) increases with the angle between surface and point of origins while the mean number of steps needed for a collision (-----) decreases.

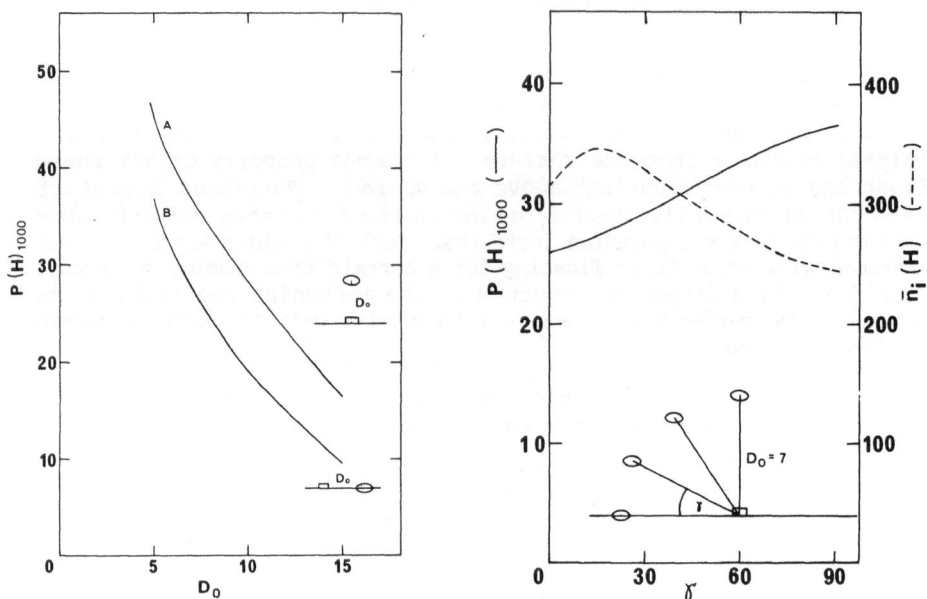

Figure 10
The effect of the relative position of a particle on the probability of encounter with a surface target. Ordinate probability of collision within 1000 diffusive steps, abcissa the initial separation distance between particle and target. Line A) The particles are generated in the bulk, above the target. Line B) The particles are generated on the surface.

Figure 11
The effect on the relative position on probability of encounter between particle and target. Ordinate probability of encounter within 1000 diffusive steps, abcissa the azimutal angle between the surface plane and point of particle origin. The computations were carried out for a constant separation distance $D_0=7$ step length. (———) probability of encounter. (----) average number of steps taken by a particle which collided with the target.

These calculations demonstrate that the mere existence of donor and acceptor on the same surface, even at very high proximity, does not ensure that the particle diffusing between them will stick to the surface. On the contrary the most probable trajectory will be to diffuse first to the bulk and then to collide with the target.

Concluding Remarks

Both kinetic measurements and theoretical consideration clearly imply that under physiologic conditions a proton dissociated from a photogenic enzyme will assume the electrochemical potential of the bulk phase within a sub microsecond time frame. Thus there is no physical evidence for the presumptions that protons may be confined to the membrane surface for a period comparable to enzyme turnover.

If one is ready to accept these consequences of the local chemiosamotic model, he will have to substitute the "local proton" by another force or mechanism functioning in parallel with the classical proton motive force (7).

On the other hand one should also question the applicability of the simple steady state formalism used for analysing bioenergetic systems. The general presentation of the couple fluxes uses the analogy to DC electrical currents. Yet Kirchoof rules governing such circuits are not applicable for short electric pulses. In the same way the discharge of a proton at the membrane interface and its rapid equilibration consist of a pulse and not a DC current. Thus one should consider whether the flux analysis currently employed should not be replaced by more precise formalism founded on relaxation dynamics.

References

1. M. Gutman (1984). Methods in Biochemical Analysis. $\underline{30}$. 1-103.
2. M. Gutman (1985). Methods in enzymology. Biomembranes: protons and water (L. Parker, ed.) in press.
3. Gutman, M., Nachliel, E. and Gershon, E. (1985) Biochemistry $\underline{24}$, No. 12.
4. Gutman, M. and Nachliel, E. (1985) Biochemistry $\underline{24}$, No. 12.
5. E. Nachliel and M. Gutman (1984). Eur. J. Biochem. $\underline{143}$, 83-88.
6. O.G. Berg and P.H. Von Hippel (1985). Ann. Rev. Biophys. Chem. $\underline{14}$, 131-160.
7. H. Rottenberg. This volume.

CONTROL OF MITOCHONDRIAL RESPIRATION BY THE PROTONMOTIVE FORCE

K. van Dam, H. Woelders and A.M.A.F. Colen

Laboratory of Biochemistry, University of Amsterdam
Plantage Muidergracht 12, 1018 TV Amsterdam
The Netherlands

INTRODUCTION

Since the development of the chemiosmotic hypothesis by Mitchell (1) a central role in energy conservation has been attributed to protons (2). The experiments, designed to verify this role, have been very successful in demonstrating that movement of protons generally accompanies the process of energy transduction in mitochondria, chloroplasts and bacteria. However, up to this moment, there are a number of observations that are difficult to reconcile with the original scheme proposed by Mitchell. A discussion of these anomalies can be found in a review by Westerhoff et al. (3).

To resolve the possible modifications of the chemiosmotic hypothesis that are able to explain the different anomalies, a quantitative description of energy transduction is needed. We developed such a description and called it mosaic non-equilibrium thermodynamics (4). It allows us to make quantitative predictions about the relation between rates of processes and the free energy differences that drive these processes. The predictions are related to the underlying mechanisms and, therefore, certain mechanisms can be falsified by experiment.

Here we report the predicted relations between rates of oxidation, rate of phosphorylation and $\Delta\tilde{\mu}_H$ under different conditions, especially for the case of mitochondria. The predictions are compared to experimental results.

THE MODEL

The model that we use is based on the original hypothesis for chemiosmotic energy transduction in mitochondria. Additionally, it is proposed that there may be a certain barrier for the protons to leave the energized complexes into the bulk aqueous phase and that there is a finite chance that the energized complexes react directly. Thus, there is an apparent intramembrane pathway for the protons (cf. 5-9). It must be stressed that the model as proposed retains all the elements of the classical chemiosmotic concept and leaves it to the comparison with experimental results to see in how far extra barriers for the protons are significant. If these resistances turn out to be negligible, the original scheme can be retained. For simplicity of discussion, the model can be represented in the form of a analogous electrical circuit (Fig.1). The 'batteries' are the respiratory chain and the ATP synthase, each with its own characteristic resistance that depends (among others) on the number of active complexes. The resistance for protons to escape into the bulk aqueous phase is given by R_e, the resistance for transmembrane

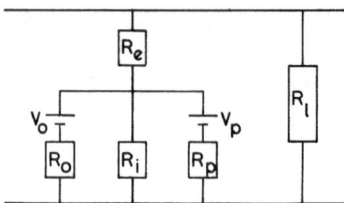

Fig.1. The electric analogon of the chemiosmotic model for energy transduction, including a resistance for protons to escape from the membrane space.

proton movement by R_1 and the resistance for backflow of protons within the membrane by R_i.

Although the description developed here is based on the proportional laws that rule in electricity, it is not too difficult to transform the equations into the (more realistic) equations based on flow-force relations of enzyme kinetics. The general conclusions remain valid in that case.

RESULTS

i) State 4

In State 4 the movement of protons through the phosphorylating branch of the circuit has come to a halt. The net current through that branch will be zero and the following relation will hold (the calculations were carried out as described in the appendix to the paper by Westerhoff et al., ref.3):

$$\Delta G_p / \Delta \tilde{\mu}_H = n_p \cdot (1 + R_e/R_1)$$

with n_p representing the number of protons pumped per ATP hydrolyzed. Formerly, it was concluded on the basis of this equation (then derived in a slightly different way, ref.4) that the original chemiosmotic hypothesis could not be correct, since experimental results showed that the ratio $\Delta G_p / \Delta \tilde{\mu}_H$ did vary if $\Delta \tilde{\mu}_H$ was varied by the addition of protonophore, i.e. by decreasing R_1. However, more recently we found that the earlier results may have been obscured by some artefacts (10). In Fig.2

106

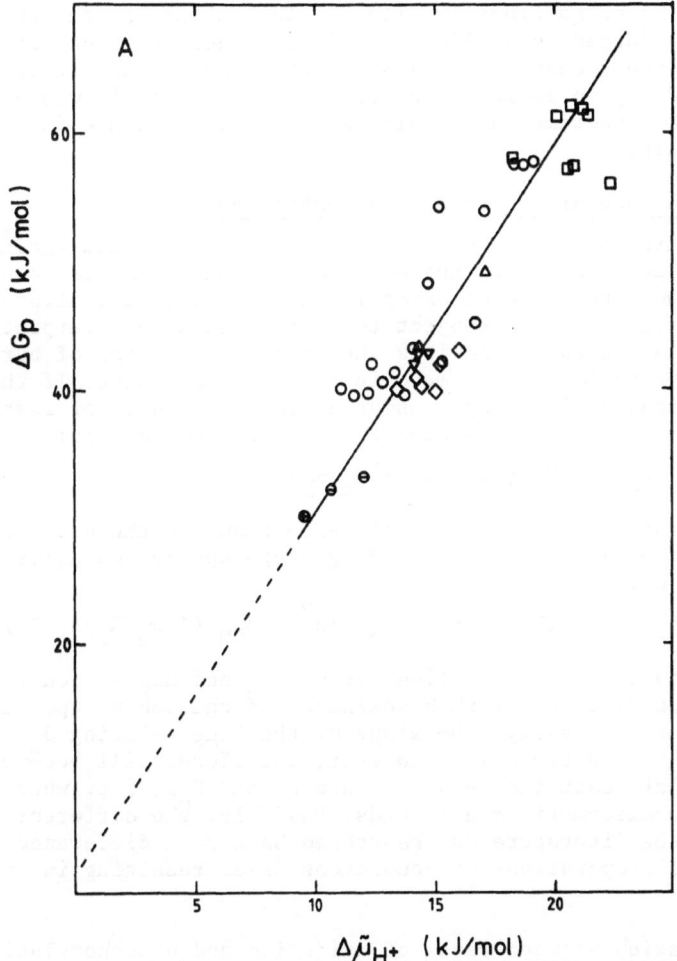

Fig.2. The relation between ΔG_p and $\Delta\tilde{\mu}_H$ in mitochondria in State 4. The magnitude of $\Delta\tilde{\mu}_H$ was varied by adding uncouplers or inhibitors of respiration. For experimental details see ref.10.

we illustrate the results of experiments in which these artefacts were avoided as much as possible. It is clear that, within the limits of our experimental observation, there exists a linear and proportional relation between ΔG_p and $\Delta\tilde{\mu}_H$. Therefore, we must conclude that in the above equation the value of R_e^H/R_1 must be constant. There are two possible ways in which this could be the case: either R_e is much smaller than R_1 or R_e varies proportionally with R_1. The latter of the two possibilities is considered to be less likely: although it is conceivable that an uncoupler would change R_e and R_1 by the same factor, it is not easy to understand why an inhibitor of the respiratory chain would have such an effect. The experimental results with uncouplers and inhibitors fall on the same line.

ii) The dependence of the rate of oxidation on $\Delta\tilde{\mu}_H$

In the literature, one of the experimental arguments against the chemiosmotic model has been the observation that there is no unique relation between the rate of oxidation and $\Delta\tilde{\mu}_H$ (11,12). Although this observation itself may be subject to difficulties in interpretation, we derive an equation that describes the relation in terms of our present model. The derivation can be made in two different ways. If the variation in respiration is brought about by changing ΔG_o, for instance by adding hexokinase, we may eliminate this force and obtain:

$$J_o = \Delta G_o/n_o^2 . R_o - \Delta\tilde{\mu}_H (1 + R_e/R_1)/n_o R_o$$

On the other hand, if we perform the experiment in the absence of phosphorylation, and vary R_1 by adding protonophore, we eliminate this constant and obtain:

$$J_o = \Delta G_o(R_e/R_i - 1)/(R_e/R_i + R_e/R_o - 1)n_o^2 - \Delta\tilde{\mu}_H/(1 - R_e/R_i - R_e/R_o)n_o R_o$$

Both equations give us a relation between J_o and $\Delta\tilde{\mu}_H$ at constant ΔG_o, the first upon titration with hexokinase and the second upon titration with uncoupler. Evidently, the slope of the line relating J_o with $\Delta\tilde{\mu}_H$ is different in the two cases. However, the slopes will become equal if $R_e = 0$ (or at least much smaller than R_i and R_o), i.e. when the pure delocalised chemiosmotic model holds. Possibly, the different findings reported in the literature can be traced back to a difference in the mitochondrial preparations or conditions used, resulting in differences in R_e.

iii) The relation between rates of oxidation and phosphorylation

In the steady state of oxidative phosphorylation, when the net flux of protons across the membrane has become zero, different relations between the rate of oxidation and the rate of phosphorylation can be derived. The usefulness of each of these relations depends on the experimental variables. If one eliminates $\Delta\tilde{\mu}_H$ and R_p, the following equation is obtained:

$$-J_p = n_o/n_p . (1 + [R_o/R_i + R_o/(R_e + R_1)])J_o - \Delta G_o(1/R_i + 1/[R_e + R_1])/n_o n_p$$

At constant ΔG_o there will be a linear relation between the two rates. The slope of the line, relating the two parameters will be steeper if the membrane is made more permeable to protons, i.e. by decreasing R_1. This prediction is confirmed by experiment (13). Another relation between rates of oxidation and phosphorylation can be derived by eliminating $\Delta\tilde{\mu}_H$ and R_o:

$$-J_p = n_o/n_p . (1 + [R_p/R_i + R_p/(R_e + R_1)])^{-1} J_o - \Delta G_p(R_p + [1/R_i + 1/(R_e + R_1)]^{-1})^{-1}/n_p^2$$

The predicted relation between the parameters is again linear. In this case, the variation in rates can be brought about by varying the rate of oxidation (at constant ΔG_p), for instance by adding inhibitors of the respiratory chain. Also the effect of adding an uncoupler on the position of the lines is clear: the intersection point with the abscissa should increase and the slope should decrease with decreasing R_1. Experimentally, the first of the two predictions turns out to be correct (14). However, the lines are parallel in the reported experiments. One possible explanation for this apparent anomaly could be that R_p is relatively small, i.e. much smaller than R_i and $(R_e + R_1)$.

iv) Double-inhibitor titrations

In so-called 'double-inhibitor titrations' (15,16) the rate of phosphorylation as a function of inhibitors of both respiration and phosphorylation is tested. To obtain an equation that can be applied to these conditions, we eliminate $\Delta\tilde{\mu}_H$. We find the following equation:

$$ J_p = \frac{(\Delta G_p/n_p - \Delta G_o/n_o)/R_o + \Delta G_p[1/R_i + 1/(R_1 + R_e)]/n_p}{n_p R_p/R_o + n_p R_p[1/R_i + 1/R_p + 1/(R_1 + R_e)]} $$

The rate of phosphorylation is described here as a function of the two forces ΔG_p and ΔG_o, and of the different resistances. To evaluate the effect of experimental variation in the resistances, one has to fix the other variables. In practice, this has not been consciously done. Therefore, the published data cannot be inserted with complete confidence in the derived equation. It is, however, possible to make some clear predictions concerning the results of double-inhibitor titrations under the appropriate experimental conditions.

If the derivative of the above function with respect to R_o is taken, it is easily seen that the effect of inhibition of the respiratory chain on J_p will depend on the values of the different resistances. At very high R_i and R_1 the effect of inhibiting the respiratory chain will be less when the ATP synthase is also partially inhibited (R_p larger). This is equivalent to the statement that under these conditions the protons show 'pool behaviour' (3). On the other hand, if the value of R_i is small compared to the value of R_p, we can see that the slope of the line relating J_p and R_o will decrease approximately proportional to the decrease in J_p. This is equivalent to the statement that under these conditions the protons show 'localised behaviour' (3).

It must be emphasized that for a proper interpretation of the data all parameters that are not explicitly accounted for must be kept experimentally constant.

CONCLUSIONS

On the basis of a relatively simple model we can derive equations for the processes of energy transduction by protons in mitochondria that are compatible with many observations reported in the literature. The extra assumption, compared with the original chemiosmotic hypothesis, is the presence of a resistance for protons for equilibration with the bulk aqueous phase. It appears that application of the equations to the available data can give a quantitative estimate for the magnitude of the extra resistance. This resistance is small relative to the resistance of the membrane for passive diffusion of protons.

This work was supported by grants from the Netherlands Organization for the Advancement of Pure Research (Z.W.O.) under the auspices of the Netherlands Foundation for Chemical Research (S.O.N.).

References

1. Mitchell, P., Nature (London) 191, 144-148 (1961).
2. Boyer, P.D., B.Chance, L.Ernster, E.Racker and E.C.Slater, Ann.Rev. Biochem. 46, 955-1026 (1977).
3. Westerhoff, H.V., B.A.Melandri, G.Venturoli, G.F.Azzone and D.B.Kell, Biochim.Biophys.Acta 768, 257-292 (1984).
4. Van Dam, K. and H.V.Westerhoff, in: Bioenergetics, L.Ernster, ed., Elsevier, Amsterdam, p.1-27 (1984).
5. Williams, R.J.P., Biochim.Biophys.Acta 505, 1-44 (1978).
6. Rottenberg, H., in: Frontiers in biological energetics, Vol.1, P.L.Dutton, J.S.Leigh and S.Scarpa, eds., Academic Press, New York, pp.403-412 (1978).
7. Kell, D.B. and Morris, J.G., in:Vectorial reactions in electron and ion transport in mitochondria and bacteria, F.Palmieri, E.Quagliariello, N.Siliprandi and E.C.Slater, eds., Elsevier, Amsterdam, pp.339-347 (1981).
8. Van Dam, K., H.Woelders, A.M.A.F.Colen and H.V.Westerhoff, Biochem. Soc.Trans. 12, 401-402 (1984).
9. Berden, J.A., M.A.Herweijer and J.B.J.W.Cornelissen, in: H^+-ATPase, S.Papa, K.Altendorf, L.Ernster and L.Packer, eds., Adriatica Editrice, Bari, pp.339-348 (1984).
10. Woelders, H., W.J.van der Zande, A.M.A.F.Colen, R.J.A.Wanders and K.van Dam, FEBS Letters 179, 278-282 (1985).
11. Padan, E. and H.Rottenberg, Eur.J.Biochem. 40, 431-437 (1973).
12. Azzone, G.F., T.Pozzan, S.Massari and M.Bragadin, Biochim.Biophys. Acta 501, 296-306 (1978).
13. Van Dam, K., H.V.Westerhoff, K.Krab, R.van der Meer and J.C.Arents, Biochim.Biophys.Acta 591, 240-250.
14. Tsou, C.S., Mitochondrial membrane processes, Ph.D.Thesis, University of Amsterdam, Mondeel Offset Drukkerij, Amsterdam (1970).
15. Baum, H., H.S.Hall, J.Nalder and R.B.Beechey, in: Energy transduction in respiration and photosynthesis,E.Quagliariello, S.Papa and G.S.Rossi, eds., Adriatica Editrice, Bari, pp.747-755 (1971).
16. Hitchens, G.D. and D.B.Kell, Biochem.J. 206, 351-357 (1982).

NONLINEARITY OF THE FLUX/FORCE RELATIONSHIP IN RESPIRING MITOCHONDRIA AS A POSSIBLE CONSEQUENCE OF HETEROGENEITY OF MITOCHONDRIAL PREPARATIONS

Lech Wojtczak, Jerzy Duszyński, Małgorzata Puka and
Anna Żółkiewska

Nencki Institute of Experimental Biology
Pasteura 3, 02-093 Warsaw, Poland

It is well recognized since early studies on oxidative phosphoryla-tion[1,2] that ADP greatly stimulates mitochondrial respiration. This reflects a tight coupling between electron transport and ATP synthesis. Nevertheless, in the absence of added ADP or after its complete phosphory-lation to ATP, mitochondrial respiration does not come to a stop but continues at a low, though measurable, level designated as state 4[2] or the resting state. This respiration may be partly attributed to external ATPase and/or recycling of Ca^{2+}. However, even if production of ATP or its exit from mitochondria is blocked and the uptake of Ca^{2+} is prevented, there still remains a significant oxygen uptake. In terms of the chemi-osmotic theory of energy coupling,[3] this resting state respiration can be interpreted as compensating the proton leak through the inner mitochondrial membrane.

In experiments with respiratory inhibitors, it has been found[4] that the rate of the resting state respiration (thus equivalent to the proton leak) is not proportional to the magnitude of the protonmotive force throughout its whole range but sharply increases at high values of the protonmotive force. An example taken from our studies is shown in Fig. 1. It has to be noted that in this figure, as well as in other experiments presented here, only the main component of the protonmotive force, viz. the membrane potential ($\Delta\psi$), was measured. However, since its ratio to the total protonmotive force remains practically constant under these experi-mental conditions,[5] such comparison is acceptable.

The nonlinearity between the respiration rate and the protonmotive force has been explained by postulating non-ohmic characteristics of the

111

inner mitochondrial membrane.[4] More recently, however, it has been proposed[6] that the resting state respiration and its nonlinearity are due, to a large degree, to "slipping" of proton pumps in the respiratory chain. In the present contribution we wish to propose another explanation for the

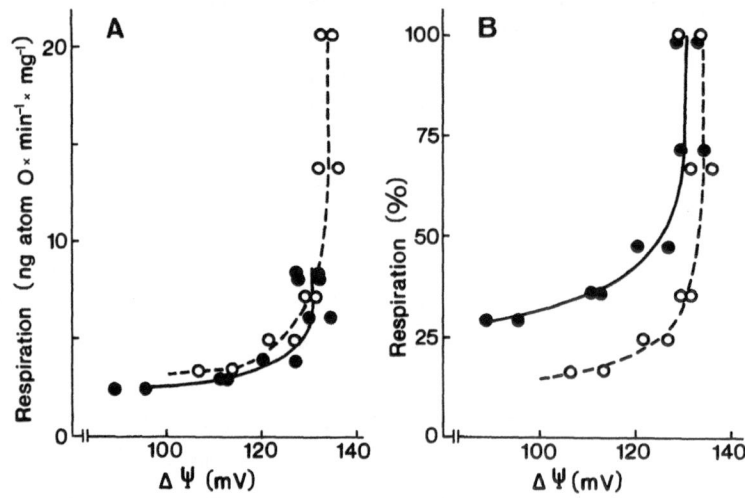

Fig. 1. Relation between the rate of resting state respiration and the membrane potential ($\Delta\psi$) in rat liver mitochondria. Oxygen uptake was measured with a Clark type electrode. Membrane potential was calculated from the distribution of [³H]triphenylmethylphosphonium.[5] ○, The respiratory substrate was succinate (+ rotenone) and the inhibitor was malonate; ●, the respiratory substrate was glutamate + malate and the inhibitor was rotenone. Temperature, 25°C.

nonlinearity of the flux/force relationship in mitochondria respiring under resting state conditions. It is based on (1) the concept of the control strength, as applied to mitochondrial respiration by Tager et al.,[7] and (2) the assumption that populations of isolated mitochondria are heterogenous concerning the degree of coupling. An expanded theoretical approach to this problem has recently been published elsewhere.[8]

For the sake of simplicity, let us assume that a mitochondrial population is composed of two subpopulations: a coupled one and a completely uncoupled one. The specific respiration rate of the uncoupled subpopulation is higher than that of the coupled subpopulation by a factor equal to the respiratory control ratio. Therefore, the respiration of the uncoupled subpopulation may have a considerable share in the overall oxygen uptake, even if this subpopulation constitutes a small proportion of the total population. On the other hand, the respiration of the uncoupled subpopulation is mainly controlled by the respiratory chain, whereas the main control strength in the respiration of the coupled subpopulation is exerted by the proton leak. Thus, when such a heterogenous mitochondrial population is titrated with a respiratory inhibitor, the respiration of uncoupled particles is strongly decreased, whereas the protonmotive force, related exclusively to coupled mitochondria, is hardly affected. Only at high concentrations of the inhibitor, the respiration of the coupled subpopulation becomes appreciably inhibited and, as result, the protonmotive force is also decreased.

The relation between the protonmotive force and the rate of respiration in tightly coupled mitochondria can be expressed by the following exponential function[9,10]

$$V = V_{max}\left\{1 - \exp\left[(\mu - \mu_{max})\,\frac{\epsilon}{kT}\right]\right\}$$

(1)

where V is the electron flow rate, V_{max} is the maximum flow capacity of the respiratory chain (practically equal to the electron flow rate under uncoupled conditions), μ is the protonmotive force, and μ_{max} is the highest attainable protonomotive force at which the electron flow completely stops (in ideally coupled mitochondria), ϵ is the elementary charge, k is the Boltzmann constant, and T is the absolute temperature.

Assuming that the proportion of the coupled subpopulation is B (that of the uncoupled one being, of course, 1-B), its respiratory control ratio is R, and the proton leak in the coupled subpopulation is _proportional_ to the protonmotive force, the following formula can be deduced from the relation between the resultant respiration rate (total respiration rate, V_{total}) and the resultant protonmotive force (apparent protonmotive force, μ_{app}) when a heterogenous mitochondrial population is titrated with a respiratory inhibitor[8]

$$V_{total} = \frac{V_{max}\left(\mu_{app} - \dfrac{kT}{\epsilon}\ln B\right)\left\{1 - B\exp\left[(\mu_{app} - \mu_{max})\,\dfrac{\epsilon}{kT} - \ln B\right]\right\}}{R\left(\mu_{max} + \dfrac{kT}{\epsilon}\ln\dfrac{R-1}{R}\right)\left\{1 - \exp\left[(\mu_{app} - \mu_{max})\,\dfrac{\epsilon}{kT} - \ln B\right]\right\}}$$

(2)

Mitochondria of the uncoupled subpopulation may fall into the following categories: (1) "open" membrane fragments, (2) "closed" submitochondrial particles, (3) mitochondria whose inner membrane is permeable to small molecules (including substrates, coenzymes, and protons), (4) mitochondria leaky to protons only. The second category may contain coupled inside-out particles. However, because of their reversed polarity, they do not accumulate lipophilic cations and therefore are regarded as uncoupled in the present experimental approach. If the hypothesis presented here is correct, then exclusion of any of these categories of uncoupled mitochondria, or their fragments, would result in making the flux/force relationship of the total population more linear.

An attempt to verify our hypothesis along this line is presented in Fig. 1. Succinate is oxidized by mitochondria and their fragments no matter how damaged or leaky they are. In contrast, the oxidation of glutamate requires the presence of nicotinamide nucleotides and therefore will not occur in membrane fragments, inversed submitochondrial particles, and mitochondria leaky to small molecules. Comparing the relationship between respiration and $\Delta\psi$ with various substrates, we found that with succinate it was more "curved" than with glutamate, if the respiration rate was expressed in percentage of the resting state respiration without the inhibitor (Fig. 1B). Fig. 1A shows another interesting feature. Both substrates energize mitochondria to the same membrane potential. However, the resting state respiration with succinate is 2.5 times higher than with glutamate plus malate, which means that proton pumping is 2.5 times less efficient. A similar or even higher difference was observed when resting state respiration of succinate and ß-hydroxybutyrate was compared. Although the $H^+/2e^-$ stoichiometry at various coupling sites may be a matter of debate, it would not account for such a high difference of pumping efficiency between succinate and NAD-dependent substrates. It could be, therefore, concluded that the mitochondrial population contains uncoupled organelles (or their fragments) classified to either category 1, 2, or 3, as listed above.

In another series of experiments mitochondria were titrated with inhibitors of either the substrate carrier or the respective dehydrogenase. As shown in Fig. 2, titration with phenylsuccinate, the inhibitor of dicarboxylate carrier, resulted in somewhat less steep plot than titration with malonate, the inhibitor of succinate dehydrogenase. In the former case only the respiration of mitochondria which were intact and not leaky to small molecules was affected.

Another support for the hypothesis presented here was provided by comparing relationships computed from Eqn. 2 with experimental results.

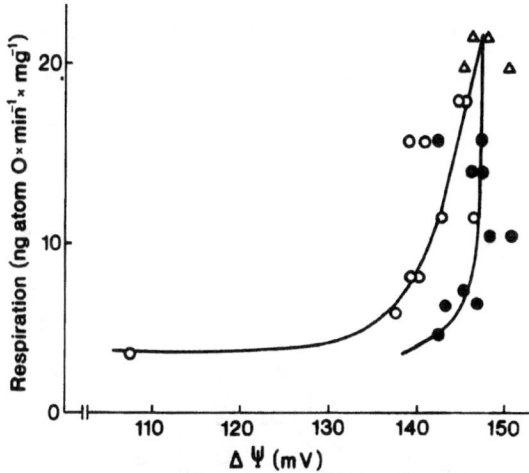

Fig. 2.   Relation between the rate of respiration
and the membrane potential in rat liver
mitochondria respiring with succinate
(+ rotenone). △ , No inhibitor;
● , titration with malonate;   o , titration with phenylsuccinate. Other conditions
and methods as in Fig. 1.

Plotting Eqn. 2 for various proportions between the coupled and the uncoupled subpopulations, one obtains a series of curves shown in Fig. 3A.
For lower values of μ_{app} they approach straight lines and bend sharply
upwards only at higher values of μ_{app}. The higher is the proportion of
the uncoupled subpopulation the larger is the deviation of the curve from
linearity. For comparison, Fig. 3B illustrates results of an experiment
with rat liver mitochondria respiring with succinate and titrated with
malonate (see also Fig. 1).

A similar relationship for a fixed proportion of coupled subpopulation
but various degree of its coupling is presented in Fig. 4A, whereas Fig. 4B
shows a series of curves for rat liver mitochondria whose respiratory
control was decreased by small amounts of carbonyl cyanide m-chlorophenylhydrazone.

Considerations presented here can explain the apparent nonlinearity
of the dependence between the rate of respiration and the protonmotive force
under resting state conditions without the necessity of introducing the

concept of non-ohmic proton leak or the molecular slip. Eqn. 2 is based on
the simplest assumption that a heterogenous population is composed of only
two subpopulations, one of which is coupled, the other one completely uncoupled. In reality, isolated mitochondria contain, most likely, particles

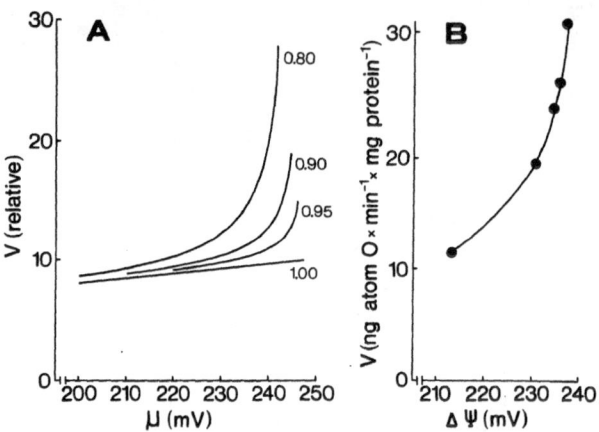

Fig. 3. Dependence between the resting state respiration and the protonmotive force in mitochondria titrated with a respiratory inhibitor. A, Theoretical curves computed from Eqn. 2 for a mitochondrial preparation containing a mixture of coupled organelles
exhibiting the respiratory control ratio (R) of 10 and totally
uncoupled particles. Proportion of the coupled subpopulation (B) is
indicated by the numbers at the curves. The value for V_{max} was
taken as 100 and that for μ_{max} as 250 mV. B, Experimentally
obtained relationship for liver mitochondria respiring with succinate (+ rotenone) and titrated with malonate. Membrane potential
was measured with a tetraphenylphosphonium-sensitive electrode.[11]
From Duszyński and Wojtczak.[8]

exhibiting a broad spectrum of the degree of coupling. This makes a mathematical description of the events more difficult, if at all possible, but
does not change the main feature of such a population, namely the apparent
nonlinearity between the resultant respiration and the protonmotive force.
For this reason, one should not expect a perfect fitting of experimental
results with computed curves. The comparison of theoretical relationships
with experimental findings, as illustrated in Figs 3 and 4, is only aimed
to stress a general similarity of the shape of respective curves.

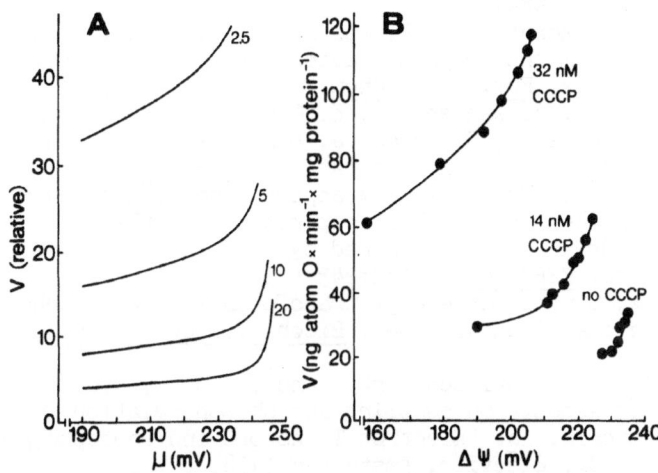

Fig. 4. Dependence between mitochondrial respiration and the protonmotive force at various degree of coupling. A, Theoretical curves calculated according to Eqn. 2 for a mitochondrial population containing 90% (B = 0.90) of particles of various degree of coupling, as characterized by their respiratory control ratios (R) of 20, 10, 5, and 2.5, as indicated. B, Experimental results with rat liver mitochondria in the absence or presence of two concentrations of carbonyl cyanide m-chlorophenylhydrazone (CCCP), as indicated. Values for V_{max} and μ_{max} and experimental conditions are the same as in Fig. 3. From Duszyński and Wojtczak.[8]

The aim of the present considerations is neither to question non-ohmic characteristics of mitochondrial membranes nor to deny the existence of slips in proton pumps. We simply want to propose another factor which may perhaps, at least partly, account for the observed nonlinear flux/force relationships in resting state mitochondria.

References

1. B. Chance and G. R. Williams, Respiratory enzymes in oxidative phosphorylation. 1. Kinetics of oxygen utilization, J. Biol. Chem., 217:383 (1955).
2. B. Chance and G. R. Williams, Respiratory enzymes in oxidative phosphorylation. 3. The steady state, J. Biol. Chem. 217:409 (1955).
3. P. Mitchell, "Chemiosmotic Coupling and Energy Transduction," Glynn Research Laboratory, Bodmin (1968).
4. D. G. Nicholls, The influence of respiration and ATP hydrolysis on the proton-electrochemical gradient across the inner membrane of rat liver mitochondria as determined by ion distribution, Eur. J. Biochem., 50:305 (1974).

5. J. Duszyński, K. Bogucka, and L. Wojtczak, Homeostasis of the proton-motive force in phosphorylating mitochondria, _Biochim. Biophys. Acta_, 767:540 (1984).
6. D. Pietrobon, M. Zoratti, G. F. Azzone, J. W. Stucki, and D. Walz, Non-equilibrium thermodynamic assessment of redox-driven H^+-pumps in mitochondria, _Eur. J. Biochem._, 127:483 (1982).
7. J. M. Tager, R. J. A. Wanders, A. K. Groen, W. Kunz, R. Bohnensack, U. Küster, G. Letko, G. Böhme, J. Duszyński, and L. Wojtczak, Control of mitochondrial respiration, _FEBS Lett._, 151:1 (1983).
8. J. Duszyński and L. Wojtczak, The apparent non-linearity of the relationship between the rate of respiration and the protonmotive force of mitochondria can be explained by heterogeneity of mitochondrial preparations, _FEBS Lett._, 182:243 (1985).
9. R. Bohnensack, Control of energy transformation in mitochondria. Analysis by a quantitative model, _Biochim. Biophys. Acta_, 634:203 (1981).
10. M. Baltscheffsky, H. Baltscheffsky, and J. Boork, Evolutionary and mechanistic aspects on coupling and phosphorylation in photosynthetic bacteria, _in_: "Electron Transport and Photophosphorylation," J. Barber, ed., Elsevier, Amsterdam (1982).
11. N. Kamo, M. Muratsugu, R. Hongoh, and Y. Kobatake, Membrane potential of mitochondria measured with an electrode sensitive to tetraphenyl-phosphonium, _J. Membrane Biol._, 49:107 (1979).

PROTON CURRENTS AND LOCAL ENERGY COUPLING IN THYLAKOIDS : A SURVEY

Yaroslav de Kouchkovsky, Claude Sigalat, Francis Haraux
and Suong Phung Nhu Hung

Laboratoire de Photosynthèse, C.N.R.S.; B.P. 1
F-91190 Gif-sur-Yvette (France)

SUMMARY

A survey of the experimental and conceptual basis of a microchemiosmotic interpretation of energy-dependent processes in thylakoids is presented. It assumes that protons circulate from their active transport points
to their backflow ports in a heterogeneous medium, having a resistivity
not only transversal, from membrane to isopotential bulk phase, but also
lateral, in or on the membrane, an idea central to our hypothesis. Consequently, the proton electrochemical potential $\Delta\mu H$ at the H^+-generators
(redox carriers) is higher than $\Delta\mu H$ at H^+-leaks (coupling factors and membrane pores), and the measured $\Delta\overline{\mu H}$ is an average of local values. However,
some flexibility in this microscopic coupling is possible, and a delocalized
behaviour may sometimes be obtained by increasing the lumen volume (osmolarity decrease) and its conductivity (ionicity increase). A one-to-one link
between primary and secondary pumps, as advocated by mosaic chemiosmosis,
seems therefore improbable here.
    The experimental procedure used in this investigation consisted in
correlating electron flow, proton gradient (ΔpH, with or without $\Delta\Psi$), and
phosphorylation rate in several conditions. To modulate $\Delta\mu H$, H^+ influx or
efflux were adjusted by light or ionophore changes: contrary to the prediction of classical chemiosmosis, these two factors have no identical effects.
Also, an increase of the distance between H^+-translocators and coupling
factors (SII chain vs. SI) lowers the phosphorylation efficiency of a given
mean $\Delta\overline{\mu H}$, which is easily explained by a ΔpH drop along the lateral resistance mentioned above. Another approach was to compare redox control and
phosphorylation in heavy water media, where proton circulation is hindered.
Additional information given concerns possible occurrence of "scalar ATP-
ases" and a quantitative estimate, at variable ΔpH, of proton flux across
coupling factor, phosphorylating or not.

Abbreviations. $\Delta\mu H$ ($\equiv \Delta\tilde{\mu} H^+$), ΔpH, $\Delta\Psi$: differences in proton electrochemical
potential, pH, and electric potential; with a bar on top: mean values, with
superscripts E or C: local values, at electron-transfer chain and coupling-
factor (CF); subscripts e, i: external (stroma or medium) or internal (lumen) phases. SI, SII: systems I and II; Chl: chlorophyll, PQ: plastoquinone;
DCPIP: 2,6-dichlorophenolindophenol, DMQ: 2,5-dimethylquinone, FeCy: ferricyanide, MV: methylviologen, P: inorganic phosphate (Pi) or "high-energy"
phosphate bond, PYO: pyocyanin, 9A: 9-aminoacridine. V_e-, V_P: rates of electron flow and phosphorylation; RC = redox control .

INTRODUCTION

Respiratory and photosynthetic membranes, which form closed structures separating a small internal compartment from the exterior — the suspending medium in vitro —, have a topographical arrangement making that protons are translocated in one direction during active redox transport and that ATP is synthesized thanks to the use of this potential energy, when H^+ flow back through the enzymatic "coupling factors". According to Mitchell's chemiosmotic hypothesis[1], the resulting transmembrane $\Delta\mu H$ concerns only solvated protons, delocalized in their respective bulk phases, whereas Williams initial proposal — which has changed to some degree[2] — was that anhydrous protons might be directly channelled from the redox chains to coupling factors, from where they would be expelled in a hydrated state. Somewhat similar by the idea of a "direct coupling" mechanism is the recent "mosaic chemiosmosis", developed by a group of investigators working on mitochondria and bacteria[3]. In this hypothesis, proton primary (redox) and secondary (phosphorylating) pumps are strictly connected, which necessarily implies structural constraints, rather improbable in the case of thylakoids if one considers the heterogeneous distribution of Systems I and II with respect to coupling factors (Fig. 1A).

Our own conception[4-6] conciliates both electrochemical basis of Mitchell's chemiosmosis and the experimental evidence that local events occur in energy transduction. We propose that protons circulate, from their input to their output points (CF, membrane leaks), in an anisotropic structure of a resistivity not only transversal, from membrane to bulk phase, as it was already proposed[7], but also lateral. This causes a lateral drop (ΔpH_L) of proton gradient, and the ΔpH, hence $\Delta\mu H$, is higher at H^+-source than at its sink. That is, the experimentally measurable $\Delta\overline{\mu H}$ is a mean value, averaged not only over the vesicle population, but also within each of its component. For instance, considering only plastoquinol reoxidation-site as a H^+ source and coupling factor as a leak, one may write[8]: $\Delta pH (PQ) = \Delta\overline{pH} + \Delta pH_L$ and $\Delta pH (CF) = \Delta\overline{pH} - \Delta pH_L$ (the same would be true for $\Delta\mu H$). Depending on the relative value of the various resistances, the level of different local $\Delta\mu H$ may vary; more precisely, that at coupling factors may be below — which probably is the general case —, near, and even above, mean $\Delta\overline{\mu H}$ (Fig. 1B, right). In the following paragraphs, we shall summarize our main results supporting this hypothesis and present some other data relevant to the subject.

EXPERIMENTAL PROCEDURE

Chloroplasts were extracted by grinding Lettuce leaves in an isotonic buffer, with an intermediate low osmolarity shock to rupture the envelope and liberate thylakoids[9]; chlorophyll was measured according to Mackinney[10].

The "standard conditions" mentioned in figure captions, were essentially: thylakoids at 15 μM Chl in 0.2 M sorbitol + 10 mM Tricine + 10 mM Hepes + 10 mM KCl + 6 mM $MgCl_2$ + 2 mM K_2HPO_4, + 0.5 mM ADP in case of phosphorylation, pH 7.8 (in heavy water, pH was corrected according to the relation[11]: p^2H = pH glass electrode + 0.4); stirred aerobic suspension at 20° C. The broadband red actinic light was at its maximum $\sim$ 0.5 kW m^{-2} and even above ("strong light"), but could be lowered with calibrated neutral filters; it was always long enough (> 1 mn) to reach steady-state of the phenomena. The narrowband analytic light, about 1 Wm^{-2}, was generally set at 420 nm to excite 9-aminoacridine fluorescence, collected around 520 nm, and at the same time to measure ferricyanide reduction (other wavelengths were used, for instance, to study $\Delta\Psi$-induced electrochromic bandshift of endogenous pigments). When necessary, a small membrane O_2-electrode or glass pH-electrode was introduced in the cuvette, to follow electron flow also by O_2 evolution (uptake with methylviologen) or to determine vectorial H^+-translocation

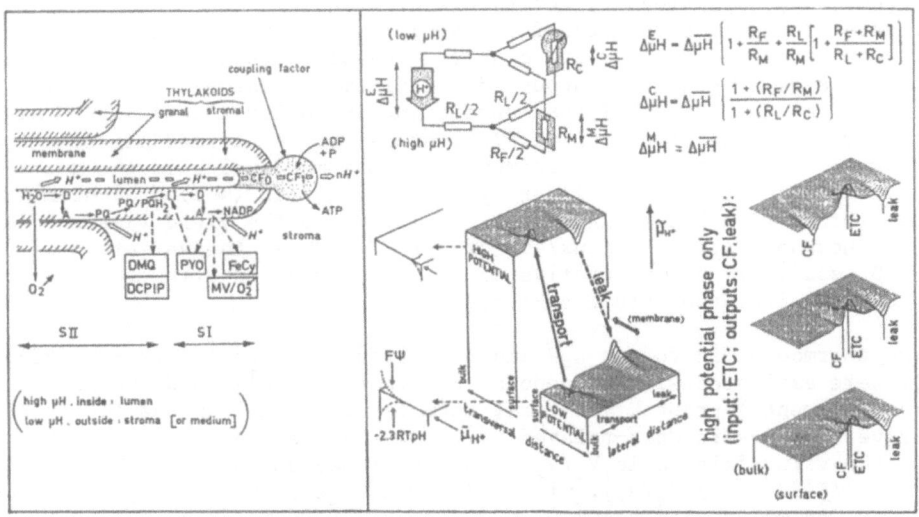

Fig. 1. Structural and electrochemical basis of energy transduction in thylakoids. Left (A): thylakoid organization, with appressed granal regions — containing almost all SII, but no CF — and free stromal lamellae — carrying all CF and most SI, but essentially devoid of SII. D and A: primary electron donors and acceptors of SI and SII, []: Fe-S protein, cytochromes b_6-f, and plastocyanin ; in boxes : exogenous redox carriers. Right (B), top: electrical analogy of proton circuit, where thick arrow is an electron-transfer chain (ETC) generator, injecting H^+ in the high potential compartment (lumen), from where they flow back (to stroma) through membrane pores, of resistance R_M, or through coupling factors (CF), of resistance R_C (which decreases when $\Delta\mu H$ rises); a lateral resistance R_L, in an interfacial and/or intramembrane phase, links the redox sources, of potential $\Delta_\mu^E H$, to the coupling factors, of potential $\Delta_\mu^C H$ (R_L increases with the distance ETC-CF); these anisotropic phases are separated from delocalized bulk phases by some resistances ($R_F/2$ on one side, connected here to the middle of R_L for the sake of simplicity; a portion of R_L is therefore shared by all H^+ pathways). Assuming a symmetric situation inside and outside, simple relations are obtained between local and mean $\overline{\Delta\mu H}$, the latter being considered close to $\Delta_\mu^M H$ across membrane leak (this slight overestimation has no influence on the reasoning[6]). In Mitchell's delocalized chemiosmosis, only R_M and R_C exist: $R_L = R_F = 0$, hence $\Delta_\mu^E H \equiv \Delta_\mu^C H \equiv \overline{\Delta\mu H}$; Williams (early) direct coupling assumes that ETC and CF form "coupling units", isolated from the bulk phase: $R_L = 0$, and $R_F \gg R_C$ and R_M, hence $\Delta_\mu^E H = \Delta_\mu^C H \gg \overline{\Delta\mu H}$. Bottom: illustration of the μH_2O shape near the membrane; its wavy lateral profile — enhanced in H_2O — results from the potential drop in R_L; on the other hand, μH decays from surface to bulk, by definition isopotential. Though on the average, i.e. except at entry and exit H^+ ports, μH is constant perpendicular to the membrane, it is not the case for its pH and Ψ components, because membrane is negatively charged (less inside in the energized state). Three situations are described on the right: 1) top: $_\mu^C H < \overline{\mu H}$, because CF and ETC are largely separated (as in SII case); 2) bottom: $_\mu^C H > \overline{\mu H}$, because CF is close to ETC; 3) middle: intermediate situation, where $_\mu^C H \approx \overline{\mu H}$.

and phosphorylation-linked scalar H^+-uptake [12]. More routinely, ATP was quantified by the luciferase luminescent method[13] on aliquots of the suspension. Two functions may be derived from these direct fluxes (V_{e^-}: electron flow, and V_p: phosphorylation rate). The redox control, RC, reflects the magnitude of the actual $\Delta\mu H$ (i.e. $\Delta_{\mu}^{E}H$: Fig. 1B) which exerts a negative feedback effect on a significantly reversible H^+-translocating steps — not H_2O oxidation — of the electron-transfer chain; it is expressed here by RC = 1 - (coupled/uncoupled V_{e^-}), where coupled = basal or phosphorylating V_{e^-}, in presence of a given $\Delta\mu H$, and uncoupled = maximum V_{e^-}, at same energy input (light) but in the absence of $\Delta\mu H$ (sufficient uncoupler added: 1-2 µM nigericin here). The other function is P/e^-, number of ATP formed per electron transferred; it lies between a minimum and a maximum, which is the quotient $(P/e^-)_{max}$ of H^+/e^- (normally, 1 per System) and H^+/ATP ($\sim$ 3) stoichiometries. In other words, $(P/e^-)_{measured} = \rho (P/e^-)_{max}$, where the yield factor ρ is the ratio of phosphorylating H^+-flux/sum of all H^+ fluxes; the variation of ρ, between 0 and 1, is a quantitative expression of the coupling factor activation by ΔpH, i.e. the actual $\Delta_{\mu}^{C}H$: Fig. 1B.

The thermodynamic forces ΔpH and $\Delta\Psi$, giving rise to $\Delta\mu H$ (= -2.3 $RT\Delta pH$ + $F\Delta\Psi$), were estimated with exogenous (9-aminoacridine[15]) and endogenous (membrane pigments) probes. The first term is computed from the fluorescence levels of 9-A in dark-relaxed, $F^\bullet$, and light energized states, F°, knowing the volumetric ratio V_e/V_i of external medium to internal lumens: $\Delta pH = \log\left[(F^\bullet/F^\circ - 1) V_e/V_i\right]$, with $V_e/V_i \approx 2 \Omega/\left[Chl\right]$[16], Ω being the medium osmolarity, in molar units as $\left[Chl\right]$. The second term may be established using previous calibrations[17]: for 1 cm optical length, $\Delta\Psi(mV) \approx 0.255 \times \Delta A_{515}/\left[Chl\right]$, where ΔA_{515} is the absorbance change at 515 nm (electrochromic peak). However most experiments were made with valinomycin (50-100 nM) added to K^+-containing media, to cancel out $\Delta\Psi$, but similar results were obtained in its absence, showing the limited weight of $\Delta\Psi$ in overall $\Delta\mu H$ at steady-state.

RESULTS

Modulation of $\Delta\mu H$ by H^+ input and output

If measured $\Delta\overline{\mu H}$ were identical to actual $\Delta\mu H$ at the site where a phenomenon takes place under its control, a univocal relation would link this $\Delta\overline{\mu H}$ to that phenomenon (such as V_{e^-}, V_p, and their derived functions RC, P/e^-). In other words, to adjust $\Delta\overline{\mu H}$ to a same value by a change of H^+ input or output should have exactly the same effect on a given process. This in fact was not observed in the flux(V_p)-force($\Delta\mu H$) correlations established on mitochondria [18], bacterial chromatophores[19], and, recently, thylakoids[6,9]. Fig. 2 is a good illustration of this. On its left side (A), it is clear that nigericin addition (increased H^+-efflux) is less efficient than light reduction (decreased H^+-influx) in diminishing V_p with $\Delta\overline{\mu H}$, in SI and SII modes. On its right side (B), one sees an extension of this type of experiment to the redox-control of a SI + SII chain involving the $\Delta\mu H$ (i.e. pH_i)-sensitive plastoquinone link, in basal and phosphorylating conditions. The insert in Fig. 2B additionally shows that redox control, at given $\Delta\overline{pH}$, is higher when coupling factor permeability is enhanced by phosphorylation activity. These observations fit nicely with our hypothesis, and they may be more directly evaluated considering the relations given in Fig. 1B. Thus, relative to $\Delta\mu H$, $\Delta_{\mu}^{C}H$ ($\to V_p$) and $\Delta_{\mu}^{E}H$ ($\to$ RC) raise when R_M drops (nigericin), and the same is true for $\Delta_{\mu}^{E}H$ when R_C is lowered by ADP addition, in this case only because the lateral resistance $R_L \neq 0$, a key point in our scheme. Therefore, that no univocal relationships are found between fluxes and forces, depending on how they are modulated, is due to the non-identity of local and mean values of $\Delta\mu H$.

As illustrated Fig. 1A, the two photosystems and the coupling factors are heterogeneously distributed, at least in granal chloroplasts[20], which is the case here. Thus, the average distance between SII and CF is quite long (> 100 nm), whereas it is short between CF and SI. The lateral resistance R_L must increase with the distance — and so the lateral pH drops — that is, $\Delta_L^C \mu \overline{H}/\Delta \mu \overline{H}$ should decrease (Fig. 1B). It was therefore predictable that for a same mean $\Delta \mu \overline{H}$, SII would be less efficient than SI in ATP synthesis[9,21]. This is represented in Fig. 3 for dimethylquinone (SII) and pyocyanin (SI) chains. Its left side displays a result for $\Delta p\overline{H}$ (qualitatively unchanged by valinomycin addition[9]), and its right side shows the same for $\Delta \Psi$. In this latter case, the discrimination between the two curves — not expected as strong as in $\Delta p\overline{H}$ representation — doesn't mean that $\Delta \Psi$ is localized, since ΔpH was still there the predominant component of $\Delta \mu \overline{H}$. Other chains were tested, such as ferricyanide + dibromothymoquinone, DBMIB (SII) vs. ferricyanide alone (SI + SII)[21] or pyocyanin (SI) vs. methylviologen (SI + SII)[9]. The same conclusion as above was reached, but SII reactions were always more clearly distinguishable than SI reactions from the SI + SII ones. This is understandable, since SI proton-pumps are more efficient than SII, as already stated, their weight predominates in the full SI + SII chain.

Some investigators[22] reported, however, that both systems behave similarly. We thought that one of the reason of this discrepancy might have been that the media used by these authors were different from ours. Indeed, the distribution of protons between the two branches we propose, one lateral (in or on the membrane) and the other via bulk phase, should depend on their relative conductance and on the value of the resistance (R_F) which connects them: these factors are likely to be highly variable with the physico-chemical conditions. Thus, to increase the bulk phase volume, by decreasing medium osmolarity, and in addition to increase its conductivity, by raising its ionicity, should favour the delocalized pathway: a confirmation of this is given by Fig. 3B. However, one should notice that generally, hypotonicity only helps to bring nearer to each other SI and SII curves — a merging was always observed in pyocyanin and methylviologen case[9] — and that their superimposition exhibited on the right side of

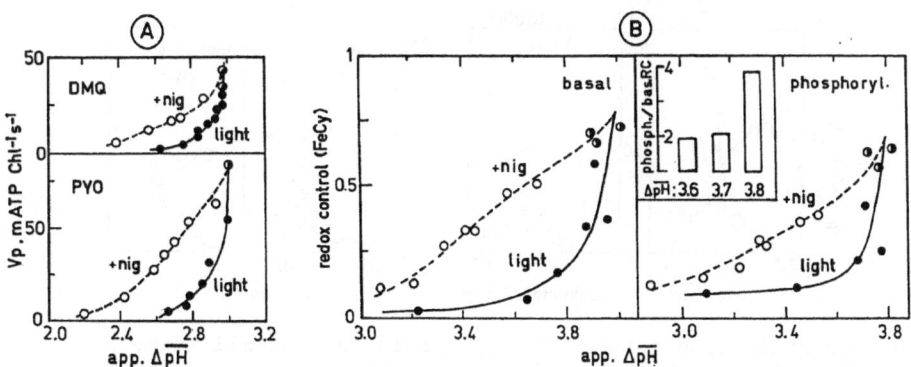

Fig. 2. Non-equivalency of adjusting proton gradient ($\Delta p\overline{H}$, here: valinomycin present) to a given value by H^+ redox-influx (light) or by H^+ leak (nigericin). Left (A): phosphorylation rate of SI (pyocyanin) or SII (dimethylquinone) chains. Right (B): redox control of SI + SII chain (ferricyanide); insert: indication that redox control (RC) is stronger in phosphorylating (low R_C: see figure 1) than in basal (high R_C) states, in a $\Delta p\overline{H}$-dependent manner. Standard conditions.

Fig. 3B is rarely so perfect. Thus, there must be more than one factor which determine the observed behaviour.

Evaluation of the proton gradient

One could think that the way $\Delta\mu H$ is evaluated leads to conclude to a microchemiosmotic type of coupling. We shall not discuss here the measurement of $\Delta\Psi$ by electrochromism — this was done elsewhere[6] — since we obtained similar results with or without valinomycin. We thus shall restrict ourselves to $\Delta\overline{pH}$. Many side reactions muddle up probe responses: electrostatic effects[23], partitioning in the membrane[6], and binding[24]. As a result, $\Delta\overline{pH}$ is overestimated, by perhaps 1 unit or more in the case of 9-aminoacridine[6]. McCarty's group, who criticizes 9-A, recognizes valid only the results obtained with radioactive amines in spite of their condition of use. We have already remarked that all these probes actually share similar reservations[6,25], but the main point is that in the kind of studies presented here, it is not important that $\Delta\overline{pH}$ scale cannot be calibrated in absolute units and even is not linear, provided the different phenomena are always compared at same values of apparent $\Delta\overline{pH}$ (this is why the word "apparent" is given on the figures).

It is nevertheless interesting to have a completely different way to estimate $\Delta\overline{pH}$, namely by the amount of H^+ translocated across the membrane. This is easily obtained from the external pH shift, ΔpH_e, measured with the glass-electrode, because ΔpH_e is proportional to $\Delta\overline{pH}$: see Fig. 4A for details. At the same time, phosphorylation rate may be computed from the scalar H^+ uptake linked to ATP synthesis[12] according to the equation given on this figure. As demonstrated Fig. 4B, 9-aminoacridine fluorescence and pH data, simultaneously recorded on the same sample, give very similar results. Thus, even if one may discuss what are the molecular mechanisms involved in one or an other probe response, and how to calibrate it, this method remains a valuable tool for comparative investigations.

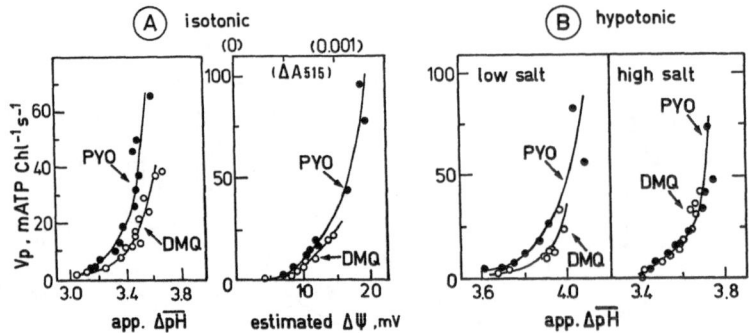

Fig. 3. Discrimination between SI (pyocyanin, ●) and SII (dimethylquinone, o) chains in their phosphorylation efficiency (Vp) for given, mean $\Delta\overline{pH}$ (i.e. $\Delta\overline{\mu H}$). Left (A): standard isotonic medium for tracing Vp vs. $\Delta\overline{pH}$ and $\Delta\Psi$ curves, $\Delta\Psi$ being estimated by the electrochromic band-shift of endogenous pigments at 515 nm. Right (B): similar medium, but without sorbitol and containing either 5 mM (left) or 50 mM (right) KCl, both cases in presence of 100 nM valinomycin. Standard conditions (except medium in case B); variable light to decrease $\Delta\overline{pH}$ and $\Delta\Psi$, thus Vp.

Hydrogen isotope exchange

These experiments were the first which we designed to analyze proton currents[4,8,22] and some additional data are presented below. The recording on Fig. 5A shows that a smaller light-induced 9-A fluorescence quenching is obtained in heavy-water — no direct effect on the probe itself was detected — which slowly relaxes; however, the first-order character of the $\Delta\overline{pH}$-decay, occuring after some apparent lag (often shorter) is preserved. The next four graphs, of Fig. 5 reveal the effect of an increasing proton replacement by the slower deuteron. In Fig. 5B, the trend towards a lessening of $\Delta\overline{pH}$ is confirmed (top), despite a marked decrease of membrane permeability (bottom); that is, H^+ influx, and therefore coupled electron flow, are significantly inhibited. Phosphorylation is even more sensitive[26] than electron flow, and P/e^- thus sharply drops, whereas the redox control rises: Fig. 5C. As demonstrated Fig. 6A, the latter result is clearly due to a larger efficiency of isotope substitution on coupled than on uncoupled redox rate, at least in strong light (the $\sim 30\%$ inhibition of uncoupled V_e- likely reflects trivial solvent effect, since all redox chains studied were similarly reduced[4]). In limited light, the inhibition vanishes, and Fig. 6B verifies that the quantum yield is unaffected, as could be expected, inasmuch as no significant membrane change is brought about by the present experimental conditions. Fig. 7A shows that though membrane permeability rises with pH in both isotopes, heavy water has little effect near neutrality, which suggests that at pH < 7, other factors than proton mobility, to mention only the most obvious, control membrane permeability to this ion. On the other hand, Fig. 7B points out to the fact that when H^+-conductance is mainly under the dependence of an external factor such as an ionophore shuttle, the isotope effect disappears (yet, at 30 nM nigericin, a significant inhibition of electron flow and of phosphorylation is maintained, together with some lowering of $\Delta\mu\overline{H}$, as expected if lateral R_L rises in ^2H_2O).

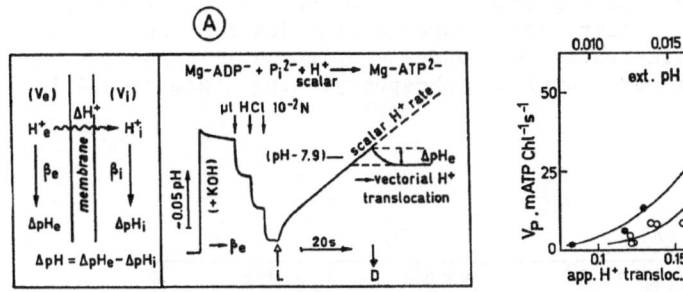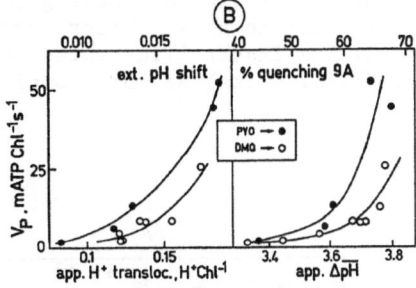

Fig. 4. Differences in SI (pyocyanin, ●) and SII (dimethylquinone, o) phosphorylation efficiency for a given mean proton gradient, established by two independent methods run in parallel: external pH change and 9-aminoacridine fluorescence quenching. Left (A): principle of measuring simultaneously phosphorylation rate, by scalar H^+ consumption[12], and proton gradient, by vectorial H^+ translocation. The amount ΔH^+ of protons transported from the external medium, of volume V_e, to internal lumens, of overall volume V_i, is related to external and internal pH-shifts ($\Delta pH_e > 0$, $\Delta pH_i < 0$) by the conservative equation: $\Delta H^+ = V_e \beta_e \Delta pH_e = -V_i \beta_i \Delta pH_i$, where β_e and β_i are the mean buffering powers (< 0) for the external and the internal pH ranges, respectively, covered by these ΔpH_e and ΔpH_i; on the other hand, by definition, $\Delta pH = \Delta pH_e - \Delta pH_i$ (= $pH_e - pH_i$), hence $\Delta pH = [1 + (V_e\beta_e/V_i\beta_i)] \times \Delta pH_e$ (ΔpH is actually here the mean $\Delta\overline{pH}$). The pH-meter recording shows the different steps. Right (B): actual data obtained by varying light intensity; standard conditions (except buffers omitted).

That ATP formation is strongly reduced in ^2H_2O might be ambiguous, if its limiting step involves protons or if the enzyme turn-over time is sufficiently decreased to determine the overall rate of the reaction. Actually, the V_p vs. $\Delta\mu\overline{H}$ curves were far from reaching saturation, and P/e^- variation, i.e. that of the yield factor ρ, which expresses coupling factor activation by $\Delta\mu H$, confirms that the key role in the phosphorylation falling-down was played by the proton gradient existing at the CF level. That is, assuming that the replacement of ^1H by ^2H significantly increases the lateral resistance, and therefore the lateral ΔpH ($\Delta\mu H$) drop, one gets, relative to the measured $\Delta\mu H$, $\Delta\mu^2H < \Delta\mu^1H$ at coupling factors (hence lower V_p and P/e^-). Conversely, one has $\Delta\mu^2H > \Delta\mu^1H$ at H^+-generators, hence the stronger redox control of coupled V_{e^-}, although an additional isotope effect on redox potentials cannot be excluded.

Membrane proton conductance

At more that one place, a good deal of attention was paid to membrane H^+ permeability, this latter word being loosely defined here. Actually, two main types of endogenous leaks exist: natural "pores" or membrane defects, induced by thylakoid preparation, and coupling factors. Whereas the former are essentially static leaks, the latter have a dynamic character, that is, their conductance steeply increases with $\Delta\mu H$ (activation[14]) and is stimulated by phosphorylation (state 3). When ATP synthesis and hydrolysis — both involving, but in opposite direction, vectorial H^+ transport — equilibrate each other (state 4), no net proton current crosses CF, and H^+-efflux can only occur through the membrane static leaks. In basal and phosphorylating conditions, membrane permeability is thus $\Delta\mu H$-dependent, and since phosphorylation "consumes" part of this $\Delta\mu H$[27], membrane conductance and proton gradient are different in these two conditions, and so is the balance between phosphorylating and non-phosphorylating proton pathways. Consequently, to calculate P/e^-_{max} simply by the ratio $V_p/(V_{e^-}$,total phosphorylating $-V_e$, basal), as classically made[28], is not rigorous, because the two electron flow rates should be taken at identical $\Delta\mu H$, not mentioning the problem of local gradients. There are two recent reports[29,30] related in part to this subject, and a more general description is given in Fig. 8. Its insert illustrates how, under continuous illumination, a partial (basal state) or more complete (phosphorylating state 3) opening

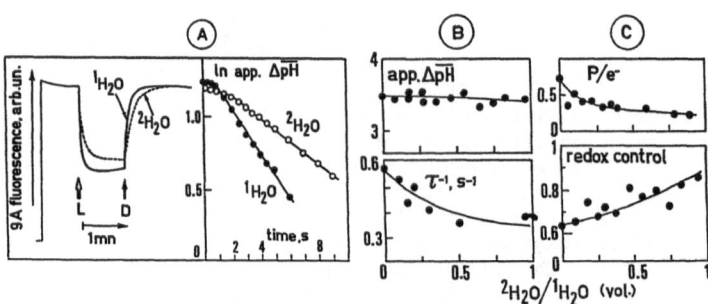

Fig. 5. Effect of ^1H_2O substitution by ^2H_2O. Left (A): recorder trace of 9-aminoacridine light(L)-induced fluorescence quenching and semilogarithmic plot of the dark(D)-relaxation of corresponding $\Delta\overline{pH}$; pure normal or heavy water media (standard). Right: variation of $\Delta\overline{pH}$ and of half-time reciprocal of its decay (B), and changes of P/e^- ratio and of redox control (C) with the enrichment in ^2H_2O; note that the latter increases (and the former decreases) despite some $\Delta\overline{pH}$-lowering. Standard conditions, strong light (FeCy chain).

of CF lowers $\Delta\overline{pH}$, its highest value being obtained with closed CF (static state 4); reduction of thiol groups by dithiothreitol does not significantly affect the observed levels of $\Delta\overline{pH}$, in agreement with the idea that in energized states, thiol-modulated CF behaves as original CF. Fig. 8A, which was traced at variable light, indicates how varies with $\Delta\overline{pH}$ (i.e. $\Delta\mu\overline{H}$, since valinomycin was added) the rate of ferricyanide reduction, therefore that of H^+ efflux. Indeed, at steady state, proton efflux and influx are equal, and H^+ influx $= (H^+/e^-) \times$ electron flow. Substraction of static rate from phosphorylating and basal rates, at same $\Delta\overline{pH}$, gives net H^+-fluxes through CF alone, in active and inactive states; their ratio represents the variable stimulation by ADP of H^+ flux through CF (Fig. 8B, bottom). Finally, the two curves on top of Fig. 8B express the increased opening of CF, in inactive and active conditions, which in the latter case is nothing else than the variation with $\Delta\mu H$ of the yield factor $\rho = (P/e^-)$ measured/ (P/e^-) theoretical maximum.

DISCUSSION

Leaving apart some speculations, such as a mixed electron and proton coupling mechanism[31], two extreme concepts bound the field of bioenergetics, namely the fully delocalized[1] and the strictly localized[3] chemiosmosis, presented in a way as irreconcilable. Our approach shows that actually a continuum of situations may exist, both theoretically (by adjusting the protonic resistances on Fig. 1A) and experimentally (Fig. 3). However, at variance with the above hypotheses, we have to conclude that the proton electrochemical potentials at the redox sources and at the coupling factors are different, because a ΔpH ($\Delta\mu H$) drop along a lateral resistance lowers the latter with respect to the former (see equations Fig. 1A). An illustration of this is given by the antiparallel behaviour of P/e^- and redox control when ^1H_2O is replaced by ^2H_2O (Fig. 4) or by the poor efficiency of SII compared to SI in the phosphorylating ability at a given $\Delta\mu H$ (Fig. 3).

One of the stumbling block for Mitchell's theory is the observation that the $\Delta Gp/\Delta\mu H$ ratio rises when $\Delta\mu H$ is decreased (ΔGp is here the free enthalpy — "phosphorylation potential" — of the ATP-forming reaction, quoted in Fig. 4). Actually, these results are now seriously questioned[32], and one must remark that besides some bias which may be introduced in this

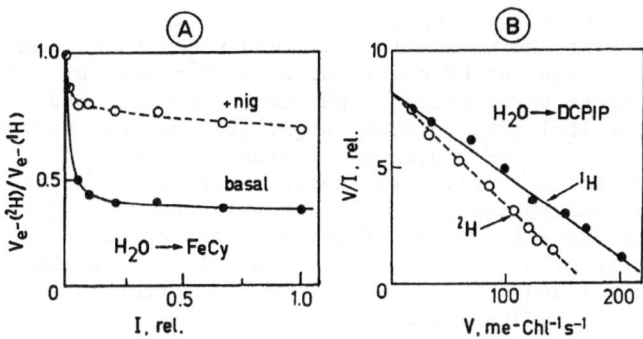

Fig. 6. Light-intensity curves of the heavy-water effect on the redox chains. Left (A): ratio of the electron flow rates, showing a stronger inhibition of the basal than of the uncoupled chain, involving plastoquinone (ferricyanide reduction). Right (B): confirmation that the quantum yield — extrapolation to zero V_{e-} — is unaffected (uncoupled DCPIP chain). Standard conditions (pure ^1H_2O or ^2H_2O buffers); I = light intensity.

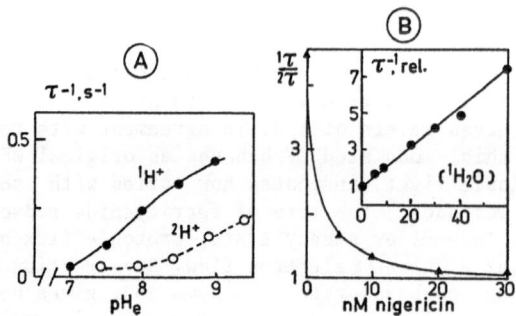

Fig. 7. Effect of heavy water on membrane permeability. Left (A): variation of the half-time reciprocal of the ΔpH decay with external
pH (corrected[11] in ^2H_2O). Right (B): isotope-independence of this
permeability when H^+-backflow is essentially determined by an
ionophore shuttle; insert: regular increase of proton-leakage
with nigericin concentration, meaning that the thylakoid suspension behaves as a homogeneous population (see Discussion).
Ferricyanide chain in strong light; standard conditions.

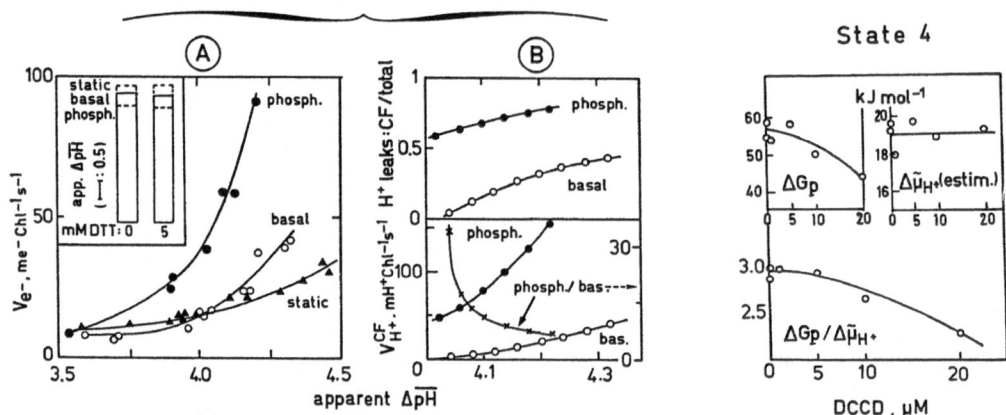

Fig. 8. Proton-leakage and coupling-factor state. Left (A): variation of
the rate of ferricyanide reduction with $\Delta\overline{pH}$, modulated by light, in
basal (b) state, phosphorylating (p) state 3, and static (s) state
4; insert: other indication, but under constant strong light and
with pyocyanin, of CF-contribution to H^+-leaks by comparison of
$\Delta\overline{pH}$ in these three states ($\Delta\overline{pH}$ static > basal), and absence of
dithiothreitol (DTT) effect. Right (B), bottom: variation of H^+
flux with $\Delta\overline{pH}$ ($\Delta\overline{\mu H}$: valinomycin present) with (partly) open CF,
inactive (o) or active (•), these rates being respectively b-s
and p-s, multiplied by H^+/e^- (2, here); the ratio (p-s)/(b-s) represents the stimulating effect of phosphorylation on leakage (x).
Top: ratio of H^+ flux via CF over total flux in these conditions,
(b-s)/b (o) and (p-s)/p (•), where the latter is the yield
factor $\rho = (P/e^-)$measured/(P/e^-)theoretical. Standard conditions.

Fig. 9. Possible occurrence of "scalar ATPases", suggested by a drop of the
phosphorylation potential ΔG_p without $\Delta\overline{\mu H}$ change. Curves traced
by increasing the concentration of dicyclohexylcarbodiimide (DCCD),
a CF_0 inhibitor; $\Delta\overline{\mu H}$ is computed from $\Delta\overline{pH}$ (valinomycin present),
corrected according to fig. 1 in ref. 6. Standard conditions in
strong light, pyocyanin chain in static state 4.

apparent variation of H^+/ATP stoichiometry by a systematic underestimation of $\Delta\mu H$ due to probes used, another source of "error" may concern ΔG_p, if ATP is hydrolysed without a vectorial backtransport of H^+ across CF. Fig. 9 shows indeed that, in state 4, ΔG_p may be reduced by addition of dicyclo-hexylcarbodiimide (DCCD) without changing $\Delta\mu H$, which suggests that this phosphorylation potential drop is due to some ATP hydrolysis by "scalar ATPases", since DCCD leaves in principle catalytic CF_1 operational whilst it blocks CF_0 proton channel.

Recently, the problem of population heterogeneity in flux-force type of experiments was raised[33]. Although this in fact is not relevant to the discussion of chemiosmosis, because a linearity between the two phenomena is by no means a prerequisite in this theory, it may indeed introduce apparent distortions open to erroneous interpretations. For instance, in the H^+-efflux (nigericin) vs. influx (light) experiments, if some thylakoids are uncoupled and some not, one may expect that nigericin curves would run over those traced by light reduction, as in Fig. 2. But if this were true,

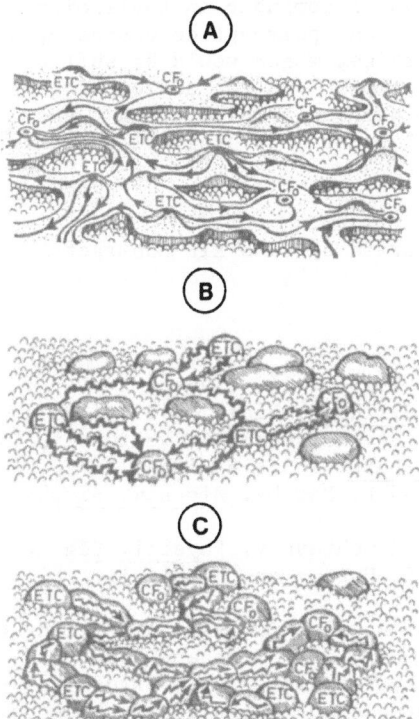

Fig. 10 Hypothetical proton lateral pathways, from electron transfer chain generators (ETC) to coupling-factors leaks, restricted here to their intramembraneous part (CF_0). Proteins (B, C) are drawn as irregular blocks scattered on the lipid layer, depicted by small balls (A, B, C). Top (A): H^+ flow in few irregular layers of "structured" water, covering hydrophilic parts of the membrane. Middle (B): H^+ move on top of polar lipid heads. Bottom (C): part of H^+ is channelled along (more or less shielded) polypeptides chains. In A and C, proton circulation avoids in part the apolar membrane core, hence its relative resistance — and that of phos-phorylation — to lipophilic nigericin; in all cases, a fraction of H^+ escapes into adjacent bulk phase. Though both membrane sides may be similar, the $\Delta\mu H$ heterogeneity may essentially depend on the restricted internal domain.

the ΔpH-relaxation, normally monophasic (see Fig. 5A) would become at least biphasic, which was never observed and, for instance, the insert in Fig. 7B shows a simple linear increase of the half-time reciprocal of the ΔpH-decay with nigericin concentration.

To end up, one may try to figure out what kind of structure, besides normal "bulk" water, would be able to conduct protons from their input to their output points. The comparison of Arrhenius plots of coupled electron flows (ln V_e-, basal, vs. 1/T) in ^{1}H and ^{2}H media[8] suggested to us that a few layers of ice-type "structured" water, bound on the membrane, may play this role (Fig. 10A). However, recent experiments on model systems[34] cast some doubt on the possibility to prevent a fast H^+ exchange between surface and bulk water. A second possibility is offered by the phospholipid polar heads (Fig. 10B), which have been found to efficiently transport H^+ ions on a monolayer surface, yet in direct contact with water[35]. But one may wonder then why lipophilic nigericin does not uncouple phosphorylation from $\Delta\overline{\mu H}$ more efficiently than shown in Fig. 2, not only in the case of SI, for which H^+ translocation may indeed occur quite close to CF, but also in case of SII: it is unlikely that lipids are absent over this long distance which separates SII from CF. In addition, neutral galactolipids and not charged phospholipids predominate in thylakoids. Thus, the most likely remains a proton hoping along polypeptide chains[36], buried in the membrane or within extrinsic proteins, which would be sufficiently shielded to avoid an excessive escape of H^+ in the surrounding medium (Fig. 10C). Though attractive, this proposal, which doesn't exclude other pathways, still needs experimental proof in biological systems.

Acknowledgments

This work was supported by CNRS-ATP contracts.

REFERENCES

1. P. Mitchell, Nature 191:144-148 (1961).
2. R. J. P. Williams, J. Theor. Biol. 1:1-17 (1961).
3. H. V. Westerhoff, B. A. Melandri, G. Venturoli, G. F. Azzone and D. B. Kell, FEBS Lett. 165:1-5 (1984).
4. Y. de Kouchkovsky and F. Haraux, Biochim. Biophys. Res. Commun. 99: 205-212 (1981).
5. F. Haraux and Y. de Kouchkovsky, Physiol. Vég. 21:563-576 (1983).
6. Y. de Kouchkovsky, F. Haraux and C. Sigalat, Bioelectrochem. Bioenerg. 13:143-162 (1984).
7. K. van Dam, A. H. C. A. Wiechmann, K. J. Hellingwerf, J. C. Arents and H. V. Westerhoff, in: "Membrane Proteins", P. Nicholls et al., eds., pp. 121-132, Pergamon Press, Oxford (1978).
8. F. Haraux and Y. de Kouchkovsky, Biochim. Biophys. Acta 679:235-247 (1982).
9. C. Sigalat, F. Haraux, F. de Kouchkovsky, S. Phung nhu Hung and Y. de Kouchkovsky, submitted to Biochim. Biophys. Acta.
10. G. Mackinney, J. Biol. Chem. 140:315-322 (1941).
11. P. K. Glasoe and F. A. Long, J. Phys. Chem. 64:188-190 (1960).
12. M. Nishimura, T. Ito and B. Chance, Biochim. Biophys. Acta 59:177-182 (1982).
13. J. L. Lemasters and C. R. Hackenbrock, in "Methods in Enzymology", S. P. Colowick and N. O. Kaplan, eds., 57, pp. 36-50, Acad. Press., New York (1978).
14. P. Gräber, U. Junesch and G. H. Schatz, Ber. Bunsenges. Phys. Chem. 88:599-608 (1984).
15. S. Schuldiner, H. Rottenberg and M. Avron, Eur. J. Biochem. 25:64-70 (1972).

16. F. Haraux and Y. de Kouchkovsky, Biochim. Biophys. Acta 546:455-471 (1979).
17. A. H. C. M. Schapendonk and W. J. Vredenberg, Biochim. Biophys. Acta 462:613-621 (1977).
18. G. F. Azzone, T. Pozzan, S. Massari and M. Bragadin, Biochim. Biophys. Acta 501:296-306 (1978).
19. A. Baccarini-Melandri, R. Casadio and B. A. Melandri, Eur. J. Biochem. 78:389-402 (1977).
20. B. Andersson and W. Haehnel, FEBS Lett. 146:13-17 (1982).
21. F. Haraux, C. Sigalat, A. Moreau and Y. de Kouchkovsky, FEBS Lett. 155:248-252 (1983).
22. J. W. Davenport and R. E. McCarty, Biochim. Biophys. Acta 766:363-374 (1984.
23. J. Barber, Biochim. Biophys. Acta 594:253-308 (1980).
24. A. B. Hope and D. B. Matthews, Aust. J. Plant Physiol. 12:9-20 (1985).
25. F. Haraux and Y. de Kouchkovsky, Biochim. Biophys. Acta 592:153-168 (1980).
26. Y. de Kouchkovsky, F. Haraux and C. Sigalat, FEBS Lett. 139:245-249 (1982).
27. U. Pick, H. Rottenberg and M. Avron, FEBS Lett. 32:91-94 (1973).
28. S. Izawa and N. E. Good, Biochim. Biophys. Acta 162:380-391 (1968).
29. M. Rathenow and B. Rumberg, Ber. Bunsenges. Phys. Chem. 84:1059-1062 (1980).
30. M. Schönfeld and H. Schickler, FEBS Lett. 167:231-234 (1984).
31. V. V. Lemeshko, Biofizika 27:420-424 (Engl. transl.: Biophysics 27: 429-434) (1982).
32. H. Woelders, W. J. van der Zande, A.-M. A. F. Colen, R. J. A. Wanders and K. van Dam, FEBS Lett. 179:278-282 (1985).
33. J. Duszynski and L. Wojtczak, FEBS Lett. 182:243-248 (1985).
34. E. Nachliel and M. Gutman, Eur. J. Bioch. 143:83-88 (1984).
35. J. Teissié, M. Prats, P. Soucaille and J.-F. Tocanne, Proc. Natl. Acad. Sci. USA 82:3217-3221 (1985).
36. J. F. Nagle and H. J. Morowitz, Biophysics 75:298-302 (1978).

THE PROTONMOTIVE ACTIVITY OF ENERGY
TRANSFER PROTEINS OF MITOCHONDRIA

Sergio Papa

Institute of Medical Biochemistry and Chemistry
University of Bari
70124 Bari, Italy

INTRODUCTION

Proton flow represents a major device for energy transfer by membrane proteins. The molecular mechanism by which the protonmotive force is generated, transmitted and utilized is, however, matter of debate. The available knowledge supports the concept of Mitchell [1,2] that vectorial organization of primary protolytic reactions at the catalytic sites is central to energy transfer. The postulate [1] that vectoriality derives simply from anisotropic diffusion of the same chemical groups involved in primary catalysis doesn't appear to be equally satisfactory [3].

A number of observations point out to a critical role of ion interactions in the polypeptides of energy transfer systems, possibly co-operatively linked to the primary chemical events [3-5]. These ionic interactions may be directly involved in determining the anisotropy of the primary chemical event, the specific ions that are transported and the stoicheiometric factors. A striking example is provided by bacterial rhodopsins, where different apoproteins complement the light-driven stereochemical transition of retinal to generate a proton or a chloride pump [6]. An analogous situation might hold for the redox-driven Na^+ pump of certain bacteria [7] which could result from selective ion interactions associated by specific proteins to the primary proteolytic process.

The present paper deals with the role that ionic interactions in polypeptides and specific aminoacid residues can play in the protonmotive activity of the cytochrome system and H^+-ATP synthase of mitochondria.

THE CYTOCHROME SYSTEM

The cytochrome system of the mitochondrial membrane is composed of twenty or more polypeptides which are in excess of the metal redox centers amounting to nine [3]. These proteins, besides providing regulatory binding sites for the redox prosthetic groups and facilitating mobility of organic hydrogen carriers [2], can constitute conducting pathways mediating H^+-translocation from the aqueous phase to hydrogen carriers in the membrane [5]. Even more it could just be the specific aminoacids in the vicinity of

133

the primary redox centers and their redox-linked pK shifts to determine asimmetriy of protonation/deprotonation of the hydrogen carriers |3|.

Electron flow along the cytochrome chain results in the effective translocation of protons from the matrix (N) space to the external (P) space. The flow of two reducing equivalents from ubiquinol to cytochrome c results in the ejection from mitochondria of 4 protons.Two derive formally from scalar oxidation of hydrogenated ubiquinol by electron carriers, the other two are transported from the matrix to the outer space |2-5,8,9|.

It is disputed whether $\Delta\mu H^+$ generation by cytochrome oxidase derives only from anisotropic arrangement of the reduction of O_2 to H_2O, whereby protons are consumed from the matrix (N) space |10-12|, or, if in addition to this energy is also conserved by proton pumping from the N to the P space |9|.

Two general mechanisms have been proposed for the protonmotive activity of cytochrome c reductase. Direct-ligand conduction mechanism, whose most detailed version is provided by the quinone cycle of Mitchell |2|(Fig. 1a) and co-operative redox-linked proton translocation by hemoproteins or

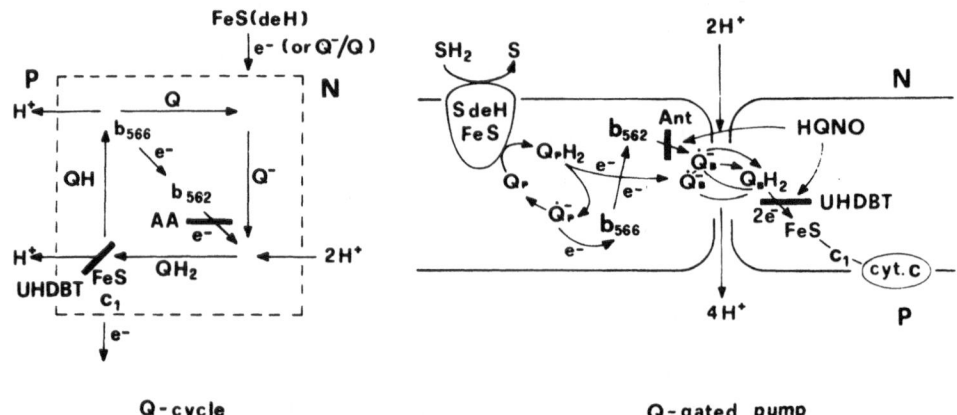

Q-cycle   Q-gated pump

Fig.1. Models for proton translocation in cytochrome c reduc-
tase.Ubiquinone cycle |2|,Q-gated proton pump |3,12|.

metalloproteins |3|. A model of the latter type was originally proposed by Papa et al.,|4,13| for b cytochromes. More detailed models of proton trans location by b cytochromes have been later developed by Von Jagow et al.|14| and Wikström et al. |9|.

Difficulties to explain the observations available, purely in terms of vectoriality or, alternatively, of co-operativity, have led us to consider the possibility that H^+ translocation derives from combination of these two types of mechanisms |3,5|.

We favour a model (Fig.1b) |3,12,15| where ubiquinol of the pool delivers 1 e^- to protein-bound semiquinone, through an antimycin-insensitive reaction. The anionic ring of the semiquinone is attracted at the P side of the membrane and is oxidized by b_{566} which then transfers electrons to b_{562} at the N side and from this to bound semiquinone through an antimycin sensitive reaction. The two electrons reunite on bound quinone and from this pass to the Fe-S protein. The b cytochrome shunt explains the well known oxidant-induced reduction of b cytochromes |3,4,16|.

134

Protonation of the UQ̊/UQH$_2$ couple from the N side of the membrane upon reduction and H$^+$ release,upon oxidation,at the P side,both provided by specific H$^+$ conduction pathways in the proteins can result in transmembrane H$^+$ translocation. Since the semiquinone has a pK around 5-6 this can transfer 2 H$^+$ per 1 e$^-$ at physiological pH's.

Evidence for a role of polypeptides in the protonmotive activity of cytochrome c reductase is provided by experiments based on chemical modification of subunits or their digestion by proteolytic enzymes [17,18].

SDS electrophoresis resolves the purified reductase in eight major bands(Fig.2). When the soluble reductase was treated with ^{14}C-dicyclohexyl-carbodiimide (DCCD)a large binding of the reagent to the 8 kDa band was observed [cf.19].

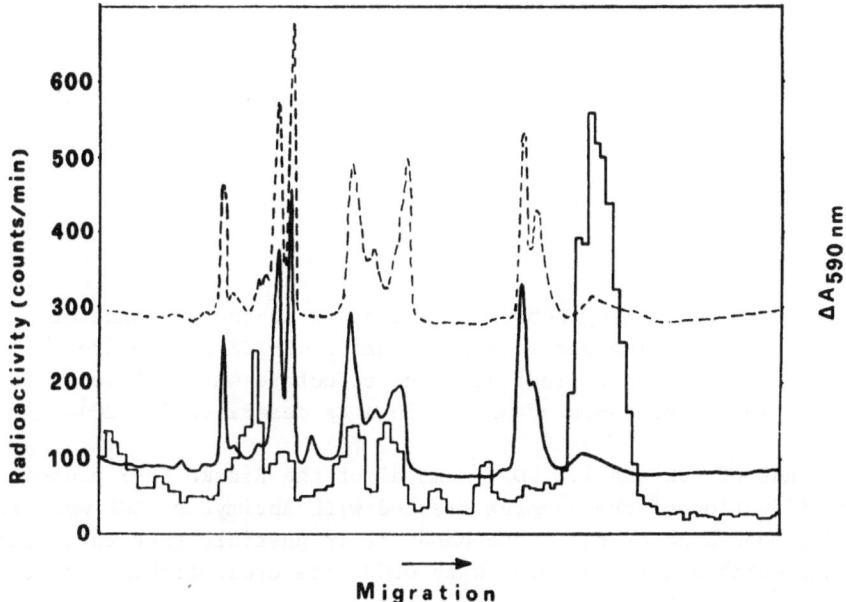

Fig.2 Densitometric profiles and ^{14}C-DCCD distribution of SDS-PAGE of cytochrome c reductase treated with ^{14}C-DCCD.The gels stained with Coomassie blue were scanned at 590 nm.The solid profile refers to the reductase incubated for 30 min at room temperature with 200 nmol.^{14}C-DCCD/mg protein.The dotted profile is of the control enzyme. From ref. [17].

Treatment with DCCD caused attenuation of the 8 kDa band.The 12 kDa band and the Fe-S protein were also reduced after treatment with DCCD and,concomitantly,a new band with no radioactivity and an apparent molecular weight of 40 kDa appeared.

It is likely that a polypeptide in the 8 and or 12 kDa band, after reaction with ^{14}C-DCCD,gets cross-linked with the Fe-S protein with release of urea derivative of ^{14}C-DCCD.

The gels in Fig.3, show that urea-treatment resolves better the two core proteins, the b and c$_1$ cytochromes and the Fe-S protein. It resolves the low M.W. bands into 5 polypeptides. The 8 kDa diffuse band is resolved

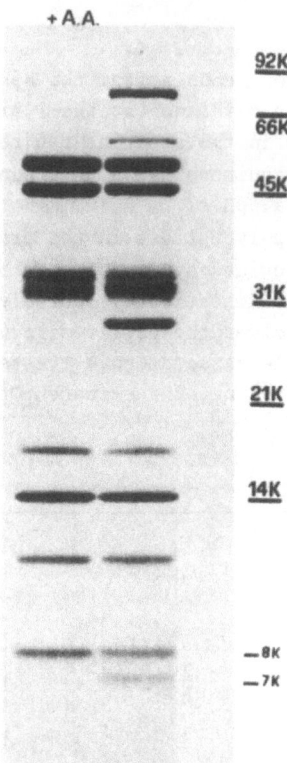

+ A.A.

92K
66K
45K
31K
21K
14K
8K
7K

Fig.3. SDS-PAGE of cytochrome c reductase in the presence of
6 M urea.The separating gel was a 12-20% acrylamide
gradient.Where indicated the reductase was isolated
in the presence of antimycin A as described in |20|.

into two bands of 8.0 and 7.0 kDa. Removal of the Rieske Fe-S protein affec-
ted by purification of the complex treated with antimycin |20| was accompa-
nied by disappearance of the 7 kDa band. It is possible that this component
is the same which after treatment with DCCD gets cross-linked with the Fe-S
protein.

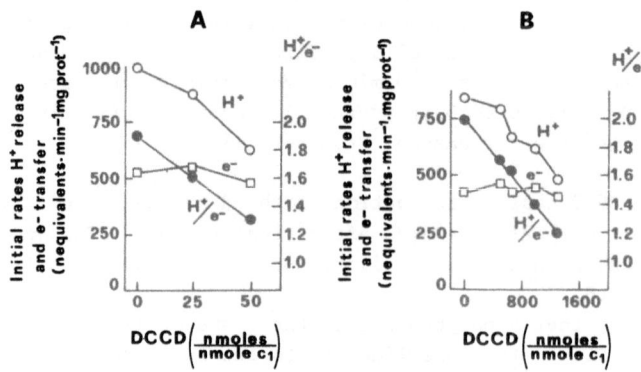

Fig.4. Effect of DCCD on redox-linked proton translocation in
cytochrome c reductase vesicles.In (A) the reductase
was treated with DCCD prior to reconstitution with phos-
pholipids;in (B) after reconstitution |from 17|.

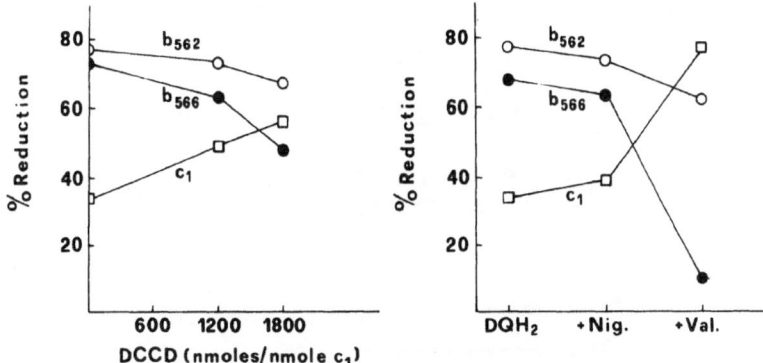

Fig.5. Effect of DCCD and ionophores on steady-state redox
levels of b and c cytochromes in turning-over cytochrome c reductase vesicles.Where indicated the vesicles were incubated with DCCD for 30 min at 25°C.The
vesicles (1 μM cytochrome c reductase)were supplemented with 0.2 μM cytochrome c and 0.1 μM soluble cytochrome oxidase. Steady-state oxygen consumption was
supported by 200 μM duroquinol.Where indicated 1 μg
valinomycin and 1 μg nigericin were added.For other
details see |15|.

Treatment of the $b-c_1$ complex with DCCD, either before or after reconstitution with liposomes depressed H^+ release elicited by the addition of
duroquinol to cytochrome c supplemented $b-c_1$ vesicles.The H^+/e^- ratio was
lowered from 2 to about 1 (Fig.4).

The experiments of Fig.5 identify the site in the reductase at which
DCCD decouples electron flow. $b-c_1$ Vesicles were supplemented with a small
amount of cytochrome c and cytochrome oxidase. The addition of duroquinol
provided steady-state electron flow to oxygen. The ensuing transmembrane
$\Delta\mu H^+$ exerted back pressure on electron flow from b cytochromes to cytochrome c_1.The steady-state reduction of b cytochromes amounted to 80% whilst
that of c cytochromes was not larger than 30%.The inhibition of electron
flow was released by collapse of $\Delta\mu H^+$ with valinomycin(plus nigericin)which
resulted in large oxidation of b_{566} and reduction of cytochrome c. Cytochrome b_{562} was not significantly affected.

Treatment with DCCD resulted in oxidation of b cytochromes, in particular b_{566} and reduction of c cytochromes. Thus the chemical modification
specifically abolished the crossover effect exerted by $\Delta\mu H^+$ between cytochrome b_{562} and cytochrome c_1.

Fig.6 shows the effect of papain-treatment on the polypeptides of the
$b-c_1$ complex |18|. The first lane represents the SDS-PAGE of the control,
the second refers to the enzyme in the presence of papain, whose activity
was blocked at zero time by TCA. The other lanes show the reductase after
various intervals of treatment with papain. This resulted in partial dige-
sion of core protein II, which was already completed in 5 min and corre-
sponded to a loss of a 3 kDa fragment. Papain caused also enhanced migration of the 15 kDa band. Furthermore there occurred digestion of the Fe-S
protein.

A time course of the effect of papain digestion on electron transfer
and H^+ translocation in $b-c_1$ vesicles shows (Fig.7) that whilst the inhibi-

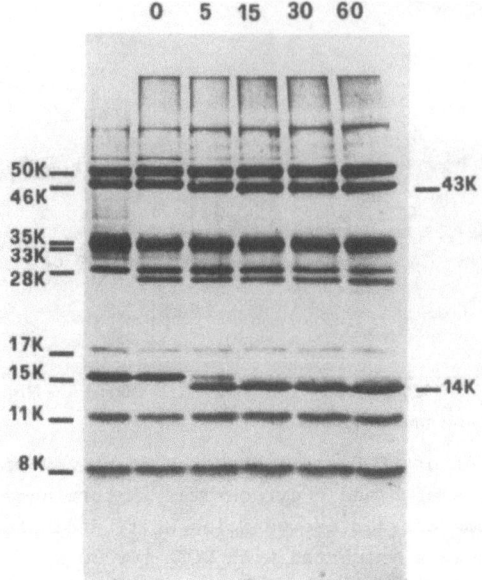

Fig.6. Coomassie blue stained gels of b-c$_1$ complex;effect
of papain digestion b-c$_1$ Complex was treated with
papain 1 mg/20 mg reductase protein at room
temperature. From |18|.

tion of the reductase activity paralleled the digestion of the Fe-S protein,
H$^+$ pumping was already suppressed by 60% after 5 min incubation, when the
Fe-S protein digestion and the inhibition of electron flow were negligible,

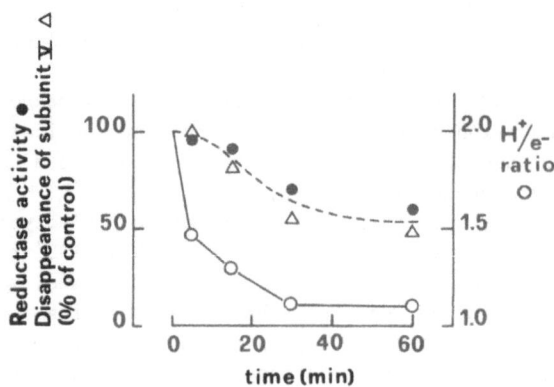

Fig.7. Time course of the effect of papain digestion on polypeptides and enzymatic activity of b-c$_1$ complex. The band
corresponding to subunit V (△—△) was integrated and the
integrals are reported as percentage of the control band.

138

but digestion of core protein II and of the 15 kDa band were almost completed.Thus these proteins,or at least one of the two,in addition to the 8 and 12 kDa proteins modified by DCCD appear to be involved in the pumping activity of the complex.

It thus appears that coupling between electron flow and H^+ translocation in the reductase depends on the ordered arrangement and interaction of various polypeptides.

Polypeptides can bind and stabilize the semiquinone species, thus preventing uncoupled dismutation. They can also provide pathways for H^+ conduction between aqueous phases and bound quinone.

Salt-bridged basic and acidic residues, co-operatively linked to protolytic redox catalysis at the quinone center,could result in the proper adjustement in space and in time of the proton donating and proton accepting capabilities so to result in asymmetric protonation-deprotonation of the quinone system (Fig.8).

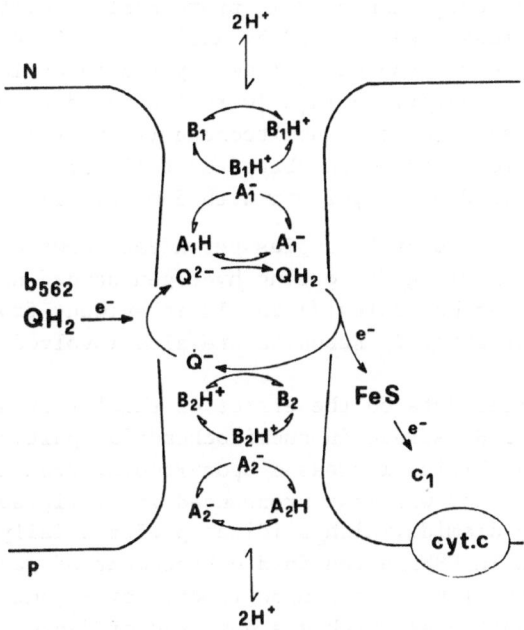

Fig.8. Possible involvement of co-operative H^+ translocating
network of salt-bridges in the proton pumping activity
of cytochrome c reductase.

The H^+-ATP synthase

The proton translocating system $(F_o)$of H^+-ATP synthase can be conceived|21-25|as consisting of two functional regions:(i)the channel proper, which mediates proton translocating across the lipid bilyer, and (ii) the gate which controls the direction of the proton current and its coupling to the hydro-dehydration catalysis in F_1.The gate is constituted by F_o and F_1 subunits apparently associated by weak electrostatic interactions |21|.

The 8 kDa hydrophobic protein of F_o is an essential component of the channel|21-23|.A critical role for proton conduction is contributed by a conserved glutammic(or aspartic)residue in the carboxyl-terminal hydrophobic segment of this protein|23|.Other polar residues,in particular hydroxyl and basic residues,located in the two hydrophobic stretches of the 8 kDa

Table 1. Effect of thiol reagents on H^+ release from submitochondrial particles and F_o-liposomes induced by valinomycin mediated K^+-influx.

	Rate H^+ release $1/t_{1/2}$ (sec^{-1})									
	ESMP (F_o,F_1)		USMP ($-F_1$)		ASMP ($-F_1;-OSCP$)		AESMP ($-F_B$)		F_o-liposomes	
	sec^{-1}	%	sec^{-1}	%	sec^{-1}	%	sec^{-1}	%	sec^{-1}	%
—	0.5		1.54		0.62		0.92		2.50	
DCCD 50 µM	0.15	−70	0.71	−54	0.20	−68	0.26	−72	0.80	−68
MalNet 5 mM	0.36	−28	1.00	−35	0.37	−40	0.29	−68	0.60	−76
DTP 2 mM	0.35	−30	0.59	−62	0.18	−71	0.32	−65	0.50	−80
Diamide 5 mM	6.67	+1234	5.0	+225	0.66	+7	0.75	−19	1.88	−25

Submitochondrial particles prepared as described in |26|,were incubated at a concentration of 3 mg protein/ml in 0.15 M KCL at 25°C.Then H^+ release was started by addition of 2 µg valinomycin/ mg protein.Proton release was followed potentiometrically.F_o-liposomes prepared as described in |25| were incubated under the same conditions as described for submitochondrial particles.Where indicated the vesicles were preincubated as follows:15 min with 50 µM DCCD, 5 mM NEM or 2 mM DTP, 2 min with 5 mM diamide.

protein ,which exists in 6 or 10 copies per ATPase complex |22-24|,could contribute to proton channeling by forming hydrogen bonded networks,possibly with inclusion of water molecules |21,25|.It is evident from the work on bacterial H^+-ATPase that other F_o subunits are also involved in the proton channel |24|.

Table 1 summarizes data on the effect of thiol reagents on passive proton conduction in the H^+-ATPase in submitochondrial particles and F_o-liposomes.The extent of inhibition increased upon removal from the particles of F_1, OSCP and Factor B and was most pronounced in F_o-liposomes.

On the contrary diamide,which oxidizes preferentially vicinal dithiols to disulfides,caused in ESMP a ten fold enhancement of passive H^+conduction in particles which still persisted upon removal of F_1,but caused inhibition of H^+conduction in particles which are deprived of Factor B and F_o liposomes. Thus the stimulation of H^+conduction effected by diamide results from modification of thiol residues different from those involved in the inhibition caused by NEM and DTP.

It should be mentioned that the inhibitory effects of NEM and DTP were additive with the inhibition exerted by DCCD or oligomycin and the stimulation by diamide was abolished by these reagents.This shows that we are, indeed,dealing with an effect of thiol reagents on proton conductivity of the ATPase complex.

The thiol groups modified by NEM were identified by electrophoretic analysis of the H^+-ATPase components after treatment of the particles with ^{14}C-NEM.Among the various bands reacting with ^{14}C-NEM significant binding of the reagent to the 25 and 8 kDa proteins of the F_o sector occurred (Fig.9).

Carboxyatractylate,which potentiated the inhibitory action of NEM on H^+translocation, enhanced significantly the binding of the reagent to the 25 and 8 kDa bands,had no effect on the binding to the 31 kDa band and γ

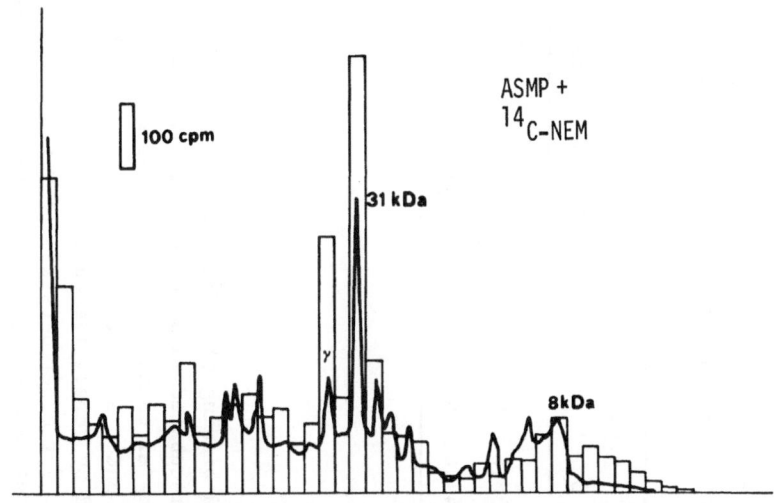

Fig.9. Densitometric profile and distribution of ^{14}C-NEM of Coomassie blue stained electrophoretic slabs of submitochondrial particles from beef-heart. The particles,prepared
by sonication of mitochondria in the presence of EDTA fol-
lowed by treatment with urea and ammonia salts|26|were in-
cubated at (3 mg/ml)with 3 mM ^{14}C-NEM,at room temperature.
Electrophoresis was performed on slabs of linear gradient
polyacrylamide(12-20%)gel.For electrophoresis,samples of
50 µg proteins of the particles were washed with acetone
and suspended in 3% SDS, 10% glycerol and 60 mM Tris-
HCl, pH 6.8. After Coomassie blue staining and densitometric analysis(solid line)the gel was sliced and protein-
bound radioactivity was estimated as described in|27|.

subunit of F_1 and depressed the binding to high M.W.bands (Fig.10).
    Table 2. shows that both in particles and F_O-liposomes DTP depressed
the binding of ^{14}C-DCCD to the 8 kDa protein. Thus the inhibition exerted
by monofunctional reagents on proton conduction in F_O seems to result from
reaction with thiol groups in the 8 kDa band.
    It is known that DTP reacts preferentially with thiol groups located
in proximity of acidic residues|28|. The observation that DTP was signifi-
cantly more powerful than MalNet in inhibiting proton conduction in submitochondrial particles and that it caused specific depression of the binding
of ^{14}C-DCCD to the 8 kDa protein, favours the possibility that the inhibition of proton conduction derives from reaction with the only cysteine residue present in the 8 kDa DCCD binding protein mitochondria at position
64, thus only six residues apart from the critical glutammic residue|23|.
On the other hand if critical thiol residues in the 25 kDa protein are in-
bolved these should be located close to the DCCD reactive glutammic residue
in the 8 kDa protein.
    As regard to the mechanism by which thiol groups are involved in pro-
ton conduction, it could be possible that protons move directly on these
residues.It is alternatively possible that the reaction of thiol residues
with monofunctional reagents causes alterations of the structure of F_O polypeptides, which results indirectly in depression of proton conduction.

Binding of ^{14}C-NEM to various protein bands resolved
by SDS-PAGE of ESMP

Reagents	ESMP					
	90 — 50 kDa	β	γ	31 kDa	25 kDa	8 kDa
^{14}C-NEM	403.2	56.6	67.7	130.0	22.1	27.2
C.A.+^{14}C-NEM	181.4	41.6	90.4	125.6	53.1	61.5

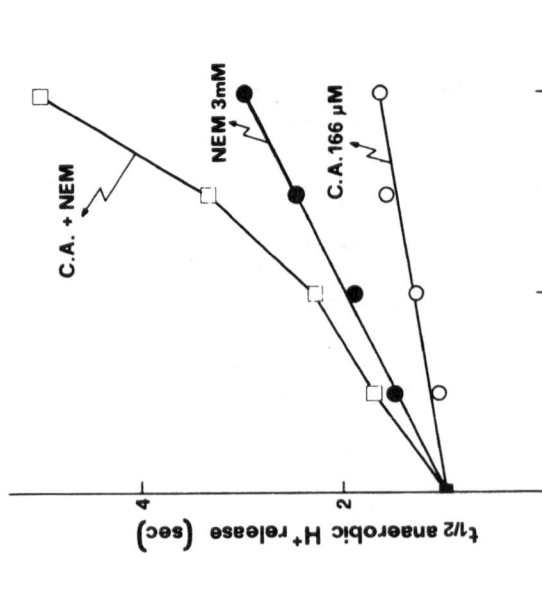

Fig.10. Effect of atractyloside on inhibition of passive H$^+$conduction by NEM and binding of ^{14}C-NEM to polypeptides in submitochondrial particles. For H$^+$translocation ESMP (3 mg/ml) were incubated in a reaction mixture containing: 200 mM sucrose, 30 mM KCl, 0.5 μg valinomycin/mg protein, 0.2 mg/ml catalase and 20 mM succinate (K$^+$ salt), pH 7.5. Incubation was carried out in a stirred glass vessel under a constant stream of N$_2$ at 25°C. After anaerobiosis respiration driven proton translocation was activated by repetitive pulses of 1-3% H$_2$O$_2$ (5 μl/ml) and the pH of the suspension was monitored potentiometrically. For other details see ref.|26|. ESMP were incubated with 3 mM cold or ^{14}C-NEM for 20'. Where indicated ESMP were preincubated with 160 μM carboxyatractylate (C.A.) for 5' before the addition of NEM. After densitometric scan gel slabs were sliced and protein radioactivity was estimated as described in|27|. The values reported in the Table refer to pmoles of ^{14}C-NEM per electrophoretic band.

Table 2. Effect of DTP on the binding of ^{14}C-DCCD to the 8 kDa band in ESMP and purified F_o

Reagents	ESMP		F_o	
	cpm	%	cmp	%
[14]C-DCCD	765		426	
DTP + ^{14}C-DCCD	403	-47	329	-23

ESMP and purified F_o (3 mg/ml) were incubated with the reagents as follows: 20 min with 50 μM ^{14}C-DCCD; 15 min with 0.8 mM DTP followed by 20 min with 50 μM ^{14}C-DCCD. After densitometric analysis gel strips were sliced and solubilized for counting as described in |27|.

In addition to diamide other aminoacid reagents have been found to enhance passive proton conduction in the H^+-ATPase. These are butanedione|29| which reacts with arginine or other basic residues and ethoxyformic anhydride, a modifier of hystidyl residues |25|(Fig.11).

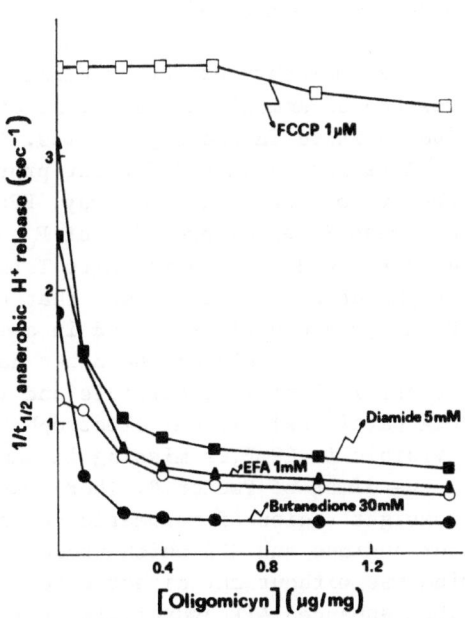

Fig.11. Effect of aminoacid modifiers on the rate of anaerobic H^+ release and its inhibition by oligomycin in ESMP. For experimental conditions see legend to Fig.10 and |26| and |27|. After addition of the reagents, particles were incubated in anaerobiosis for 2 min, then oligomycin was added at the concentration reported in the figure and after 10 min respiratory pulses were elicited by addition of oxygen.

As observed for diamide also the stimulatory effects of butanedione and EFA were completely reversed by oligomycin.

A comparison of the stimulatory effect of the three reagents in various types of particles (Table 3) showed that whilst the stimulatory effect of butanedione and diamide were retained when F_1 was removed from the membrane, the effect of EFA was lost. Furthermore the activity of F_1 obtained from particles treated with EFA was found to be severely inhibited.

Table 3. Effect of diamide,ethoxy-formic anhydride and butanedione on anaerobic release of respiratory $\Delta\mu H^+$ in sonic submitochondrial particles.

Additions	ESMP (F_O,F_1) $1/t\frac{1}{2}$ (sec^{-1})	USMP ($-F_1$) $1/t\frac{1}{2}$ (sec^{-1})	AESMP ($-F_B$) $1/t\frac{1}{2}$ (sec^{-1})
—	1	2.0	1.8
Diamide	7.7	6.6	1.5
EFA	7.1	1.7	4.5
Butanedione	2.0	5.0	12.0

Submitochondrial particles were incubated in anaerobiosis with 5 mM diamide, 1 mM EFA or 10 moles butanedione per mg protein for 2 min at room temperature.For other experimental details see legend to Fig.10 and |26 and 28|.

As previously reported the stimulatory effect of diamide was lost in particles prepared in the presence of ammonia and EDTA.The effect of butanedione was,on the other hand, potentiated in these particles.

Thus the three modifiers reacts with different proteins apparently involved in the gate of the proton conducting pathway. EFA acts on F_1 polypeptides, butanedione on membrane integral proteins of F_O, diamide on a peripheral protein of F_O which is removed by ammonia salts.This component cannot be the 30 kDa band.Electrophoretic analysis showed that this band which is heavily labelled by ^{14}C-NEM, was clearly retained in ammonia-EDTA particles. The components modified by diamide could,on the other hand,be Factor B|30|. This protein has in fact critical vicinal dithiols and is reported to be removed from ammonia-EDTA particles|30|. Gel electrophoresis revealed that a 23 kDa band, clearly visible in ESMP,is missing in ammonia-EDTA particles. This band may represent the dimer of Factor B. There was also a decrease of a band in the molecular weight region of monomeric Factor B|31|.

At difference of butanedione and EFA which produced inhibition of ATP hydrolysis by F_1, diamide was without any direct effect on the catalytic activity of soluble F_1,but enhanced ATP hydrolysis by membrane bound ATPase |31|. Interesting enough the diamide-stimulated ATPase activity exhibited the same sensitivity to oligomycin as the control|31|.

Treatment of submitochondrial particles with diamide under conditions, resulting in oligomycin-sensitive enhancement of passive proton conductivity and ATPase activity,caused partial depression of the [32]Pi-ATP exchange and of the energy-linked reversed electron flow from succinate to NAD$^+$|31|.

It is conceivable that oxidation of dithiol(s) to difulfide in peripheral protein(s) of F_O induces an open configuration of the gate thus enhancing the flow of protons from the catalytic site in F_1 at the M side to the oligomycin-sensitive proton conduction pathway in F_o. Proton conduction by the F_o channel in the reverse direction from the C to the M side,is even more markedly enhanced and partly decoupled from hydro-dehydration catalysis by diamide treatment.

The thiols modified by diamide are apparently located in superficial proteins of F_o which together with F_1 components constitute the gate of the H^+-ATPase. Hystidine residues in F_1 subunits and basic residues in membrane integral components of F_o, apparently participate in the gate function and may play a critical role in the coupling between proton translocation and hydro-dehydration catalysis.

REFERENCES

1. P.Mitchell, Chemiosmotic coupling in oxidative and photosynthetic phosphorylation, Glynn Research Bodmin,(1966).
2. P.Mitchell, Possible molecular mechanism of the protonmotive function of cytochrome system, J.Theor.Biol. 62:327,(1976).
3. S.Papa and M.Lorusso, The cytochrome chain of mitochondria:Electron transfer reactions and transmembrane proton translocation,in:"Biomembranes"R.M.Burton and F.Carcalho Guerra eds.,Plenum Pub.Corp.N.Y.(1984).
4. S.Papa, Proton translocating reaction in the respiratory chain,Biochim. Biophys.Acta 456:39, (1976).
5. S.Papa, Molecular mechanism of proton translocation by the cytochrome system and the ATPase of mitochondria.Role of proteins. J.Bioenerg. Biomembr. 14/69, (1982).
6. J.Tittor, P.Hegemann and D.Oesterhelt, Retinal as a molecular switch in ion pumps,in:"Ion interactions in energy transport systems",G.Papageorgiou,J.Barber,M.Karajannis,S.Papa eds.,Plenum Publ.Corp.N.Y.(1985)
7. H.Tokuda,M.Sugasawa and T.Unemoto,Roles of Na^+ and K^+ in α-aminoisobutyric acid transport by the marine bacterium Vibrio alginolyticus, J.Biol.Chem.257:788,(1982)
8. S.Papa,M.Lorusso and F.Guerrieri,Mechanism of respiration driven proton translocation in the inner mitochondrial membrane.Analysis of proton translocation associated with oxidation of endogenous ubiquinol, Biochim.Biophys.Acta 387:425, (1975)
9. M.Wikström, K.Krab and M.Saraste,Proton-translocating cytochrome complexes, Ann.Rev.Biochem. 50:623,(1981).
10.P.Mitchell, Protonmotive cytochrome system of mitochondria, Ann.N.Y. Acad.Sci.341:564, (1980)
11.S.Papa, F.Guerrieri and M.Lorusso,Mechanism of respiration-driven proton translocation in the inner mitochondrial membrane.Analysis of proton translocation associated to oxido-reductions of the oxygenterminal respiratory carriers, Biochim.Biophys.Acta 357:181, (1974).
12.S.Papa,Mechanism of active proton translocation of cytochrome systems,in: "Membranes and Transport"A.N.Martonosi ed.Plenum Publ.Corp.N.Y. (1982).
13.S.Papa,F.Guerrieri,M.Lorusso and S.Simone,Proton translocation and energy transduction in mitochondria,Biochimie 55:703 (1973).
14.G.Von Jagow,W.D.Engel and H.Schägger, On the mechanism of proton translocation linked to electron transfer at energy conversion site 2, in: "Vectorial Reactions in Electron and Ion Transport in Mitochondria and Bacteria",F.Palmieri,E.Quagliariello,N.Siliprandi,E.C.Slater,eds. Elsevier/North Holland Biomedical Press, Amsterdam, (1981).

15. S.Papa,M.Lorusso,D.Boffoli and E.Bellomo,Redox-linked proton transloca-
    tion in the b-c$_1$ complex from beef-heart mitochondria reconstituted
    into phospholipid vesicles.General characteristics and control of
    electron flow by $\Delta\mu H^+$, Eur.J.Biochem.137:405 (1983).
16. M. Wikström, and J. Berden,Oxidoreduction of cytochrome c in the presen-
    ce of antimycin, Biochim.Biophys.Acta 283:403, (1972).
17. M.Lorusso,D.Gatti, D.Boffoli, E.Bellomo and S.Papa, Redox-linked proton
    translocation in the b-c$_1$ complex from beef-heart mitochondria recons-
    tituted into phospholipid vesicles.Studies with chemical modifiers of
    aminoacid residues, Eur J.Biochem.137:413 (1983).
18. M.Lorusso,D.Gatti,M.Marzo and S.Papa,Effect of papain digestion on redox
    linked proton translocation in b-c$_1$ complex of beef-heart reconstitu-
    ted into liposomes, FEBS Lett.182:370 (1985).
19. M.Degli Esposti,E.M.Meier,J.Timoneda and G.Lenaz, Modification of the
    catalytic function of the mitochondrial cytochrome b-c$_1$ complex by
    dicyclohexylcarbodiimide,Biochim.Biophys.Acta 725:349, (1983).
20. G.Von Jagow,H.Schägger,W.D.Engel,P.Riccio,H.J.Kolb and M.Klingenberg,
    Complex III from beef-heart:Isolation by hydroxyapatite chromatography
    in Triton X-100 and characterization,in:Methods in Enzymology,Vol.53,
    S.Fleischer and L.Packer eds.,Academic Press, New York,(1978).
21. S.Papa,F.Guerrieri,Proton conduction by H^+-ATPase,in:"Chemiosmotic Pro-
    ton Circuits in Biological Membranes"V.P.Skulachev and P.C.Hinkle eds.
    Addison-Wesley Publishing Company,Inc.Reading Mass,(1981).
22. P.Pedersen,M.Amzel,Proton ATPases:Structure and Mechanism,Ann.Rev.Bio-
    chem.52:801 (1983).
23. W.Sebald,J.Hoppe and E.Wachter,Aminoacid sequence of the ATPase proteo-
    lipid from mitochondria,chloroplasts and bacteria(Wild type and mu-
    tants) in:"Function and Molecular Aspects of Biomembrane Transport"
    E.Quagliariello,F.Palmieri,S.Papa,M.Klingenberg eds., Elsevier/North
    Holland Biomedical Press, Amsterdam (1979).
24. J.Hoppe and W.Sebald,The proton conducting F$_o$-part of bacterial ATP
    synthases, Biochim.Biophys.Acta 768:1 (1984)
25. S.Papa,F.Guerrieri,F.Zanotti and R.Scarfò,Flow and interactions of pro-
    tons in the H^+-ATPase of mitochondria,in:"Information and Energy
    transduction in biological membranes" C.L.Bolis,E.J.M.Helmereich,H.
    Passow,eds.,Alan R.Liss, Inc.New York, (1984).
26. A.Pansini,F.Guerrieri and S.Papa,Control of proton conduction by the
    H^+-ATPase in the inner mitochondrial membrane,Eur.J.Biochem.92:45(1978).
27. J.Kopecky,F.Guerrieri and S.Papa,Interaction of dicyclohexylcarbodiimide
    with the proton conducting pathway of mitochondrial H^+-ATPase, Eur.J.
    Biochem. 131:17 (1983)
28. K.Brocklehurst and Little,G.,Reactions of papain and low M.W.thiols with
    some aromatic disulphides.Biochm.J. 133:67, (1973)
29. F.Guerrieri and S.Papa,Effect of chemical modifiers of aminoacid resi-
    dues on proton conduction by the H^+-ATPase of mitochondria, J.Bioenerg.
    Biomembr.13:393 (1981).
30. R.Sanadi,Mitochondrial coupling Factor B, Biochim.Biophys.Acta,683:39
    (1982).
31. F.Zanotti,F.Guerrieri,R.Scarfò,J.Berden and S.Papa, Effect of diamide
    on proton translocation by the mitochondrial H^+-ATPase, in prepara-
    tion.

PROTON TRANSPORT-COUPLED ATP SYNTHESIS CATALYZED BY THE CHLOROPLAST ATPase

Peter Gräber, Ulrike Junesch, Günter Schmidt
and Petra Fromme

Max-Volmer-Institut für Biophysikalische und
Physikalische Chemie, Technische Universität
Berlin, Strasse des 17. Juni 135, 1000 Berlin
12, FRG

INTRODUCTION

The membrane-bound ATPase in chloroplasts can catalyze ATP synthesis/hydrolysis coupled with a transmembrane proton transport. The detailed mechanism by which proton transport gives rise to the formation of a phosphate-anhydrid bond is still not known. We have investigated the kinetics of ATP synthesis with chloroplasts and with CF_oF_1 reconstituted into liposomes. From these results we want to draw conclusions about the mechanism of the reaction. The rate of ATP synthesis was measured with a rapid double-mixing system using artificially impressed transmembrane pH difference, ΔpH, electric potential difference, $\Delta \Psi$, and phosphate potential, ΔG_p (1).

METHODS

ATP synthesis was measured using energization by an acid-base jump with additional K^+/valinomycin diffusion potential. The ATP yield was determined at different reaction times using a rapid multimixing system (1). Fig. 1 shows such experiments at constant ΔpH with varying diffusion potentials. At small reaction times a linear increase of the yield is seen, while at higher reaction times the slope decreases due to a decrease of ΔpH caused by proton efflux. The rate of ATP synthesis was calculated from the slope of the initial linear range, where ΔpH is still at the initial given value. Fig. 2 shows the ATP yield as a function of reaction time at constant $\Delta \Psi = 85$ mV and ΔpH varying between 0.85 and 4.2.

The magnitude of $\Delta \Psi$ was calculated from the Goldmann-Hodgkin-Katz equation as described earlier (1), since the permeability coefficients of the ions, other than K^+, cannot be neglected under our conditions.

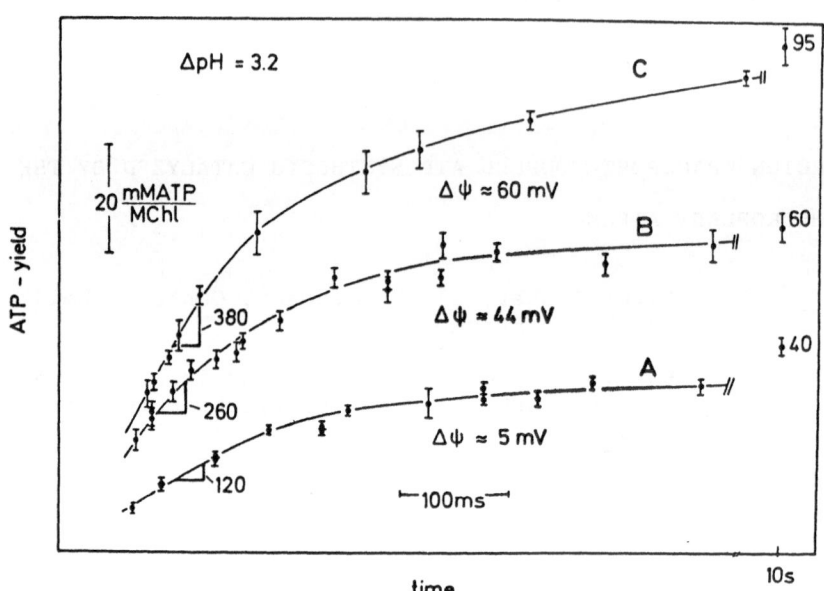

Fig. 1. ATP yield as a function of reaction time from quenched flow experiments at pH_{out} = 8.2, ΔpH = 3.2 and $\Delta\Psi$ = 6 mV, 44 mV and 60 mV. The error bars represent the standard deviation from four determinations of the ATP-concentration. The slopes are given in mM ATP/ (M Chl·s).

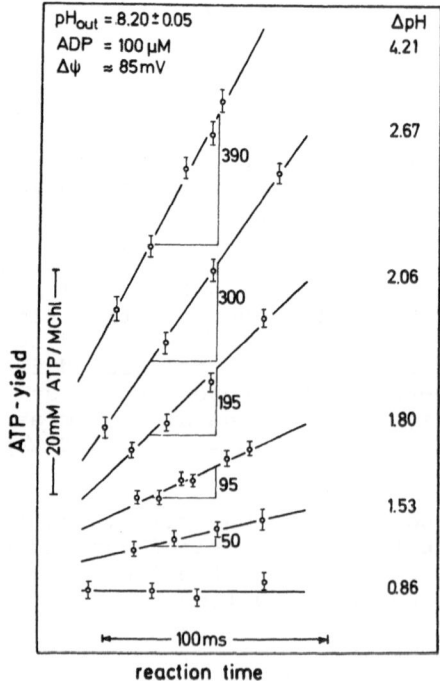

Fig. 2. ATP yield as a function of reaction time at pH_{out} = 8.2, $\Delta\Psi$ = 85 mV and ΔpH between 0.86 and 4.21. Details see Fig. 1.

RESULTS AND DISCUSSION

The rate of ATP synthesis as a function of Δ pH at three different $\Delta\psi$ is shown in Fig. 3. A sigmoidal increase of the rates with increasing ΔpH can be seen at all three curves. The maximal rate is 380 mM ATP/M Chl·s) corresponding to 410 ATP/ (CF$_o$F$_1$·s). At low ΔpH no ATP hydrolysis is observed even when the ATP concentration is increased up to 1 mM. If the data from Fig. 3 are plotted versus the electrochemical potential difference of protons, $\Delta\tilde{\mu}_H$+, all data can be described by a

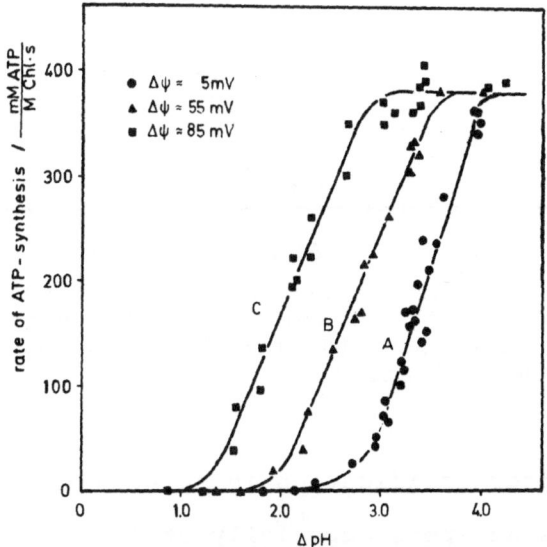

Fig. 3. The rate of ATP synthesis as a function of Δ pH and $\Delta\psi$ at pH$_{out}$ = 8.2. Data from Figs. 1&2 and similar sets of experiments.

single curve (1). This implies that the rate of ATP synthesis measured with chloroplasts isolated from dark adapted leaves is only a function of $\Delta\tilde{\mu}_H$+.

If chloroplasts are illuminated in the presence of DTT, the ATPases become modified. It is known that this DTT-treatment results in a reduction of an -S-S-group to HS-groups in the γ-subunit (2-5). Futhermore, these DTT-modified ATPases can catalyze proton transport-coupled ATP hydrolysis (6-10). Usually, ATP hydrolysis was investigated using DTT-modified chloroplasts and ATP synthesis was measured with non-modified chloroplasts. Therefore, we have measured the functional dependence of the rate of ATP synthesis on ΔpH with DTT-modified and non-modified chloroplasts (see Fig. 4). The rate of ATP hydrolysis of these modified chloroplasts is 100 mM ATP/(M Chl·s)(11).

It can be seen that the functional dependence of the rate of ATP synthesis on ΔpH is different for DTT-modified and non-modified ATPases:

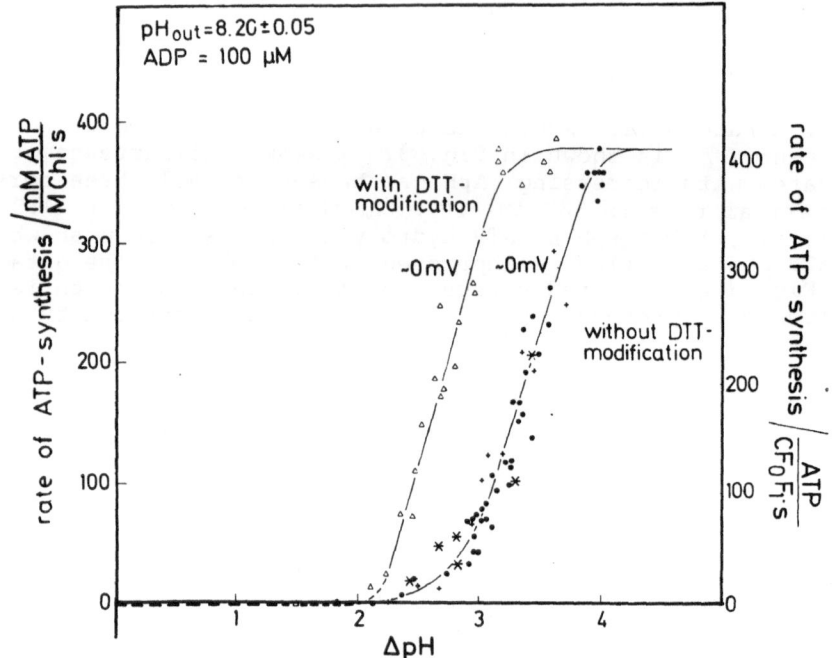

Fig. 4. The rate of ATP synthesis as a function of
Δ pH at $\Delta\Psi$ = 0 and pH_{out} = 8.2 using chloroplasts without DTT modification and after
DTT modification.

(1) The curve after modification is shifted to lower Δ pH-
    values.
(2) The slope is steeper, especially at low rates.
(3) At low Δ pH ATP hydrolysis is observed.
(4) At high Δ pH the same maximal rate of ATP synthesis is
    found with and without modification.

These results can be explained by the scheme shown in
Fig. 5 (12). The inactive ATPase, E_i, is transformed into
an active form, E_a, by energization ($\uparrow\Delta$ pH and/or $\Delta\Psi$). This
event is accompanied by the release of tightly bound ADP (13);
two H^+ are necessary for the activation, and the pK values of
the protonizable groups is 5.9 (1). In the active form, DTT
can react with the ATPase, forming a "modified" active ATPase
within 2-5 min. Also, when the membrane is not energized,
the ATPase can be modified by DTT; however, it then takes
about 1-3 hours. It is not yet known whether E_i can be
directly modified to E_i^m as suggested in the scheme. Since
under deenergized conditions a small fraction of ATPases
(10^{-6})(1) is always in the state E_a, it is possible that this
small fraction is modified to E_a^m which then reacts to E_i^m. In
this way all E_i may be successively modified. Functionally,
E_i^m and E_i^m cannot catalyze ATP synthesis/hydrolysis. In the
form E_i a rather high ΔpH is necessary to transform the
ATPase to E_a (the pK value for the H^+-binding sites from inside for activation is about 5.9 (1)). This ΔpH is - at
least at low phosphate potentials - higher than the equilibrium Δ pH for ATP synthesis/hydrolysis so that under these
conditions no ATP hydrolysis is observed.

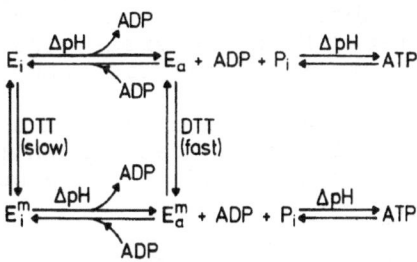

Fig. 5. Simplified scheme of activation of the
ATPase by ΔpH (and $\Delta\psi$) and modification by DTT. For details see text.

In the form E_i^m a low ΔpH is necessary to transform the
ATPase to E_a^m (the pK value for the H^+-binding sites from the
inside is higher than 5.9). It has been shown that the ΔpH
for half-maximal activation is shifted by about 0.9 units to
lower values (14). Under these conditions ATP hydrolysis
can be observed and, additionally, the ΔpH generated by ATP
hydrolysis is high enough to maintain the ATPase in the active
modified form, E_a^m. The rate of the back reaction $E_a^m \rightarrow E_i^m$,
(and presumably also that of $E_a \rightarrow E_i$) depends on different
parameters; e.g., ADP- and P_i-concentration (15-19).

ATP hydrolysis has been measured usually with ATPase in
the form E_i^m and E_a^m; ATP synthesis has been measured with
ATPase in the form E_i and E_a. In our earlier work (13,20)
it was shown that the observed rate of ATP synthesis/hydrol-
ysis at constant ATP, ADP and P_i is given by

$$V_{ATP} = W_{ATP}\ (\Delta pH, \Delta\psi)\ \frac{E_a}{E_t}\ (\Delta pH, \Delta\psi)$$

where E_a/E_t is the fraction of active ATPases and W_{ATP} is the
rate of the catalytic reaction (ATP synthesis/hydrolysis) per
active ATPase. For the modified form we obtain correspondingly:

$$V'_{ATP} = W_{ATP}\ (\Delta pH, \Delta\psi)\ \frac{E_a^m}{E_t^m}(\Delta pH, \Delta\psi)$$

Two extreme cases can now be discussed. In non-modified
ATPases the proton-binding sites for activation leading to
E_a have lower pK's than the proton binding sites for the
catalytic reaction, W_{ATP}. This implies that the observed
rate V_{ATP} reflects practically only the dependence of E_a/E_t
on ΔpH and $\Delta\psi$. If the ATPase is modified with DTT, the
pK value of the proton-binding sites for activation is increased to values higher than those for the catalytic reaction. This implies that V_{ATP} reflects mainly the dependence
of W_{ATP} on ΔpH and $\Delta\psi$.

Our interpretation of the results in Fig. 4 is, there-fore, that the curve obtained with non-modified ATPases reflects the activation, E_a/E_t, whereas, that obtained with modified ATPase reflects the catalytic reaction, W_{ATP}.

These different processes and some data are collected in Fig. 6. Biochemically it has been shown (2-5) that DTT-treatment leads to a reduction of an S-S group to -S-H groups

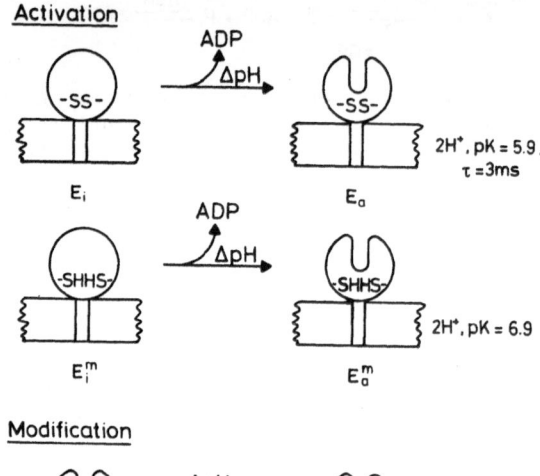

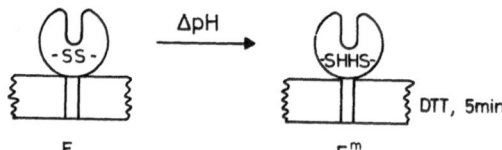

Fig. 6. Schematic illustration of activation, modification and catalytic reaction of the chloroplast ATPase.

in the γ-subunit. Furthermore, the release of tightly bound ADP is coupled with the activation of the ATPase (13) and the rebinding is coupled with the inactivation (15-19).

Based on these results, the following may be concluded: E_i contains an -S-S-group in the γ-subunit. Additionally, the enzyme contains tightly bound ADP. The groups to be proton-ized for activation have a pK of about 5.9. In E_a these groups are protonized and the tightly bound ADP is released. E_i^m con-tains a tightly bound ADP and the -S-S-group is reduced, the goups to be protonized for activation have a pK $>$ 5.9; in E_a^m these groups are protonized, the –S–S–group in the γ-subunit is reduced and the tightly bound ADP is released.

According to the chemiosmotic theory (21), a transmembrane electrochemical potential difference of protons, $\Delta\tilde{\mu}_{H^+}$, is sufficient to drive ATP synthesis by the membrane-bound ATPase, CF_oF_1, without involvement of other proteins. Thus, the rate of ATP synthesis after isolation and reconstitution of CF_oF_1 into liposomes should be as high as in the natural membrane. However, only rates of less than 1% of the maximal in vivo rate were reported (22-24). It is not clear whether these low rates are due to experimental difficulties or to fundamental reasons; namely, the fact that in vivo protons might migrate within the membrane to the coupling factor without involvement of the bulk phase. Therefore, we have measured the rate of ATP synthesis by reconstituted CF_oF_1 liposomes, using the rapid mixing technique described above (25). After optimization of the assay conditions and of the conditions for reconstitution, a rate of 200 ATP/ $(CF_oF_1 \cdot s)$ is obtained (see Fig. 7). This is half the rate

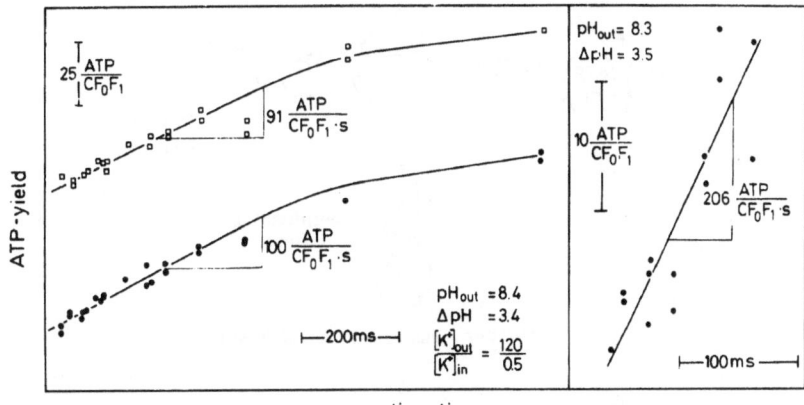

Fig. 7. ATP yield as a function of reaction time using reconstituted CF_oF_1 liposomes, pH_{out}= 8.4, $\Delta pH = 3.4$. The slopes are given in ATP/$(CF_oF_1 \cdot s)$. Left: ATP yield measured with luciferin/luciferase (lower curve), with ^{32}P (upper curve). Right: different CF_oF_1 preparations.

observed at thylakoid membranes (cf. Fig. 3). Since any correction that ought to be applied (e.g., not all the CF_oF_1 added to the reconstitution medium is inserted into liposomes, not all CF_oF_1 inserted into liposomes have the correct sidedness, CF_oF_1 may be partly denatured) would increase the rate per correctly incorporated CF_oF_1, we came to the conclusion that the activity of the reconstituted CF_oF_1 is almost the same as in the natural thylakoid membrane.

The data presented here do not answer the question, in which way protons are transferred from the electron-transport chain (source) to the ATPase (sink); i.e., via an intra-membrane pathway (26) or via the internal bulk phase (21). However, the data show that a bulk-bulk proton transfer is kinetically competent for the high rates of ATP synthesis observed in the natural thylakoid membrane. Furthermore, the high activity of reconstituted ATPase represents good starting material for further studies of the mechanism of ATP synthesis.

It has been shown that isolated F_1 can bind ATP under single site conditions and that on the enzyme a rapid equilibrium ATP $\rightleftharpoons$ ADP + P_i is established with an equilibrium constant K = 0.5 (26). Using the reconstituted CF_0F_1 liposomes, we have tried to measure this equilibrium constant on the membrane-bound CF_0F_1. This experiment has been carried out as follows (27)(see Fig. 8). The CF_0F_1 in the reconsti-

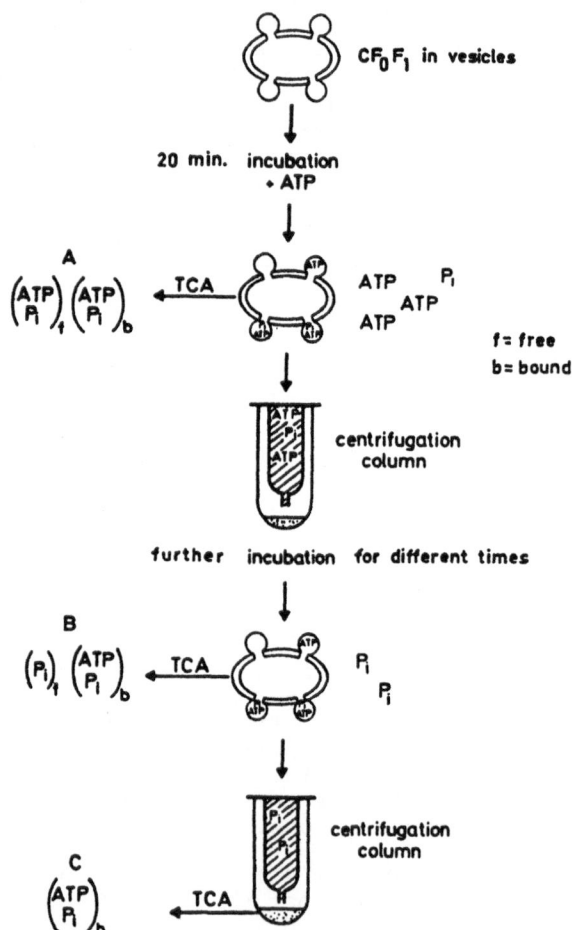

Fig. 8. Scheme of the experimental procedure for measuring the equilibrium constant between bound P_i and bound ATP at the reconstituted CF_0F_1. For details see text.

tuted liposomes were first brought into the state E_a^m (see Fig. 5) by incubating first with DTT and energizing the membrane with an acid-base jump. After decline of the energization the vesicles were incubated with γ-[^{32}P]-ATP (t = 0). After 20 min incubation, the non-bound ATP is removed by a G50 centrifugation column. The resulting preparation is then stored at room temperature and samples are taken at different times. The ^{32}P released during this incubation is removed by a second centrifugation column. After denaturation, the amount of bound ^{32}P and bound γ-[^{32}P]-ATP was measured. The result of these measurements is shown in Fig. 9.

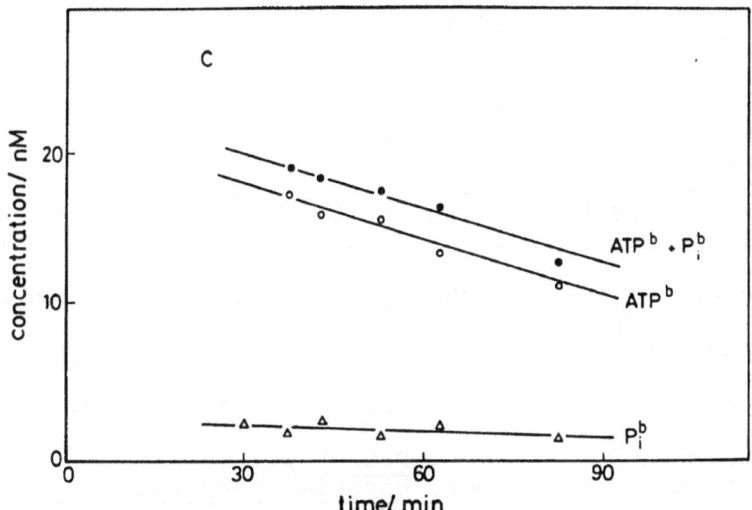

Fig. 9. Bound γ-[^{32}P]-ATP and bound ^{32}P as a function of time. From these data the equilibrium constant and the rate constant for the release of P_i is calculated.

From these data it follows that

$$K_2 = \frac{[^{32}P]_{bound}}{[\gamma\text{-}[^{32}P]\text{-}ATP]_{bound}} \approx 0.13$$

If K_2 is an equilibrium constant, it is expected that on incubation with ADP and P_i enzyme-bound ATP is formed. According-ing to a hypothesis of P. Boyer, ATP is formed on the enzyme spontaneously and the energization is required to release this tightly bound ATP (28).

Additionally, the data obtained in this type of experiment (see Fig. 8) permit an estimation of the rate constant for the P_i release under single site conditions. Provided that this reaction can be described by first order kinetics and that the back reaction can be neglected, a rate constant of $k_3 \approx 2 \cdot 10^{-4}$ s^{-1} resulted. The measurements for the isolated mitochondrial ATPase, MF_1, resulted in an equilibrium constant $K_2 = 0.5$ and a rate constant of P_i-release $k_3 = 2.7 \cdot 10^{-3}$ s^{-1} (26). This is a surprisingly good agreement between the constants, if we take into account not only the difference in the species but also the possible differences between the isolated, soluble enzyme and the membrane-bound one.

155

ACKNOWLEDGEMENTS

The financial support by the Deutsche Forschungsgemein-
schaft and the Fonds der Chemischen Industrie is gratefully
acknowledged.

REFERENCES

1. P. Gräber, U. Junesch and G.H. Schatz, Ber. Bunsenges.
   Phys. Chem. 88:599 (1984)
2. S.R. Ketcham, J.W. Davenport, K. Warncke and R.E. McCarty,
   J. Biol. Chem. 259:7286 (1984)
3. C.M. Nalin and R.E. McCarty, J. Biol. Chem. 259:7275 (1984)
4. J.V. Moroney, C.S. Fullmer and R.E. McCarty, J. Biol. Chem.
   259:7281 (1984)
5. J.L. Arna and R.H. Vallejos, J. Biol. Chem. 257:1125 (1982)
6. U. Petrack and F. Lipmann in: Light and Life, McElroy, V.D.
   and Glass, H.B., eds., pp. 621-630, Hopkins, Baltimore, MD
   (1961)
7. L. Packer and R.M. Merchant, J. Biol. Chem. 239:2061 (1963)
8. C. Carmeli, FEBS Lett. 7:297 (1970)
9. T. Bakker-Grunwald and K. van Dam, Biochim. Biophys. Acta
   292:808 (1973)
10. J.W. Davenport and R.E. McCarty, J. Biol. Chem. 256:8947
    (1981)
11. U. Junesch and P. Gräber, FEBS Lett. in press (1985)
12. U. Junesch and P. Gräber, Biochim. Biophys. Acta, in
    press (1985)
13. P. Gräber, E. Schlodder and H.T. Witt, Biochim. Biophys.
    Acta 461:426 (1977)
14. B. Rumberg and U. Becher in: H^+-ATPase/synthase; struc-
    ture, function, regulation, S. Papa et al., eds.), pp.
    421-430, Adriatica Editrice, Bari, Italy (1984)
15. C. Carmeli and Y. Lifshitz, Biochim. Biophys. Acta 267:
    86 (1972)
16. S. Bickel-Sandkötter and H. Strotmann, FEBS Lett. 126:188
    (1981)
17. J. Schumann and H. Strotmann in: Photosynthesis, Akoyu-
    noglou, G., ed., Vol. II, pp. 881-892, Balaban Intern.
    Sci. Serv., Philadelpha, PA (1981)
18. N. Shavit, C. Aflalo and D. Bar-Zvi in: Energy Coupling
    in Photosynthesis, Selman, B.R. and Selman-Reimer, S.,
    eds., pp.197-207
19. K.R. Dunham and B.R. Selman, J. Biol. Chem. 256:212
    (1981)
20. P. Gräber and E. Schlodder in: Photosynthesis, Akoyunoglou,
    G., ed., Vol. II, pp. 867-880, Balaban Intern. Sci. Serv.,
    Philadelphia, PA (1981)
21. P. Mitchell, Nature 191:144 (1961)
22. G.D. Winget, N. Kanner and E. Racker, Biochim. Biophys.
    Acta 460:490 (1977)
23. T.G. Dewey and G.G. Hammes, J. Biol. Chem. 256:8941 (1981)
24. G. Hauska, D. Samoray, G. Orlich and N. Nelson, Eur. J.
    Biochem. 111:535 (1980)
25. G. Schmidt and P. Gräber, Biochim. Biophys. Acta 808:46
    (1985)
26. G. Grubmeyer, R.L. Cross and M.S. Penefsky, J. Biol. Chem.
    257:12092 (1982)
27. P. Gräber and P. Fromme, FEBS Lett. in press (1985)
28. P. Boyer in: Bioenergetics of Membranes, C.P. Lee, ed.,
    Addison-Wesley (1979)

COMPARISON OF THE TWO RETINAL PROTEINS BACTERIORHODOPSIN AND HALORHODOPSIN

Jörg Tittor, Peter Hegemann and Dieter Oesterhelt

Max-Planck-Institut für Biochemie

D-8033 Martinsried, FRG

Two retinal proteins act as light-driven pumps in halobacteria. Bacteriorhodopsin (BR) translocates protons to the medium and halorhodopsin (HR) transports chloride into the cytoplasma. Bacteriorhodopsin's functional and structural properties are well known. Transport is mediated by a photochemical cycle of the retinal chromophore accompanied by trans to cis isomerization and a reversible deprotonation of its Schiff base. Spectroscopy in combination with the use of retinal analogue compounds demonstrated that the trans to cis isomerization is connected to the primary photochemical reaction. (1).

Halorhodopsin was isolated as a native chromoprotein of the apparent molecular weight 25,000 and functionally reconstituted into black lipid membranes as a light-driven chloride pump (2,3). Elucidation of its primary structure is presently carried out by a combination of protein chemical and gene technological methods. At present (40 % of the sequence) sequence homologies between bacteriorhodopsin and halorhodopsin are weak. Strong similarities, however, exist on the level of retinal protein interaction. Resonance Raman spectroscopy (4,5) proved the all-trans state of retinal in HR and a very similar environment in the protein compared to BR. Trans to cis isomerization occurs, but no deprotonation of the chromophore is observed as an intermediate step of the photocycle. A reversible deprotonation, however, can occur in the time range of minutes leading to slow deactivation of halorhodopsin. Photochemical reactivation is observed with blue light, thus demonstrating the photochromic property of HR. This type of light regulation might have physiological relevance (6). The de- and reprotonation of the halorhodopsin chromophore is catalyzed by azide, which allowed a detailed analysis of the photochemical cycle (7,8). In a model of ion translocation the differences of the photocycles in HR and BR can be reconciled with very similar molecular events occurring in BR and HR leading to the transport of a proton or a chloride ion (9).

References

1) H.-J. Polland, M.A. Franz, W. Zinth, W. Kaiser, P. Hegemann and D. Oesterhelt, Optical Picosecond Studies of Bacterio-

rhodopsin Containing a Sterically Fixed Retinal. BBA 767: 635 (1984).

2) E. Bamberg, P. Hegemann and D. Oesterhelt, Reconstitution of the Light-driven Electrogenic Ion Pump Halorhodopsin. BBA 773:75 (1984).

3) E. Bamberg, P. Hegemann and D. Oesterhelt, The Chromoprotein of Halorhodopsin is the Light-driven Electrogenic Chloride Pump in Halobacterium halobium. Biochemistry 23: 6216 (1984).

4) T. Alshuth, M. Stockburger, P. Hegemann and D. Oesterhelt, Structure of the Retinal Chromophore in Halorhodopsin. A Resonance Raman Study. FEBS Lett. 179: 55 (1985).

5) S. Smith, M. Marvin, R. Bogomolni and R. Mathies, Structure of the Retinal Chromophore in the HR_{578} Form of Halorhodopsin. J. Biol. Chem. 259:2326 (1984).

6) P. Hegemann, D. Oesterhelt and E. Bamberg, The Transport Activity of the Light-driven Chloride Pump Halorhodopsin is Regulated by Green and Blue Light. BBA (in press, issue Sept./Oct. 1985).

7) P. Hegemann, D. Oesterhelt and M. Steiner, The Photocycle of the Chloride Pump Halorhodopsin. I. Azide Catalyzed Deprotonation of the Chromophore Intermediates Inactivating the Pump. EMBO J. (in press, issue Sept. 1985).

8) D. Oesterhelt and P. Hegemann, The Photocycle of the Chloride Pump Halorhodopsin. II. Quantum Yields and a Kinetic Model. EMBO J. (in press, issue Sept. 1985).

9) D. Oesterhelt, P. Hegemann, P. Tavan and K. Schulten, A Model for Ion Translocation in Halorhodopsin by trans-cis Isomerization of its Retinal Moiety (in preparation).

MEASUREMENT OF PROTON/M_{412} RATIOS IN SUSPENSIONS OF PURPLE AND WHITE MEMBRANE FROM HALOBACTERIUM HALOBIUM

A.Edward Robinson, Eva Hrabeta and Lester Packer

Department of Physiology-Anatomy
University of California
Berkeley CA94720 U.S.A.
        and
Membrane Bioenergetics Group
Lawrence Berkeley Laboratory
Berkeley CA94720 U.S.A.

SUMMARY

We have studied isolated purple membrane from Halobacterium halobium strain S_9 and white membrane from the mutant strain JW-5, making careful measurements of the proton to M_{412} stoichiometry of our preparations, and also examining them by Quasi Elastic Light Scattering (QELS) and by negative staining electron microscopy (EM), to assess extent of aggregation and hydrodynamic radius.

In the past decade there have been varying reports of the stoichiometry of proton release from purple membrane in suspension and protons pumped after its incorporation into phospholipid vesicles. Interpretation of the results, especially when by a group other than that which carried out the experiments, has sometimes been clouded by a misunderstanding of the differences between the various measurements involved and also by a lack of recognition that the aggregation state of membranes will affect the measured ratios.

In an attempt to resolve these misunderstandings, we have investigated the H^+/M_{412} ratio, measured in suspensions of purple membranes by the laser flash induced response of pH indicator dyes. We have found this ratio to reflect the aggregation state of the membrane, assessed on the same preparations by QELS (and EM, see accompanying chapter: Lefort-Tran et al.). We have also found our measurements of proton release stoichiometry on retinal reconstituted white membrane suspensions to give much higher H^+/M_{412} ratios than the same measurements on purple membrane suspensions.

INTRODUCTION

Bacteriorhodopsin is a small (approx. 26kD) protein found in the plasma membrane of the bacterium Halobacterium halobium (1). Its functions is to pump protons out of the cytoplasm into the surrounding medium, using the energy provided by sunlight. This electrochemical gradient is then utilised by a number of other membrane proteins, e.g. for ATP production and solute transport. Bacteriorhodopsin is produced in large quantities by the organism when the oxygen content of the medium falls so low as to make conventional respiration growth limiting, a situation which occurs, for example, when the salt concentration in evaporating sea-water pools becomes very high. H. halobium is thus able to grow well even in saturated salt solution.

The reported stoichiometry of protons released per M_{412} intermediate formed in the photocycle is variable. It has been reported (2) that, in the presence of high salt concentrations and at low light intensity, the H^+/M_{412} ratio approaches three, but that in distilled water, the ratio is 0.3-0.5 at pH7.5. Lozier et al. (3) found the H^+/M_{412} ratio for membrane sheets in water to be 1.0 at the same pH. Using flash photolysis with pH indicator dyes, Govindjee et al. (4) reported H^+/M_{412} for purple membranes in suspension to vary with salt concentration, unlike membranes incorporated into liposomes, which have a consistent H^+/M_{412} for pumping of about 2.0, and that proteolytic cleavage of the C-terminal tail of bacteriorhodopsin by trypsin specifically lowers the H^+/M_{412} stoichiometry (5). Several laboratories (6) (7) (8) (9), however, report that the removal of the carboxyl tail region does not affect proton pumping of the purple membranes after incorporation into liposomes.

CLARIFICATION OF THE PARAMETERS

In the treatment of the figures on proton stoichiometry, it must first be understood that each of the techniques used looks at a somewhat different event.

In aqueous suspension, the laser/indicator dye technique measures the average transient release of protons from the two faces of the membrane. In this case the proton pumping cannot be detected, only net proton release from the membrane. This observed proton release is, most likely, due to a difference of pK of the two faces of the membrane (10) and as such, is probably an indicator of varying chemical nature of the two sides.

The dyes commonly used will partition into the membrane interface region due to their high hydrophobicity, and thus the figure obtained will be weighted toward the transient changes in this region and perhaps more importantly, modulated by differing surface charges of preparations due to the influence of the Gouy-Chapman double layer. These are problems in calibration of the dye response (which is carried out by adding protons to the bulk aqueous phase and measuring an equilibrium response) and means that figures cannot easily be compared between preparations having different surface charge densities, nor absolute values for H^+/M_{412} easily obtained. If a probe with a large permanent charge of the same polarity as the surface is used, then this latter problem is considerably reduced, if not eliminated.

Measurements by the pH electrode method, because of the longer time scale of the electrode response, are a response to the equilibrating proton release into the bulk aqueous phase, leading to a steady-state. This will be an even more complicated function of the titratable groups at the surface, since in the intervening time the released protons will have diffused to other areas of the surface, as well as into the bulk phase. The calibration in this case, however, is under the comparable conditions. Once again, a pumped proton gives no response.

In liposomes, proton pumping is the primary object of the measurement techniques. In the laser technique, dye is normally added to the external aqueous phase so that the measurement detects alkalinization due to protons pumped into the liposomes. As before, it is a transient response which is detected and subject to the same influence of surface charge of both bacteriorhodopsin and lipid on the dye measurement (If the dye is also present inside the liposomes, then the data obtained would reflect average proton release from the internal and external faces of the membrane). Because of the differing interior and exterior volumes and consequent difficulties of calibration, however, this is probably not a useful measurement.

In liposomal suspensions, the pH electrode measures a similar event to the laser technique, but once again responding to the pH change on a rather slower time scale to that technique.

METHODS

Membranes

Purple membranes from H. halobium S$_9$ were isolate as previously described (11). Membranes were bleached by hydroxylamine plus light (12). White membranes from the JW-5 strain were reconstituted with all-trans retinal to form functional bacteriorhodopsin (13). White and purple membranes were suspended in 10mM azide at pH7 and kept in the cold; before use the suspensions were washed into test media.

Trypsin treatment

1mg/ml of purple or white membrane, suspended in 50mM Tris-Cl, 30mM NaCl, 10mM CaCl$_2$ pH7.5 buffer solution were treated with trypsin, suspended in the same buffer. The final concentration of trypsin was 0.017mg/ml for purple membrane and 0.0334mg/ml for white membrane. Samples were incubated for four hours at 37°C, and reaction stopped by dilution with double-distilled water and centrifugation (35000 x g, 30min, 5°C). Two additional washes completed removal of the lighter cleavage product.

Laser Apparatus

Measurements were made at a controlled temperature of 20°C in a 1cm path length cell. Actinic light was provided by flashes from a Phase-R DL 1100 dye laser, using Rhodamine 575, the flash duration was 350ns and the flash artifact less than 40μs. The photo-current from a photomultiplyer was recorded using a Gould Biomation Model 4500 digital oscilloscope. Essentially noise-free traces were obtained on averaging of 16 flash responses.

Proton Release

Proton release from purple or white membrane sheet preparations was measured using the pH indicator dye, 7-hydroxycoumarin. The working concentrations were 0.312mg/ml membranes in distilled water, and 0.1mM hydroxycoumarin. The dye, dissolved in ethanol, was added in 30μl to 3ml of sample. The pH was adjusted to 7.5 between each measurement. Laser flash photolysis measurements were made at 412nm without and at 365nm with and without the dye. Transient absorption differences at 365nm were assessed at the time of maximum difference. The response of the dye (r) at 365nm to added protons was determined for each sample by careful titration with small aliquots of standard HCl, following absorption changes at 365nm using a Perkin Elmer Lambda 5 spectrophotometer. The final calculations of H$^+$/M$_{412}$ were made using the following equation.

$$\text{Ratio} = \frac{\Delta I_{365}^{dye} \times \varepsilon_{365}^{M}}{\Delta I_{365}^{prot} \times r_{365}^{dye}}$$

the extinction due to M_{412} at 365nm was calculated using:

$$\varepsilon_{365}^{M} = \frac{\varepsilon_{412}^{M} \times \Delta I_{365}^{protein}}{\Delta I_{412}^{protein}}$$

The ε_{412}(protein) used was 23,000 $M^{-1}cm^{-1}$.

Since the measurements on the sample with and without the dye are made under the same conditions and the response due to M_{412} calibrated at 365nm, all other variables involved in the calculation of absorbances from the photomultiplier current measurements cancel out. There is also the advantage that reproducibility of flashes and electronic response is only required over the short period of time necessary to make the three measurements. In fact reproducibility is excellent so long as the instrument is closely watched and calibration of the dye response is very carefully carried out for each sample. Samples of white membranes reconstituted with retinal and measured on different days, for example, gave results differing by less than 0.1 H^+/M_{412}.

QELS Studies

QELS (14) allows determination of the translational diffusion coefficient (D_t), from which the hydrodynamic radius (R) for spherical particles can be calculated, using the following equation:

$$D_t = \frac{k\,T}{6\,\pi\,\eta\,R}$$

where k is the Boltzman constant; T and η the absolute temperature and viscosity of the medium, respectively. The apparatus for QELS studies has recently been described for characterization of sarcoplasmic reticulum (15) and purple membranes (16).

RESULTS AND DISCUSSION

Proton release figures for native and trypsin treated membrane preparations are given in Fig. 1. It can be seen that, upon trypsin treatment, the proton release activity of purple membranes declines from an initial value over 1.5 H^+/M_{412} to a value below 0.5. Retinal reconstituted white membrane, on the other hand, starts with a much higher proton release stoichiometry, close to 4.0 H^+/M_{412}, and even after trypsin treatment it retains most of this activity. In summary, whilst purple membrane looses two thirds of its activity, retinal reconstituted white membrane looses less than one tenth of its activity.

Whilst the absolute H^+/M_{412} figures presented here cannot be directly compared, due to the surface charge considerations discussed earlier, the relative figures are most informative. One must remember that, on cleavage of the C-terminal tail, the surface looses considerable negative charge, which will in turn increase the amount of deprotonated hydroxycoumarin in the interface region and so cause an increase in sensitivity of the dye to protons released into this region, decreasing any lowering

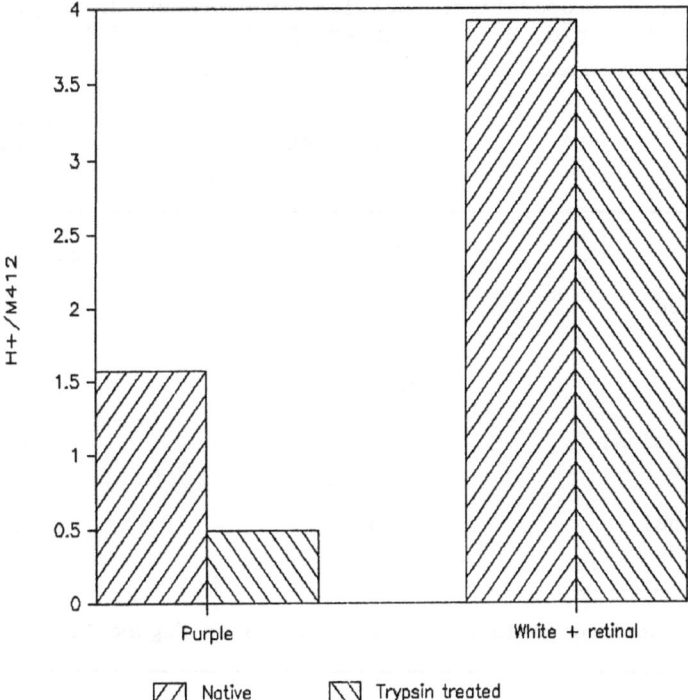

H+/M412

4

3.5

3

2.5

2

1.5

1

0.5

0

Purple   White + retinal

▨ Native   ◩ Trypsin treated

Figure 1. Proton release per M412.

of H^+/M_{412} stoichiometry upon trypsin treatment. We are confident, however, that our trypsin treatment procedure causes a similar extent of cleavage in both membrane preparations, so this increase in sensitivity should affect each of the samples equally. The fact is that the white membranes show a much smaller decline in activity than the purple membranes. What, then, is the reason for the decrease?

In Table 1, some representative results of a series of QELS experiments on the same samples are presented. There was close agreement of the experimental traces with theoretical curves, indicating that the samples had good size homogeneity. There were some deviations in the low frequency region, where some much larger particles make a contribution, but these were rather small.

Native purple membrane consists of a suspension of particles with homogeneous size distribution in a rather narrow range (around 260nm). Upon trypsin treatment the membranes become extensively aggregated, with a QELS pattern more characteristic of a gel than of a suspension of particles. The appearance of such large aggregates gives us a clue as to why the H^+/M_{412} figures decline so much on trypsin treatment. With the membranes stacked and confined, as they must then be, it will be impossible for protons to be released into a region where they can be detected by the dye, or indeed by any other measurement technique. The light flash will still, however, generate M_{412} so that the observed ratio of H^+/M_{412} must decline.

In the case of white membranes there is a quantitatively different effect. In the first place, white membrane patches are rather smaller (88nm radius), even before trypsin treatment. Following treatment, they

TABLE 1
Translocational diffusion coefficients and hydrodynamic radii of purple and white membrane preparations (Results obtained on our preparations by B. Arrio, G. Johannin and P. Volfin (17))

Membrane Preparation	Diffusion Coefficient x 10^{-9} (cm^2s^{-1})	Hydrodynamic Radius[a] (nm)
native purple	8.5	260 ± 20
trypsin treated purple	0.63	see text
white	24.4	88 ± 50
white after retinal reconstitution	24.4	88 ± 50
white, + retinal trypsin treated	4.3	515 ± 50

[a]   mean $\pm$ standard deviaton, conditions as in figure 1.

too aggregate, but to a rather lesser extent than do purple membranes (final radius 515nm, and homogeneous) and perhaps in a rather more open structure. The H^+/M_{412} ratio remains quite high, as a consequence.

Such aggregation after proteolysis has been previously observed by ourselves and others (18) and attributed to the reduction in electrostatic repulsion allowing the hydrophobic forces to induce aggregation. We have found that untreated preparations of purple membrane, over a range of concentration between 0.02 and 9mg/ml, stored for many months at $2°C$ showed no sign of aggregation by QELS. We and others (19) have also observed, however, that in some preparations of purple membrane, long term storage allows the development of protease activity, most likely arising from bacterial contamination. This activity is inhibited by benzamidine, indicating that it is trypsin-like (20). After proper washing of cell homogenates and purification of purple membrane by sucrose density gradient, little or no benzamidine inhibited protease activity is present. Thus, it is apparent that in order to get good H^+/M_{412} figures one must take great pains to exclude all bacterial contamination.

ACKNOWLEDGMENTS

Research supported by the Office of Biological Energy Research, Division of Basic Energy Sciences, U.S. Department of Energy (Contract #DEAC03 76SF00098), Centre National des Recherches Scientifique (A.T.P. "Bioenergetique" n° 3093/79) and a NATO Grant for International Collaboration in Research.

REFERENCES

1. Blaurock, A.E., Stoeckenius, W., Oesterhelt, D. and Scherphof, G.L. Structure of the cell envelope of Halobacterium halobium. J. Cell. Biol. 71, 1-22 (1970).

2. Kuschmitz, D. and Hess, B. On the ratio of the proton and photochemical cycles of bacteriorhodopsin. Biochemistry 20, 5950-5957, (1981).

3. Lozier, R.H., Niederberger, W., Bogomolni, R.N., Hwang, S. and Stoeckenius, W. Kinetics and stoichiometry of light-induced proton release and uptake from purple membrane fragments, Halobacterium halobium cell envelopes and phospholipid vesicles containing oriented purple membrane. Biochem. Biophys. Acta, 440, 545-556 (1976).

4. Govindjee, R., Ebrey, T.G. and Crofts, A.R. The quantum efficiency of proton pumping by the purple membrane of Halobacterium halobium. Biophys. J., 30, 231-242 (1980).

5. Govndjee, R., Ohno, K. and Ebrey, T.G. Effect of removal of the C-terminal region of bacteriorhodopsin on its light induced H^+ changes. Biophys. J. 38, 85-87 (1982).

6. Gerber, G.E., Wildenauer, D. and Khorana, H.G. Orientation of bacteriorhodopsin in Halobacterium halobium as studied by selective proteolysis. Proc. Natl. Acad. Sci. USA 74, 5426-5430 (1977).

7. Ovchinnkov, Y.A., Abdulaev, N.G., Feigina, M.Y., Kiselev, A.V. and Lobanov, N.A. Recent fndings in the structure-functional characteristcs of bacteriorhodopsin. FEBS Lett. 84, 1-4 (1977).

8. Liao, M.-J. and Khorana, H.G. Removal of the carboxyl-terminal peptide does not affect refolding or function of bacteriorhodopsin as a light dependent proton pump J. Biol. Chem. 259, 4194-4199 (1984).

9. Govindjee, R., Ohno, K., Chang, C.-H. and Ebrey, T.G. In: Transduction in Biological Membranes, Plenum Press, In Press.

10. Garty, H., Klemperer, G., Eisenbach, M. and Caplan, S.R. FEBS Lett. 81, 238-242 (1977).

11. Oesterhelt, D. and Stoeckenius, W. Isolation of the cell membrane of Halobacterium halobium and its fractionation into red and purple membrane, In: Methods in Enzymology (S. Fleischer and L. Packer, eds.), vol. 31, pp. 667-678, New York, Academic Press, (1974).

12. Oesterhelt, D. Reconstitution of the retinal proteins bacteriorhodopsin and halorhodopsin. In: Methods in Enzymology (L. Packer, ed.), vol. 88, pp. 10-17, New York, Academic Press, (1982).

13. Mukohata, Y., Sugiyama, Y.,Kaji, Y., Usukura, J. and Yamada, E. The white membrane of crystalline bacterioopsin in Halobacterium halobium strain R_1mW and its conversion nto purple membrane by exogenous retinal. Photochem. Photobiol. 33, 593-600, (1981).

14. Ware, B.R. and Haas, D.D. Electrophoretc light scattering. In: Fast Methods in Physical Biochemistry and Cell Biology (R.I. Sha'afi and S.M. Fernandez, eds.), pp. 174-220, Elsevier Science Publishers, (1983).

15. Arrio, B., Johannin, G., Carrette, A., Chevallier, J. and Brethes, D. Electrokinetic and hydrodynamic properties of sarcoplasmic reticulum vesicles: a study by laser Doppler electrophoresis and quasi-elastic light scattering. Arch. Biochem. Biophys. 228, 220-229, (1984).

16. Arrio, B., Johannin, G., Volfin, P. and Packer, L. Quas-elastic laser light scatterng and Doppler electrophoresis of purple membranes. Biophys. Soc. Abstr. 45, 212a, (1984).

17. Arrio, B., Johannin, G., Volfin, P., Lefort-Tran, M., Packer, L., Robinson, A.E. and Hrabeta, E. Aggregation and proton release of purple and white membranes following cleavage of the C-terminal tail of bacteriorhodopsin. Submitted to Archives of Biochemistry and Biophysics (1985).

18. Wallace, B.A. and Henderson, R. Location of the carboxyl terminus of bacteriorhodopsin in purple membrane. Biophys. J. 39, 233-239, (1982).

19. Govindgee, R., Ohno, K. and Ebrey, T.G. Effect of the removal of the COOH-terminal region of bacteriorhodopsin on its light-induced H^+ changes. Biophys. J. 38, 85-87, (1982).

20. Walsh, K.A. Trypsinogens and trypsins of various species. Methods in Enzymology (Perlmann, G.E. and Lorand, L., eds.) vol. 19, pp. 41-63 (1971).

ON THE FOLDING OF BACTERIORHODOPSIN

D.M. Engelman

Yale University
Department of Biophysics and Biochemistry
260 Whitney Avenue, P.O. Box 6666
New Haven, CT 06511, U.S.A.

INTRODUCTION

A great deal of effort has been expended in trying to understand the folding of soluble proteins, the tertiary structures of which have been determined in many cases. The problem has turned out to be very complex, and the current lines of study have met with only limited success. Surprisingly, the possibility of structural prediction may be greater in the case of membrane proteins, where no structures of the membrane spanning regions are yet known at high resolution. This circumstance arises as a consequence of the existence of topological and energetic constraints which place important limits on the range of secondary and tertiary structures expected for globular membrane proteins. In the following discussion, I discuss the energetic arguments as I presently view them, outlining the use of an energy calculation to identify membrane spanning regions from protein sequence information, and considering the covalent, polar, and packing considerations which may be important in the final folding of globular membrane proteins in lipid bilayers.

BACTERIORHODOPSIN

Bacteriorhodopsin is a protein found in the plasma membrane of the organism Halobacterium halobium (1,2,3). A number of reviews concerning the structure and function of bacteriorhodopsin have appeared (4-8). The fact that the protein exists in a two dimensional crystaline lattice in the membrane (9) and that it is a small protein which pumps protons against the electrochemical gradient using the energy from absorbed light (10) have stimulated a great growth in the number of studies of its function and structure. In fact, it has become one of the most active fields in life science research (11) as pointed out by Stoeckenius. Interest has been accelerated by improved understanding of the function of bacteriorhodopsin and, more particularly, by the fact that structural studies have led to a more detailed description of the protein's organisation than exists for any other membrane transport apparatus. Using very low electron fluxes to produce diffraction patterns and images of purple membranes, Unwin and Henderson (12,13) obtained three dimensional density maps showing that bacteriorhodopsin contains seven rods of density extending across the membrane. These are interpreted as being α-helical segments of the polypeptide.

HYDROGEN BONDING ENERGIES IMPLY HELICAL STRUCTURES

The large free energy cost of transfering an unsatisfied hydrogen bond donor or acceptor from an aqueous to a nonpolar environment (14,15) or of breaking such a bond in a nonpolar environment (5 to 6 kcal/mole) suggests that hydrogen bonds must be satisfied as proteins are inserted into a lipid bilayer. A structural motif which satisfies hydrogen bonds systematically and which is adopted by peptides in nonaqueous environments is the α-helix (16,17). Thus, it is expected that helical structures will be important motifs in the portions of membrane proteins spanning the nonpolar regions of lipid bilayers.

THE ENERGY OF PARTITIONING HELICES INTO LIPID BILAYERS CAN BE CALCULATED

We may now consider the factors which might lead to favourable partitioning of helices from the aqueous phase into a membrane bilayer. The major energetic factor favouring partitioning into the bilayer is the hydrophobic effect factors favouring its solution in the aqueous phase are interactions of polar and charged side chains with water. In order to make a quantitative estimate of the energies involved we must assign energies for hydrophobic and hydrophylic components for each amino acid, and sum the energies.

The hydrophobic energy thus measured has been shown for a wide variety of cases to correlate with the total surface area in contact with water (18). A typical energy is 25 cal/A^2, so that calculation of the total area which can be removed from contact with water leads to a value for the hydrophobic transfer free energy (19). We have used the surface area computations of Richmond and Richards (20) to obtain the surface area for each amino acid as it would be exposed in an alpha helix. The surface areas can then be converted into hydrophobic free energies, the results are shown in Table 1.

TABLE 1. Transfer Free Energies for Helical Groups (KCAL/MOL)

	ΔG_{Phobic}	$+\Delta G_{Philic}$	$=\Delta G_{Water-Oil}$		ΔG_{Phobic}	$+\Delta G_{Philic}$	$=\Delta G_{Water-Oil}$
PHE	−3.7		−3.7	SER	−1.6	1.0	−0.6
MET	−3.4		−3.4	PRO	−1.8	2.0	0.2
ILE	−3.1		−3.1	TYR	−3.7	4.0	0.7
LEU	−2.8		−2.8	HIS	−3.0	6.0	3.0
VAL	−2.6		−2.6	GLN	−2.9	7.0	4.1
CYS	−2.0		−2.0	ASN	−2.2	7.0	4.8
TRP	−4.9	3.0	−1.9	GLU	−2.6	10.8	8.2
ALA	−1.6		−1.6	LYS	−3.7	12.5	8.8
THR	−2.2	1.0	−1.2	ASP	−2.1	11.3	9.2
GLY	−1.0		−1.0	ARG	−4.4	16.7	12.3

The free energy for inserting charged groups can be considered as having two components: the energy required to produce an uncharged species by protonation or deprotonation, and the energy required to partition the uncharged but polar portions of side chains from water to nonaqueous phase. We consider that potentially charged amino acids will be transferred as the uncharged species (21). By assuming the process to occur at or near neutral pH, we can calculate the energy required. The energies are included in the hydrophylic energies listed in Table 1.

There will also be an energy cost associated with the transfer of uncharged polar groups. These energies arise principally from the participation of side chain groups in hydrogen bonds to water. It is difficult to treat the hydrogen bonding potential explicitly, and one must rely to a large extent on experimental measurements based on solubility of various compounds. Extensive reviews of these data are available (14,15).

It is important to consider the details of structural interaction within helices. Threonine and serine, for example, may participate in bifurcated hydrogen bonds with main chain carbonly groups, reducing their effective polarity. This structure is found in many protein interiors. Furthermore, strong polar interactions, possible including ion pairs, may occur between amino acids which are positioned appropriately along an α-helix. An effort has been made to include these considerations in our energy calculations.

IDENTIFICATION OF MEMBRANE SPANNING HELICAL REGIONS FROM THE AMINO ACID SEQUENCE

Using the free energies assigned in Table 1, a computational approach has been developed by Adrian Goldman (22,23). The concept of the computation is quite simple. The amino acid sequence is first considered to be an extended α-helix. The transfer free energy for successive segments of the helix is calculated as the sum of the free energies for the amino acids involved. The length of the segment is chosen as that length appropriate for spanning the nonpolar region of lipid bilayer. A plot is then obtained which describes the position of the buried segment (identified by the n-terminal amino acid for the segment) versus the free energy for insertion. The resulting curve will have minima where sections of the sequence would be most favored as an inserted helix. Figure 1 shows the result of such a computation done on bacteriorhodopsin. Seven minima are seen in the free energy diagram with varying degrees of clarity, and it is possible to use the energy diagram to make an assignment of segments of the amino acid sequence to the helices. The result of such an assignment is shown in

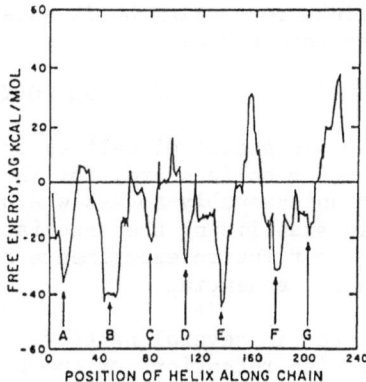

Figure 1.   Energy plot for Bacteriorhodopsin. The 7 minima permit a possible identification of the regions of the amino acid sequence which correspond to the 7 helices (22,23).

Figure 2. The results of several modification studies, principally using proteolytic enzymes and immunological methods have served to support this assignment.

WHAT HOLDS A GLOBULAR MEMBRANE PROTEIN TOGETHER?

If it is agreed that many of the membrane spanning regions of globular membrane proteins are α-helices and that such helices can be identified from the amino acid sequence, the next question concerns the stabilization of the association of helices to make a folded, compact protein in the

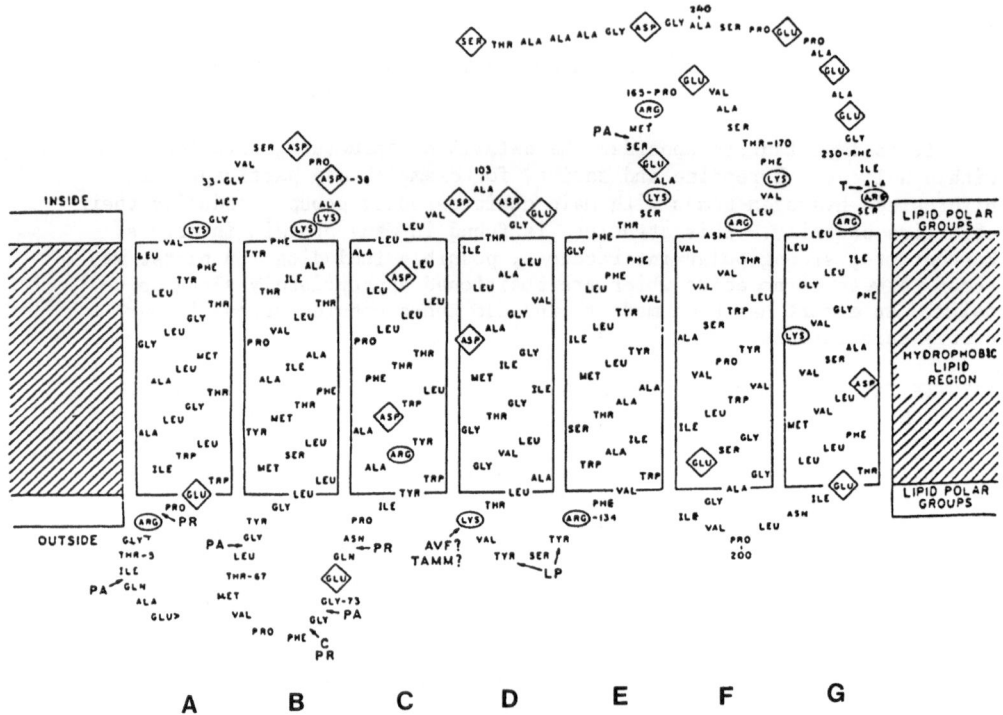

A   B   C   D   E   F   G

Figure 2.   Bacteriorhodopsin Helical Segments. Helices are located in the amino acid sequence and shown as they would span a lipid bilayer. The assignment is consistent with known sites of proteolytic cleavage and chemical modification (22).

lipid bilayer. Three factors will be considered in this discussion: stabilization by the polypeptide links connecting helix ends, polar interactions between helices, and packing influences.

STABILIZATION OF BACTERIORHODOPSIN BY CONNECTING POLYPEPTIDE LINKS

It can be seen from the assignment of helices in Figure 2 that the polypeptide stretches connecting helices are, for the most part, rather short. Therefore it is not unreasonably to ask whether these connections. may play an imporant role in stabilizing the association of the structure'. Proteolytic cleavage and reconstitution experiments suggest that many if not most of these links are not essential.

Proteolytic cleavages made in several parts of bacteriorhodopsin have been shown to be compatible with retention of the spectrum and pumping activity. In recent experiments, we have shown that molecules which were cut in the regions between helices B and C and also between E and F and from which the carboxy terminal peptide had been cleaved are still found in two dimensional lattices giving diffraction patterns which are very similar to those of the native structure. Therefore, at least these cleavages can be sustained with no radical rearrangement of the structure and the links are, consequently, not to be viewed as essential.

In another experimental approach Khorana and his colleagues (24,25) have shown that the molecule can be reconstituted in functional form from pairs of proteolytic fragments obtained either by cutting between helices A and B or between E and F. In a further development, we have shown that protein reconstituted from such fragments can reform the original purple membrane lattice to give diffraction patterns similar to those of the original structure. Thus,these linkages are not essential for the folding of the molecule from a fully denatured state in which both lipid and retinal had been removed.

POLAR INTERACTIONS MAY CONTRIBUTE TO THE STABILITY OF THE BACTERIORHODOPSIN MOLECULE

Using neutron diffraction from partially deuterated bacteriorhodopsin molecules, we have found that the orientation of polar groups in the helices tends to be towards the molecular interior where helix-helix contacts are involved rather than toward the molecular exterior where helix-lipid contacts exist. Thus, the protein appears to be inside-out when compared with the organisation of soluble proteins. This observation suggests the possibility that, in addition to the functional roles which are undoubtedly involved for many of the polar groups, their mutual interaction in the nonpolar environment may help to stabilize the structure. The situation can be understood in terms of the free energy associated with the pairing of different groups on the helix surfaces. Clearly, the role of polar groups must be taken into account as a possible factor in the stability of globular membrane proteins. An additional factor is the possible role of helix dipoles.

It is known (26) that α-helices possess a significant dipole moment arising from the alignment of peptide bonds in the helical structure. This dipole moment will tend to produce an attraction between two helices aligned in antiparallel orientation, resulting in an attractive force between them. Thus, it is anticipated that many of the helix-helix relationships in the final structure will involve an antiparallel orientation of helices. It will be of interest when the assignment of helices to positions in the structure is finally established, to see whether such a prediction is borne out.

PACKING INFLUENCES OF THE LIPID PHASE

A third important factor which must be considered is the interaction of the rough surfaces of helices with each other and with the nonpolar region of the bilayer. Formally, this amounts to a comparison of the relative fits in the case of lipid-lipid, lipid-helix, and helix-helix interactions. The relatively limited range of conformational possibilities for a polymethylene lipid chain make it impossible for the lipid bilayer to achieve a good fit between the lipid molecules and the surface of an α-helix, with the many projections on it produced by amino acid side chains. On the other hand, lipid molecules fit very well with each other and helices can pack quite well together. It is a well known observation that integral membrane proteins increase the permeability of lipid bilayers to the passage of ions and small molecules. Perhaps it is that there are packing voids in the interface between the lipid and protein surfaces. If this is the case, then one expects proteins to minimize their contact with the lipid bilayer through association of helices with each other. At present, work is under way to understand the magnitude of these effects.

CONCLUSION

We have examined the folding of polypeptide chains to form globular membrane proteins with a particular focus on bacteriorhodopsin. It appears that many transmembrane helical structures can be identified from amino acid sequence and that the interactions of helices to form a globular structure may involve both the participation of polar interactions and packing influences. If these can be understood in quantitative terms, it may be possible to derive detailed models of membrane protein structure starting from the information contained in the genetic code.

REFERENCES

1. Larsen, H. (1967) Adv. Microbiol 1, 97-132
2. Stoeckenius, W. and Rowan, R. (1967) J. Cell Biol. 34, 365-393
3. McClare, C.W.F. (1967) Nature 216, 766-771
4. Henderson, R. (1977) Ann. Rev. Biochem. Bioeng. 6, 87-109
5. Stoeckenius, W., Lozier, R., and Bogomolni, R. (1979) Biochim. Biophys. Acta 505, 215-78
6. Lanyi, J.K. (1978) Microbiol. Rev. 4, 682-706
7. Ottolenghi, M. (1980) Adv. Photochem. 12, 97-200
8. Stoeckenius, W. and Bogomolni, R. (1982) Ann. Rev. Biochem. 52, 587-616
9. Blaurock, A. and Stoeckenius, W. (1973) Nature New Biol. 233, 152-155
10. Oesterheldt, D. and Stoeckenius, W. (1973) Proc. Nat. Acad. Sci. USA 70, 2853-2857
11. Garfield, E. (1980) Curr. Contents 23 (40), 5-12
12. Unwin, P.N.T. and Henderson, R. (1975) J. Mol. Biol. 94, 425-450
13. Henderson, R. and Unwin, P.N.T. (1975) Nature 257, 28-32
14. Davis, S.S., Higuchi, T. and Rytting, S. (1974) Adv. Pharmaceut. Sci. 4, 73
15. Tanford, C. (1980) The Hydrophobic Effect (New York: J. Wily & Sons)
16. Singer, S.J. (1962) Adv. Prot. Chem. 17, 1
17. Singer, S.J. (1971) in: Structure and Function of Biological Membranes (New York: Academic Press) p. 145
18. Reynolds, J.A., Gilbert, D.B. and Tanford, C.S. (1974) Proc. Natl. Acad. Sci. USA 71, 2925
19. Richards, F. (1977) Ann. Rev. Biophys. Bioeng. 6, 151
20. Richmond, T. and Richards, F. (1978) J. Mol. Biol. 119, 537
21. Honig, B.H. and Hubbell, W. (1984) Proc. Natl. Acad. Sci. USA 81, 5412
22. Engelman, D.M., Goldman, A. and Steitz, T.A. (1982) Meth. Enzymol 88, 81
23. Engelman, D.M. and Steitz, T.A. (1984) in: The Protein Folding Problem, D. Wetlaufer, Ed. AAAS Symposia
24. Huang, K.S., Bayley, H., Liao, M.J., London, E. and Khorana, H.G. (1981) J. Biol. Chem. 256, 3802
25. Liao, M-J, Huang, K.S. and Khorana, H.G. (1984) J. Biol. Chem. 259, 4200
26. Wada, A. (1976) Adv. Biophys. 9, 1

STACKING OF PURPLE MEMBRANES IN VITRO

Marcelle Lefort-Tran, Monique Pouphile, Bernard Arrio[+],
Georges Johannin[+], Pierre Volfin[+], and Lester Packer*

Laboratoire de Cytophysiologie de la Photosynthese
Gif Sur Yvette, France

+Institut de Biochimie, Bat 432
Universite de Paris XI
91405 Orsay, France

*Membrane Bioenergetics Group
Applied Science Division, Lawrence Berkeley Laboratory
University of California
Berkeley, California, 94720, USA

ABSTRACT

Negative staining electron microscopy showed that purple membranes
isolated from Halobacterium halobium are aggregated in vitro in the form of
stacked arrays. This effect is more marked after trypsin treatment. White
membranes isolated from mutant strains do not stack and exhibit an average
size consistent with previous results of electron microscopy. White membrane fragments also do not exhibit stacking in vitro after retinal reconstitution or trypsin treatment. Quasi-elastic light scattering was also
used to characterize the size (hydrodynamic radius) of isolated purple and
white membranes before and after proteolysis. These results also show that
native purple membrane preparations are larger in size than expected and
that, following trypsin treatment, they are on average more than an order
of magnitude larger. In stacked purple preparations, cations are unable to
exchange freely with the aqueous medium. This explains why proteolysis
lowers the efficiency of proton release by illuminated bacteriorhodopsin in
purple membranes in vitro. Thus, previously reported decreases in efficiency of proton release by bacteriorhodopsin in proteolyzed purple membranes are due to the stacking effect and not per se to loss of the carboxyl terminus tail.

INTRODUCTION

Purple membranes of Halobacterium halobium have been observed by electron and neutron diffraction of intact cells to consist as "patches" in the
membrane consisting of regular hexagonal arrays of bacteriorhodopsin trimers
interspersed with lipids (1). Previous studies of purple membranes by electron microscopy reported that isolated purple membranes exist as flat membrane sheets in a size range 800-1500 Å. After oriented adsorption to

173

cationic surfaces, electron microscopy also reveals that the two surfaces differ in appearance (2), one surface smooth and the other (the extracellular surface from which light-induced proton release by bacteriorhodopsin is believed to occur) cracked in appearance. Studies of proton release from purple membrane in suspension and proton pumping after incorporation into liposomes, have shown that only the light-adapted form of bacteriorhodopsin is active in proton translocation (3).

The reported stoichiometry of protons released per M_{412} intermediate formed in the photocycle is variable. It has been reported (4) that in the presence of high salt and at low light intensity the H^+/M_{412} ratio approaches 3, but that in distilled water the ratio is 0.3-0.5 at pH 7.5. Lozier et al. (5) found that H^+/M_{412} ratio for membrane sheets in water to be 1.0 at the same pH. Using flash photolysis with pH indicator dyes, Govindjee et al. (6) reported H^+/M_{412} stoichiometry for purple membranes in suspension to vary with salt concentration, unlike membranes incorporated into liposomes, which have a consistent H^+/M_{412} for pumping of about 2, and that proteolytic cleavage of the C-terminal tail of bacteriorhodpsin by trypsin specifically lowers H^+/M_{412} stoichiometry (7). Several laboratores Gerber et al. (8), Ovchinnikov et al. (9), Liao and Khorana (10), Govindjee et al. (11) , however, report that the removal of the carboyxl tail does not effect proton pumping of purple membranes incorporated into liposomes.

By combining electron microscopy with the use of quasi-elastic light scattering (QELS), which give information about the hydrodynamic radius of particles in suspension, we have found that removal of the carboyxl terminal tail of bacteriorhodopsin caused marked changes in the structural arrangement of purple membranes in aqueous suspension, which may account for the known discrepancy of lower proton release previously reported for some of the native and proteolyzed purple membrane preparations and for the decrease in proton release in protelyzed sheets observed in a companion study (cf. Robinson et al., in these Proceedings). This explains earlier discrepancies between proton translocation after incorporation into liposomes, proton release in native and proteolyzed purple membranes in suspension, and the inconsistencies in the lower-limit values of H^+ released/M_{412}.

METHODS

Membranes

Purple membranes from H. halobium S_9 were isolated as previously described (12). Membranes were bleached by hydroxylamine plus light (13). White membanes from the R_1mW strain (gift of Y. Mukohata) were reconstituted with all-trans retinal to form functional bacteriorhodosin (14). White and purple membranes were suspended in 10 mM azide at pH 7 and kept in the cold; before use, suspensions were diluted into test media.

Trypsinization

The trypsin treatment consisted of a 5 minute incubation of the membranes in the presence of 10 micrograms/ml of trypsin at pH 8.2 and 20C of 100 mM Tris buffer, then the membranes were washed twice by centrifugation at 34,000 g for 20 minutes and resuspended in the HEPES buffer.

Electron microscopy

For electron microscopy, negative staining of membranes was performed directly on the formvar coated grids according to "drop method" (Haschemeyer and Myers, 1972). The samples were diluted, if necessary, to the appropriate concentration with low concentration buffer (10 mM HEPES, pH 7.2). Sodium silicotungstate (1%) or ammonium molybdate (1%) solution at pH 7.0-7.2 were used.

QELS studies

QELS (17) allows determination of the translational diffusion coefficient (D_t), from which the hydrodynamic radius (R) for spherical particles can be calculated according to the following equation

$$D_t = \frac{k\,T}{6\pi\,\eta\,R}$$

.where k is the Boltzman constant, T and η, the temperature and viscosity of the medium, respectively. The apparatus for QELS studies has recently been described for characterization of sarcoplasmic reticulm (18) and purple membranes (19). All the QELS size measurements were carried out in 10 mM HEPES buffer at pH 7.2 and 20°C.

RESULTS AND DISCUSSION

Negative staining electron microscopy was used to compare the appearance of native purple membranes isolated from H. halobium and white membranes isolated from the R_1mW mutant (Figure 1). Here, native purple membranes often appear as folded upon one another, whereas the white membranes at the same magnification appear to be smaller, more uniform in appearence, and not so discreet in their profiles. After trypsinization of native purple membranes, large aggregates are formed which are unsuitable for clear observation by electron microscopy. However, after mild homogenization, individual trypsinized purple membranes profiles can be observed (Figure 2, a-b). After sonication the trypsinized purple membranes can be teased apart and appear in this case as more individualized profiles of smaller size (Fig. 2c).

A comparison of the size of purple and white membranes using QELS methods is shown in Table 1. It can be seen that the white membranes are much smaller in size than purple membranes, closely correlating with results

Table 1. Measurement of the size of purple and white membranes by QELS methods

	Hydrodynamic radius nm	
	Native	Trypsinized
Purple membranes	265	>1000 (75)*
White membranes	115	155

* Value observed after sonication. The control remained constant with the same treatment.

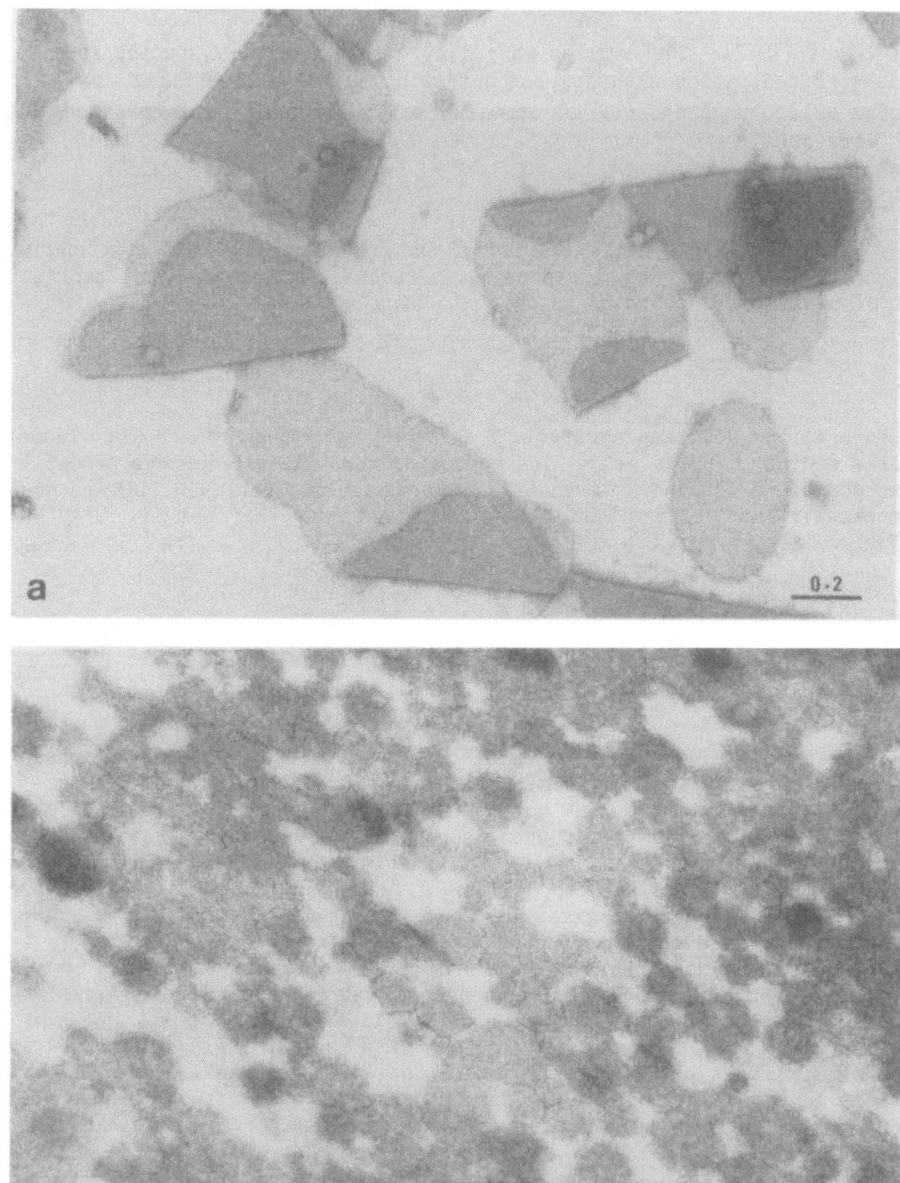

Figure 1. Comparative sizes of native folded purple membranes (a) and white membranes of mutant R 1 m W (b). x 55,500. Negative staining (silicotungstate).

of electron microscopy. The QELS data also show that the trypsinized native purple membranes are too large to obtain an accurate size measurement by this technique (which affords a measure of the hydrodynamic radius). On the other hand, after the trypsinized purple membrane preparations are son-

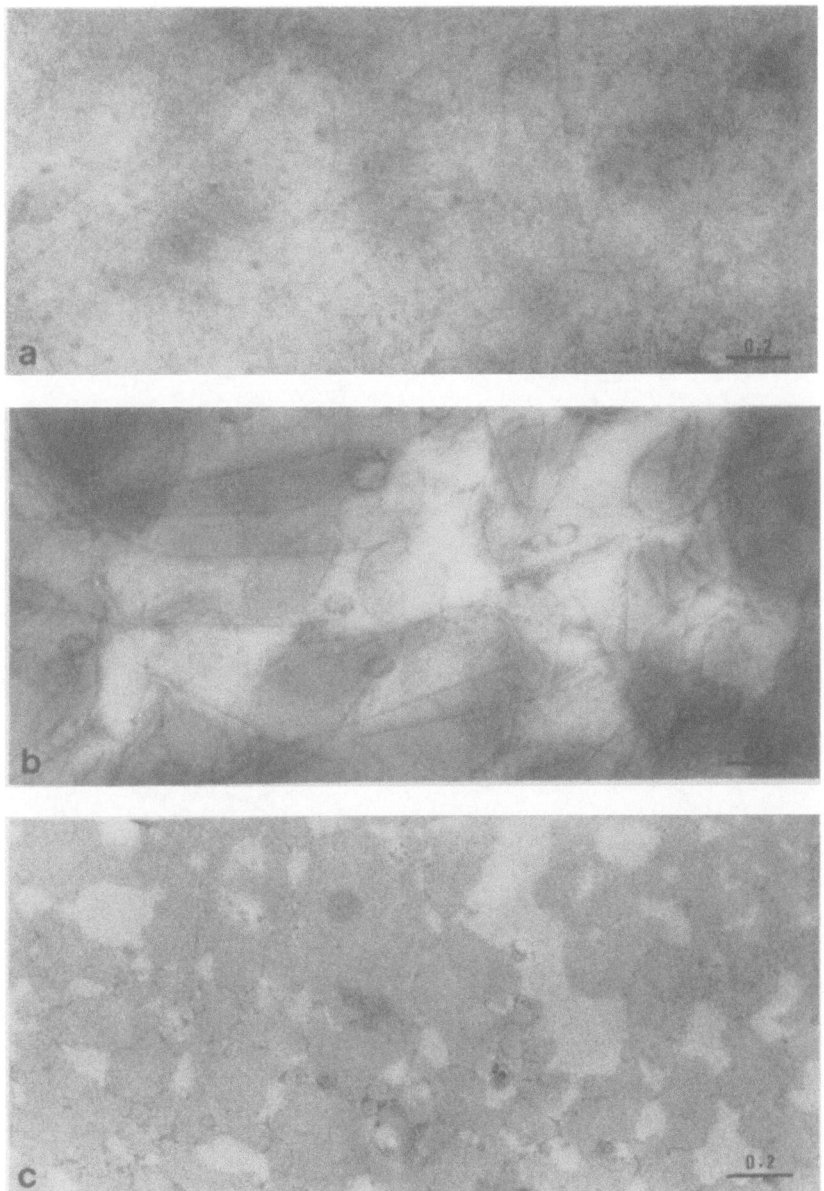

Figure 2. Native purple membranes after trypsin treatment. Negative staining with sodium silicotungstate (S.T.A.) x 55,500.

a) Trypsinized purple membranes
b) Trypsinized purple membranes after mild homogenization
c) After sonication (compare with Figure 1a)

icated, the QELS data reveal that the trypsinized preparations are of
reduced size, similar to that observed in electron microscopy. This effect
is reversable several hours after sonication. The trypsinized purple membrane preparations once again become aggregated as seen by electron microscopy and QELS (results not shown). White membranes on the other hand are

smaller in size than purple membranes and do not show an appreciable
increase in stacking or aggregation after trypsin treatment.   Again these
results closely coincide with direct structural observations by negative
staining electron microscopy.

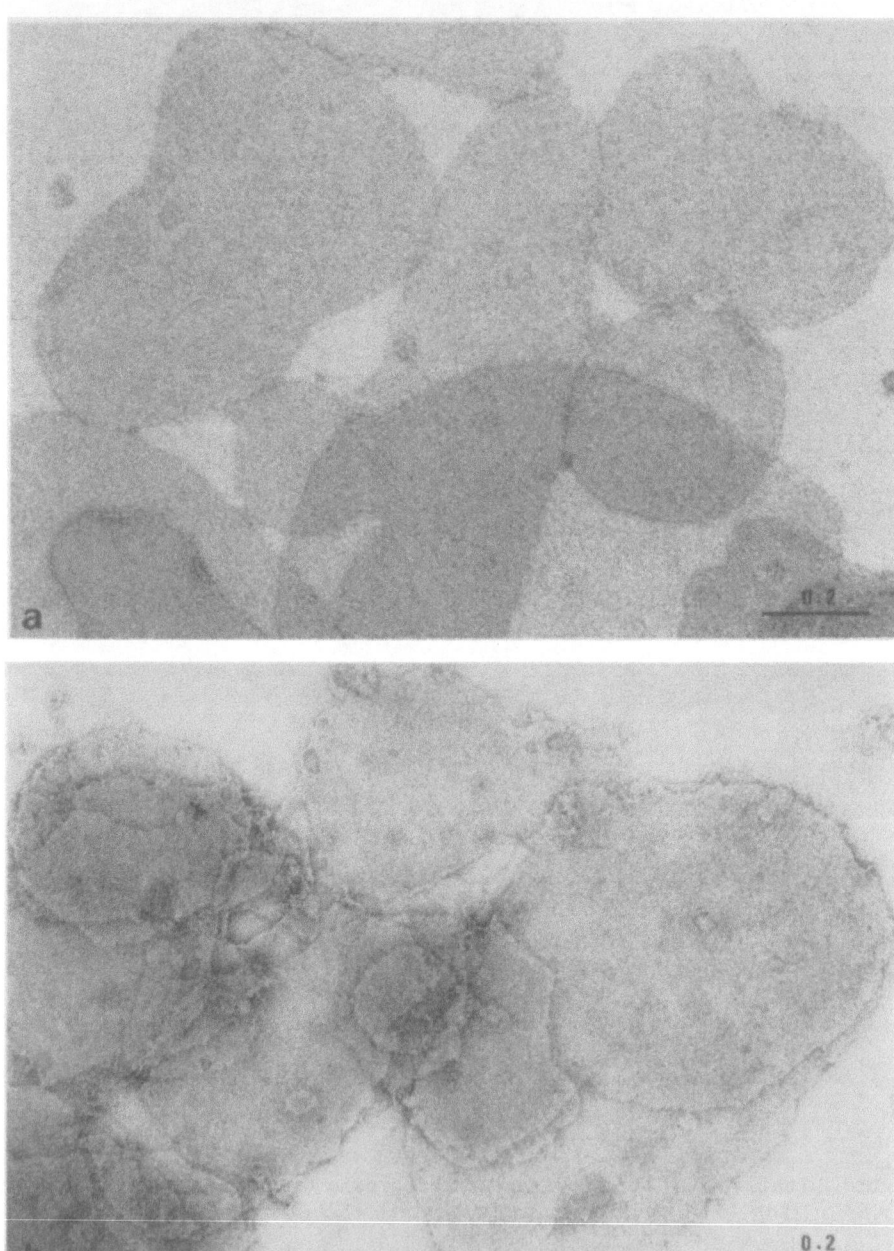

Figure 3. Comparison of native purple membrane and trypsinized purple mem-
        branes. x 83,000.

        a) Native purple membranes.   Negative staining with silico-
        tungstate.
        b) Trypsinized purple membranes.   Negative staining with ammonium
        molybdate.

The hydrodynamic behavior of purple membranes in suspension, as measured by QELS, approximates that of a suspension of spherical particles. Such behavior justifies the use of the given formula to provide an accurate indication of the population size range. In these studies we observe that the linear relation between half band width and $\sin^2 (\theta/2)$ found in angular light scattering studies of purple membranes is in better agreement with a spherical shape rather than the thin rigid disk shape seen by electron microscopy. This property appears to be due to folding of the purple membranes in suspension.

Whether or not negative staining electron microscopy was carried out with silicotungstate or ammonium (Figure 3) when native purple membranes are compared with trypsinized purple membranes there is always a persistence of the aggregation in the trypsinized preparations. With purple membranes, selective areas on the grid show individualized sheets as spread out without displaying much of the stacking effect, whereas marked aggregation or stacking is always observed in the trypsinized preparations. At higher magnifications (110,000 diameters), the profiles of native white membranes are discerned quite clearly by negative staining. They display a relatively uniform size and a more spherical pattern than purple membranes. After trypsinization, some aggregation and stacking is observed (Figure 4).

Trypsin treatment of purple membranes causes considerable size increase ($\frac{1}{2}$-1 micrometer). Such aggregation after proteolysis has been previously observed (22) and attributed to the reduction in electrostatic repulsion allowing hydrophobic forces to induce aggregation. We and others (7) have observed however that in some preparations of native purple membrane, long-term storage allows development of protease activity, most likely arising from bacterial contamination. This activity is inhibited by benzamidine, indicating that it is trypsin-like (23). These findings suggest that protease activity in preparations of purple membranes should be monitored from the beginning of the membrane preparation. After systematic washing of cell homogenates and purple membrane preparations, and purification by density gradient centrifugation, little or no benzamidine inhibited protease activity is present. The results suggest that protease activity cleaves portions of the C-terminal polypeptide of bacteriorhodopsin that lie exposed at the surfaces of the membrane. Trypsin is known to cleave off the C-terminal tail of bacteriorhodopsin in purple membranes (24), and further it causes extensive aggregation effects. This indicates that the presence of low level protease activity in purple membrane preparations during storage causes the aggregation/stacking effects observed.

As was suspected earlier (19) and as Ovchinnikov (25) and Kiselev (personal communication) report, these results have important implications regarding the stoichiometries of proton and cation release from purple membranes in suspension during photocycle activity. We find that the changes in the H^+/M_{412} ratio of membrane preparations correlate with the stacking data obtained (cf. Robinson et al. in these Proceedings). Trypsin treatment reduces the proton release of purple membrane sheets by a factor of approximately 0.3. Proton release by reconstituted white membrane sheets, however, seems to be less affected by trypsinization. Thus, the published difference between H^+/M_{412} ratios measured for trypsin treated and native purple membrane preparations appear to reflect the stacking effect observed. Stacked membrane preparations cannot exchange a full complement of cations (8) from membrane surfaces. Determination of precise stoichiometry requires purple membranes whose surfaces are freely exposed to the aqueous medium, as would be the case if the active preparation of purple membranes in suspension existed only as individual membrane sheets or in the

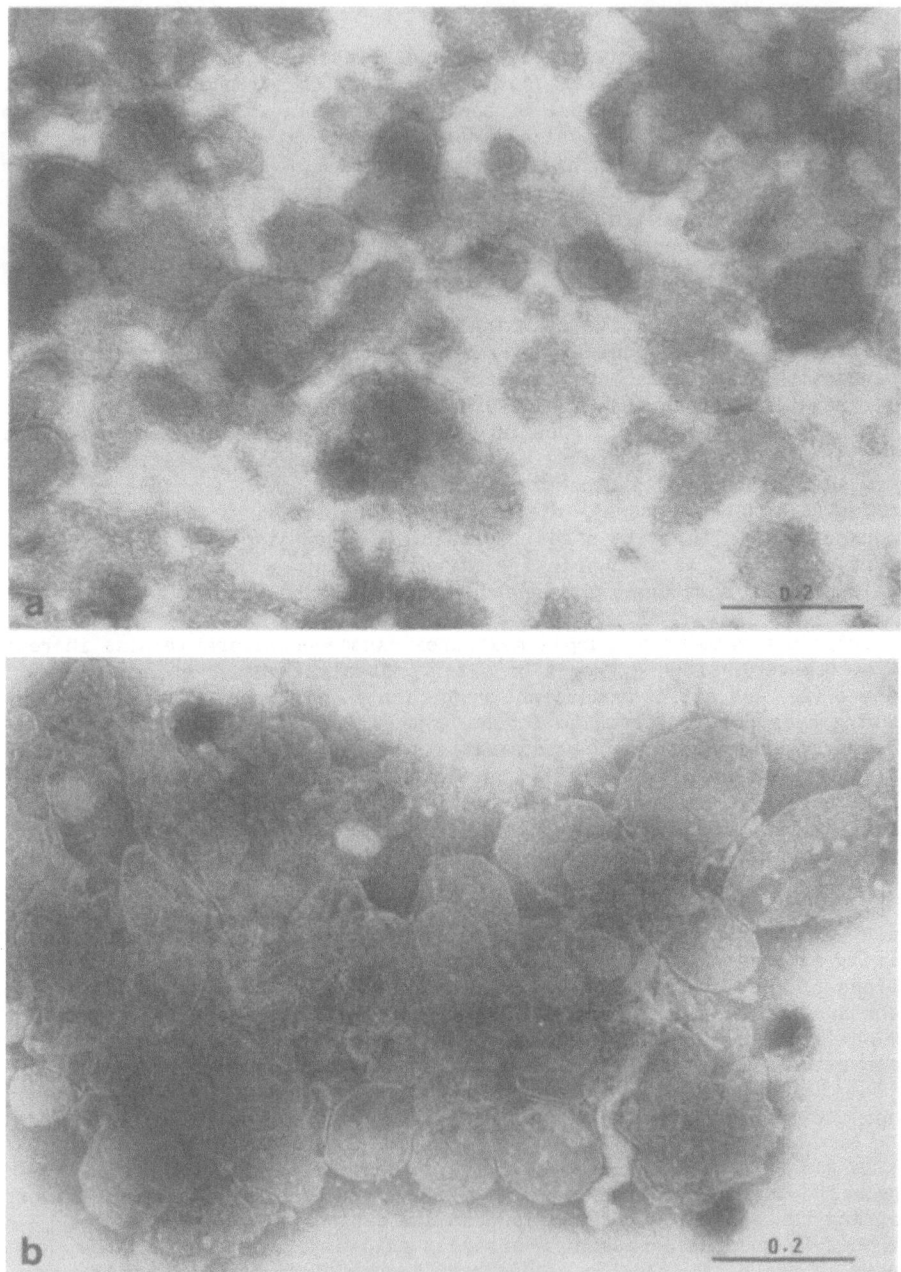

Figure 4. Comparison of native and trypsinized white membranes. x 110,000.
Negative staining; silicotungstate.

a) Native white membranes b) After trypsin treatment

reconstituted and liposome preparation in which stacking is probably greatly
reduced or eliminated. When this is not the case, normal photocycling with
lower proton release will be observed, erroneously indicating short circuit-
ing of the proton translocating function of bacteriorhodopsin.

ACKNOWLEDGMENTS

    Research support from the Centre National de les Recherche Scientifique
(A.T.P. "Bioenergetique" n° 3093/79) and the Office of Biological Energy
Research, Division of Basic Energy Sciences, U.S. Department of Energy
(Contract #DEAC03-76SF00098), and grants from the U.S. National Science
Foundation and NATO for International Collaboration in Research.

REFERENCES

1. Blaurock, A.E., Stoeckenius, W., Oesterhelt, D., and Scherphof, G.L.
    Sturcture of the cell envelope of Halobacterium halobium (1970) J.
    Cell. Biol. 71, 1-22.

2. Fisher, K., Yanagimoto, K., and Stoeckenius, W. Oriented absorption of
    purple membrane to cationic surfaces (1978) J. Cell Biol. 77, 611-621.

3. Fahr, A. and Bamburg, E. Photocurrents of dark-adapted bacteriorho-
    dopsin on black lipid membranes (1982) FEBS Lett. 140, 251-253.

4. Kuschmitz, D. and Hess, B. On the ratio of the proton and photochem-
    ical cycles in bacteriorhodopsin (1981) Biochemistry 20, 5950-5957.

5. Lozier, R.H., Niederberger, W., Bogomolni, R.N., Hwang, S., and Stoeck-
    enius, W. Kinetics and stoichiometry of light-induced proton release
    and uptake from purple membrane fragments, Halobacterium halobium cell
    envelopes, and phospholipid vesicles containing oriented purple mem-
    brane (1976) Biochim. Biophys. Acta, 440, 545-556.

6. Govindjee, R., Ebrey, T.G., and Crofts, A.R. The quantum efficiency of
    proton pumping by the purple membrane of Halobacterium halobium (1980)
    Biophys, J. 30, 231-242.

7. Govindjee, R., Ohno, K., and Ebrey, T.G. Effect of removal of the C-
    terminal region of bacteriorhodopsin on its light-induced H^+ changes
    (1982) Biophys J. 38, 85-87.

8. Gerber, G.E., Wildenauer, D., Khorana, H.G. Orientation of bacterio-
    rhodopsin generate a membrane potential (1977) Proc. Natl. Acad. Sci.
    USA 74, 5426-5430.

9. Ovchinnikov, Y.A., Abdulaev, N.G., Feigina, M.Y., Kiselev, A.V., and
    Lobanov, N.A. Recent findings in the structure-functional character-
    istics of bacteriorhodopsin (1977) FEBS Lett. 84, 104.

10. Liao, M.-J. and Khorana, H.G. Removal of the carboxyl-terminal peptide
    does not affect refolding or function of bacteriorhodopsin as a light
    dependent proton pump (1984) J. Biol. Chem. 259, 4194-4199.

11. Govindjee, R., Ohno, K., Chang, C.-H., and Ebrey, T.G. In: Transduc-
    tion in Biological Membranes, New York: Plenum Press. In press.

12. Oesterhelt, D. and Stoeckenius, W. Isolation of the cell membrane of
    Halobacterium halobium and its fractionation into red and purple mem-
    brane (1974). In: Methods in Enzymology (S. Fleischer and L. Packer,
    eds.), Vol. 31, Academic Press, pp. 667-678.

13. Oesterhelt, D. Reconstitution of the retinal proteins bacteriorho-
    dopsin and halorhodopsin (1982). In: Methods of Enzymology (L.
    Packer, ed.), Vol. 88, Academic Press, pp. 10-17.

14. Mukohata, Y., Sugiyama, Y., Kaji, Y., Usukura, J., and Yamada, E. The white membrane of crystalline bacteriorhodopsin in Halobacterium halobium strain R$_1$mW and its conversion into purple membrane by exogenous retinal (1981) Photochem. Photobiol. 33, 593-600.

15. Katsura, T., Lam, E., Packer, L., and Seltzer, S. Light dependent modification of bacteriorhodopsin by tetranitromethane. Interaction of a tyrosine and a tryptophan residue with bound retinal (1982) Biochem. Internatl. 5, 445-456.

16. Scherrer, P., Packer, L., and Seltzer, S. Effect of iodination of purple membrane on the photocycle of bacteriorhodopsin (1981) Arch. Biochem. Biophys. 202, 589-601.

17. Ware, B.R. and Haas, D.D. Electrophoretic light scattering (1983) In: Fast Methods in Physical Biochemistry and Cell Biology (R.I. Sha'afi and S.M. Fernandez, eds.), Elsevier Science Publishers, pp. 174-220.

18. Arrio, B., Johannin, G., Carrette, A., Chevallier, J., and Brethes, D. Electrokinetic and hydrodynamic properties of sarcoplasmic reticulum vesicles: a study by laser Doppler electrophoresis and quasi-elastic light scattering (1984) Arch. Biochem. Biophys. 288, 220-229.

19. Arrio, B., Johannin, G., Volfin, P., and Packer, L. Quasi-elastic laser light scattering and Doppler electrophoresis of purple membranes (1984) Biophys, Soc. Abstr. 45, 212a.

20. Haschemeyer, R.M. and Myers, R.J. (1972) In: Principles and Techniques of Electron Microscopy (M.A. Hayat, ed.), Vol. 2, Van Nostrand Reinhold Company, pp. 99-147.

21. Packer, L., Tristram, S., Herz, J., Russell, C., and Borders, C.L. Chemical modification of purple membranes: role of arginine and carboxylic acid residues in bacteriorhodopsin (1971) FEBS Lett. 108, 243-248.

22. Wallace, B.A. and Henderson, R. Location of the carboxyl terminus of bacteriorhodopsin in purple membrane (1982) Biophys. J. 39, 233-239.

23. Walsh, K.A. Trypsinogens and trypsin of various species (1971) In: Methods in Enzymology (G.E. Perlmann and L. Lorand, eds.), Vol. 19, pp. 41-63.

24. Abdulaev, N., Feigina, M., Kiselov, A., Ovchinnikov, Y., Drachev, L., Kauger, A., Knitrina, L., and Skulachev, V. Products of limited proteolysis of bacteriorhodopsin generate a membrane potential (1978) FEBS Lett. 90, 190-194.

25. Ovchinnikov, Y.A. (1984) In: EMBO Workshop on Molecular Biology of Retinal Proteins.

PLASMA MEMBRANE POTENTIAL OF ANIMAL CELLS GENERATED BY ION PUMPING, NOT

BY ION GRADIENTS

C.L. Bashford and C.A. Pasternak

Department of Biochemistry
St George's Hospital Medical School
Cranmer Terrace, London SW17 0RE

INTRODUCTION

The plasma membrane potential of animal cells is generally thought
to be generated predominantly by the diffusion of K^+ out of cells
(Williams, 1970), with only very minor contributions from the diffusion
of other ions or from an electrogenic Na^+ pump (Thomas, 1972; Lew et
al., 1979). The experiments to be described show that this view is no
longer universally applicable, and that plasma membrane potential of at
least two cell types, - a mouse tumour cell and human neutrophils, -
arises predominantly from electrogenic pumping of Na^+ and, to a lesser
extent, of H^+. In such cells electrical and osmotic stability are
preserved by the operation of anion leak and electroneutral ion
transport pathways (Bashford & Pasternak, 1984, 1985a,b).

METHODOLOGY

Lettre cells are conveniently grown as ascites fluid in mice: 1g of
cells, virtually devoid of contamination by blood cells (Mehta et al.,
1985), can be routinely obtained from 1 mouse. Human neutrophils are
obtained from fresh blood as described previously (Segal et al., 1980;
Bashford & Pasternak, 1985b). Membrane potential is measured with the
negatively-charged dye oxonol-V (Smith et al., 1976; Bashford et al.,
1979a,b; Bashford and Pasternak, 1984). This dye has the advantage over
the positively charged, cyanine dyes that it does not accumulate in
mitochondria (which have a membrane potential, negative inside, more
than double that of the plasma membrane potential). By measuring changes

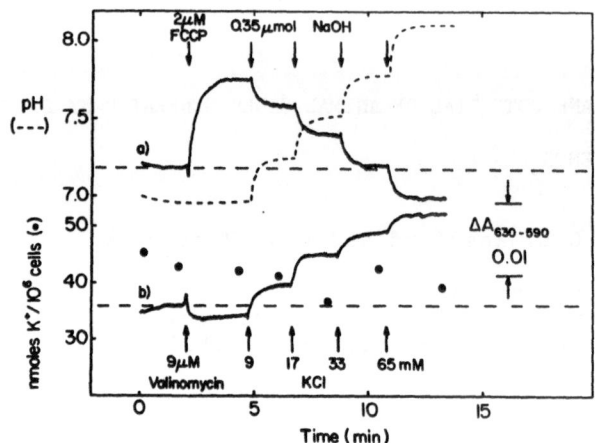

Fig. 1. Measurement of membrane potential of Lettre cells
using oxonol-V. 5×10^6 cells/ml in 150mM NaCl, 5mM
KCl, 1mM $MgCl_2$, 2μM oxonol-V at 32°C and (a) 0.5mM
Hepes, pH 7.0 or (b) 5mM Hepes, pH 7.3. FCCP, valinomycin and KCl were added to give the final concentrations
indicated. In (a) pH was recorded with a semi-micro
combination electrode (Corning, Type 19). In (b) K^+
was measured by atomic absorption spectroscopy after
pelleting cells through oil. Reprinted by permission
from Bioscience Reports 3, 631-642 copyright (C) 1983
The Biochemical Society, London.

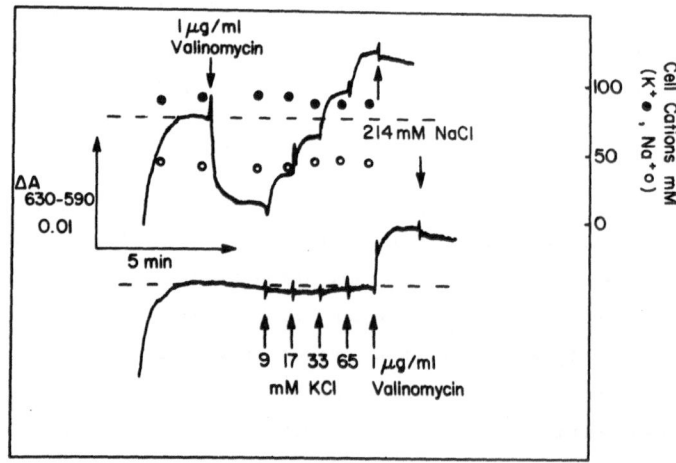

Fig. 2. Lettre cell membrane potential does not depend on the
K^+ gradient unless valinomycin is also present. 5×10^6
cells/ml in 150mM NaCl, 5mM KCl, 5mM Hepes, 1mM $MgCl_2$,
pH 7.4 (NaOH) at 37°C (HBS) containing 5mM glucose, 2mM
$CaCl_2$, 1mM Na-pyruvate and 2μM oxonol-V. Valinomycin,
KCl and NaCl were added to give the final concentrations
indicated. Reprinted from J.Membrane Biol. 79, 275-284,
1984.

184

in absorbance ($A_{630-590}$) rather than in fluorescence, potential artefacts, due to fluorescence quenching or enhancement, are avoided.

The calibration of membrane potential by oxonol V is shown in Fig. 1. The membrane potential is set either to the H^+ diffusion potential by introducing the proton ionophore carbonyl cyanide p-trifluoromethoxyphenylhydrazone (FCCP, trace a), or to the K^+ diffusion potential by introducing the K^+ ionophore valinomycin (trace b; Harris and Pressman, 1967; Henderson, McGivan and Chappell, 1969), and then titrating with H^+ (OH^-) or K^+ respectively in order to determine the 'null-point'; that is namely the concentration of H^+ or K^+ in the medium at which ionophore causes no change in the absorbance of cells treated with oxonol-V (Hoffman and Laris, 1974; Bashford et al., 1985). An increase in $A_{630-590}$ indicates depolarization. From a knowledge of cytoplasmic H^+ and K^+, the membrane potential can then be calculated using the Nernst equation:

$$V = RT/F \ln ([ion]^{null\ point}/[ion]_{cytoplasm})$$

where V is the membrane potential, T the absolute temperature, R the gas constant and F the Faraday constant. Lettre cells prove to have a potential between -50 and -60mV, and neutrophils a potential between -70 and -80mV when suspended in media containing 150mM NaCl and 5mM KCl at 37°C.

Potential not set by K^+ diffusion

Fig. 2 shows (a) that Lettre cells are hyperpolarized when valinomycin is added (upper trace see also Fig. 1), (b) that membrane potential does not alter when K^+ is added (lower trace) and (c) that in high K^+ medium valinomycin depolarizes cells. Neutrophils, which in our preparations have lower K^+ content and more negative potentials than Lettre cells, are slightly depolarized by valinomycin in media containing 5mM KCl but, like Lettre cells, do not depolarize when further K^+ is added in the absence of ionophore. These results are consistent with the notion that Lettre cells and neutrophils lack the K^+-selective channels by which cells usually generate their plasma membrane potential. That ion gradients do not contribute to potential is further indicated by the results of Fig. 3. Here Lettre cells are suspended in either the normal 150mM NaCl, 5mM KCl medium (panel A) or

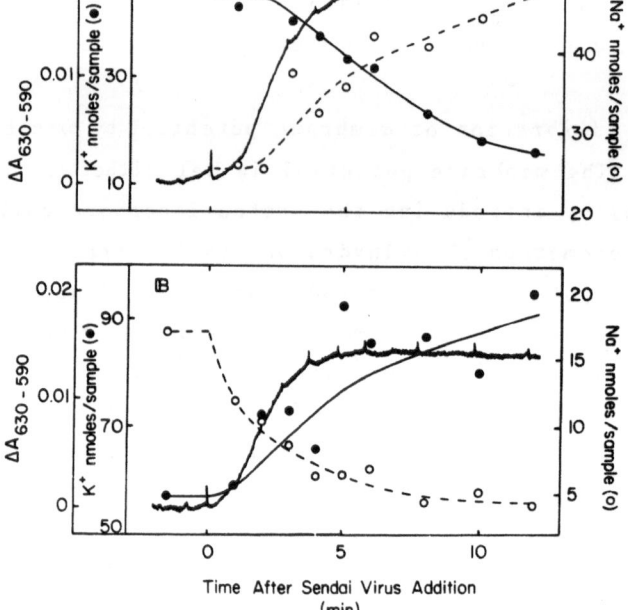

Fig. 3. Insensitivity of Lettre cell membrane potential to KCl.
5×10^6 cells/ml in 5mM Hepes, 1mM $MgCl_2$, 2μM oxonol-V,
pH 7.4 at 33°C and A 150mM NaCl, 5mM KCl or B 155mM
KCl. 2 HAU Sendai virus/ml (final concentration) was
added and cell cations determined after pelleting
through oil. Reprinted from J.Membrane Biol. 79, 275–
284, 1984.

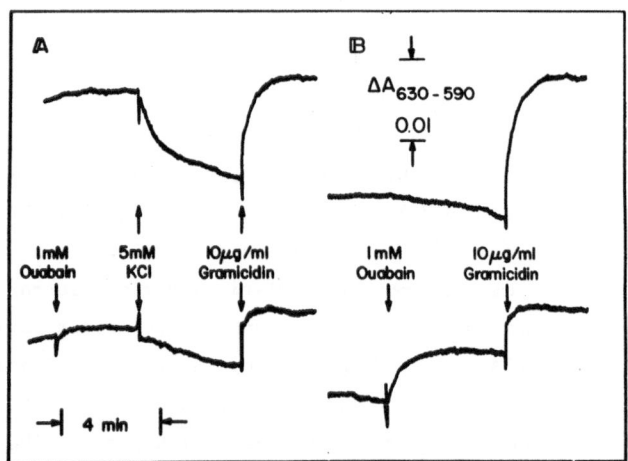

Fig. 4. The contribution of the Na^+ pump to Lettre cell
membrane potential. 2×10^6 cells/ml in HBS (see Fig.
2 legend) containing 1mM glucose, 2μM oxonol-V, pH
7.4 at 32°C (B) or a similar medium lacking KCl (A).
Ouabain, KCl and gamicidin were added to give the
final concentrations indicated. Reprinted from
J.Membrane Biol. 79, 275–284, 1984.

in 155mM KCl (panel B). Cells are then permeabilized by the addition of Sendai virus (Impraim et al., 1980). Cells depolarize in either situation (A) or (B). In A, K^+ leaks out and Na^+ leaks in; in B, the reverse is true. Hence the Na^+ and K^+ gradients must have been the opposite in the two situations, before virus was added. Yet the potential was approximately the same (-50mV) in each case. Neutrophils are depolarized by the chemotactic peptide fMetLeuPhe (Tatham et al., 1980) to a similar extent whether they are suspended in normal (5mM K^+) medium or in 155mM KCl (Bashford and Pasternak, 1985b). These results confirm that potential in Lettre cells or neutrophils cannot be set by K^+ diffusion (or, for that matter, by Na^+ diffusion).

Potential set by ion pumps

The Na^+ pump obligatorily requires intracellular Na^+ and K^+ in the suspending medium. Hence cells whose potential is dependent on an electrogenic Na^+ pump are unable to generate a potential when suspended in K^+-free medium. Fig. 4 shows this for Lettre cells: addition of 5mM KCl <u>hyperpolarizes</u> cells (A, upper trace), - in an ouabain-sensitive manner (A, lower trace). Conversely cells in 5mM KCl (Fig. 4B) are depolarized when ouabain is added; note that ouabain does not entirely depolarize cells (there is further depolarization by gramicidin): this may be due partly to the fact that the Na^+ pump of mouse cells is rather insensitive to ouabain and partly to the fact that there may in addition to the Na^+ pump, be a frank H^+ pump (see Figs. 5-7). Similar results are obtained with neutrophils (Bashford and Pasternak, 1985b). When Lettre cells are suspended in Na^+-free media the cell content of Na^+ falls and the cells are hyperpolarized by the addition of Na^+ in a ouabain-sensitive fashion (Bashford and Pasternak, 1985a). Thus the contribution of the Na^+-pump to the potential of these cells is indicated by (1) the requirement for intracellular Na^+; (2) the requirement for extracellular K^+; (3) depolarization by ouabain in the absence of changes in Na^+ or K^+ gradient across the plasma membrane. Cells whose potential is generated by K^+ diffusion, such as lymphocytes, behave differently in all these situations (Bashford and Pasternak, 1985a).

The possibility that Lettre cells, in some circumstances, may be polarized by a H^+ proton pump is suggested by the following observations:

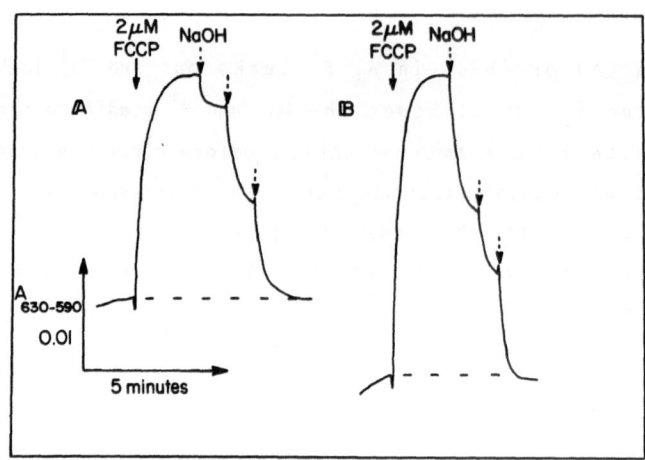

Fig. 5. Na+-dependence of Lettre cell membrane potential.
5 x 10⁶ cells/ml in 10mM Hepes, 5mM KCl, 1mM MgSO₄,
2.5μM oxonol-V, pH 7.4 (NaOH) at 36°C and 0.3M
mannitol (A) or 0.15 mannitol, 0.075M NaCl (B).
FCCP was added to give the final concentration
indicated and NaOH to restore the absorbance to its
value before the addition of FCCP. The 'null-point'
pH values were: 8.0 (A) and 8.3 (B).

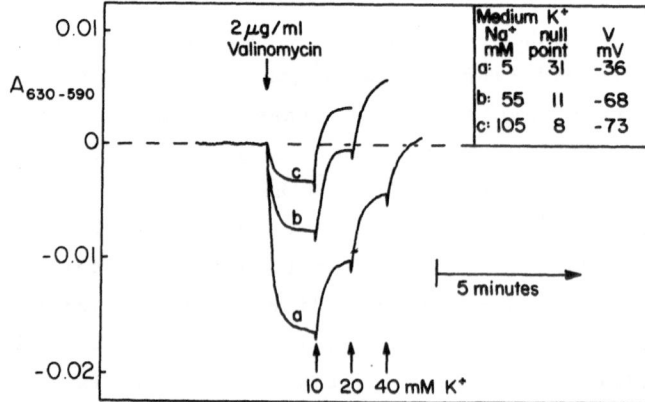

Fig. 6. Na+-dependence of Lettre cell membrane potential.
5 x 10⁶ cells/ml in 10mM Hepes, 5mM KCl, 1mM MgSO₄,
2.5μM oxonol-V, pH 7.4 (NaOH) at 37°C and 0.3M
mannitol (a), 0.2M mannitol, 0.05M NaCl (b) or 0.1M
mannitol, 0.1M NaCl (c). Valinomycin and KCl were
added to give the final concentrations indicated.

A.   Cells in a Na^+-free medium, in which the Na^+ pump is inactive
retain a significant FCCP-sensitive membrane potential (Fig. 5A),
although this is less than that found when the cells are supplemented
with Na^+ (Fig. 5B). The difference in 'null point' of 0.3pH unit
suggests that the Na^+ supplemented cells were 20mV more polarized than
the low Na^+ controls, assuming that the cells maintain similar cytoplasmic pH values in the two situations. Figure 6 shows the behaviour
of the K^+ 'null point' of cells suspended in media containing different amounts of Na^+. Trace a shows the result when Na^+ is omitted
(except for that required to adjust the pH of the medium), traces b and
c the results when Na^+ is included in the medium. For clarity of
presentation the $A_{630-590}$ value before the addition of valinomycin has
been assigned a value of zero. In this set of experiments the addition
of Na^+ polarised the cells by some 37mV.

B.   Dicyclohexyl carbodiimide (DCCD) an inhibitor of a number of
electrogenic proton pumps (Linnett & Beechey, 1979; Bashford et al.,
1976) depolarises Lettre cells suspended in normal media (Fig. 7A),
where it acts additively with ouabain and in Na^+-free media (Fig. 7B)
where ouabain is without effect. Note that in no instance do ouabain,
DCCD or their combination fully depolarize cells as judged by the
criterion of their sensitivity to FCCP. In the experiments illustrated
in Fig. 7 the dotted traces indicate that the behaviour of cells to
which neither ouabain or DCCD was added. The ability of Lettre cells,
in the absence of Na^+ pumping, to maintain a significant potential
which is partially DCCD sensitive strongly suggests the presence of
other, presumably H^+, electrogenic pumps in the plasma membrane. This
is compatible with reports regarding the acidification of endosomes
(Galloway et al., 1983; Hopkins, 1984; Yamashiro et al., 1983).

Na^+ pump acts as a H^+ pump at steady state

An electrogenic Na^+ pump (or H^+ pump) cannot maintain a
membrane potential at steady state, unless (a) electroneutral return
mechanisms and (b) a balancing current exist. Fig. 8 illustrates these
mechanisms in the case of Lettre cells. We have found evidence for a
$Na^+:H^+$ exchange mechanism, for $Na^+:K^+$ exchanges, and for
balancing HCO_3^- and lactate$^-$ leaks (Bashford and Pasternak, 1984,
1985a). Thus the membrane potential of Lettre cells is set essentially
as indicated by the bold arrow. Human neutrophils behave similarly to
Lettre cells, although in this case passive lack of Cl^- is also an

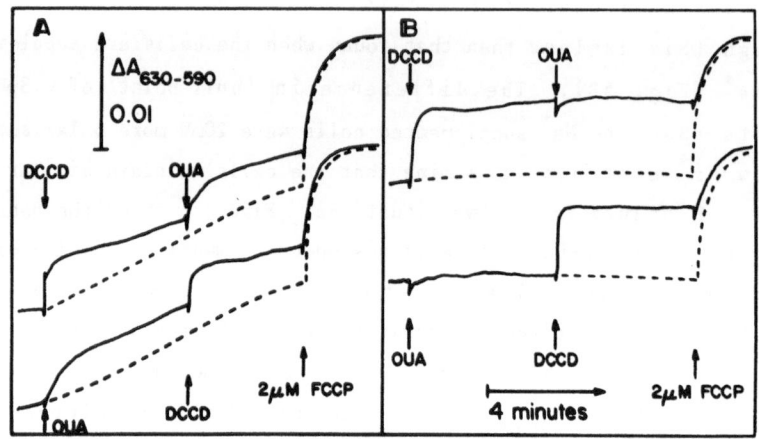

Fig. 7. Effect of dicyclohexylcarbodiimide (DCCD) and ouabain on Lettre cell membrane potential. 4×10^6 cells/ml in 10mM Hepes, 5mM KCl, 1mM MgSO$_4$, 2.5μM oxonol-V, pH 7.4 (NaOH) at 32°C and 150mM NaCl (A) or 300mM mannitol (B). Ouabain and DCCD (1mM, final concentration) and FCCP were added as indicated. The dashed lines indicate the behaviour of control cells to which ouabain and DCCD were not added.

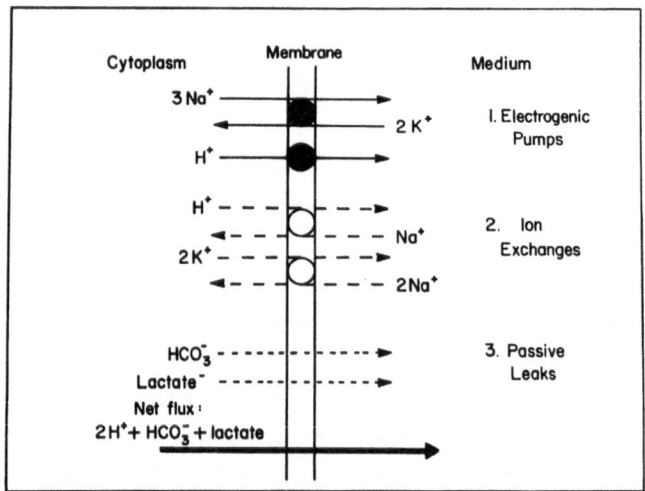

Fig. 8. Factors affecting the 'steady state' membrane potential of Lettre cells. Reprinted from J.Membrane Biol. <u>79</u>, 275-284, 1984.

190

important balancing mechanism (Bashford and Pasternak, 1985b).

We propose that any cell that lacks K^+ channels, but maintains a membrane potential near -60mV, is able to do so through the operation of an electrogenic Na^+ pump acting essentially as a H^+ pump; in addition, a frank proton pump may be present, as in Neurospora crassa plasma membrane.

CONCLUSION

The plasma membrane potential of certain animal cells is set by a mechanism more akin to that of mitochondria (H^+ pump), than to that of other animal cells (K^+ diffusion).

ACKNOWLEDGEMENTS

We thank Mrs V. Marvell and Mrs B. Bashford for preparing the paper for publication and to the Cell Surface Research Fund for financial support.

REFERENCES

Bashford, C.L., Casey, R.P., Radda, G.K. and Ritchie, G.A., 1976, Energy-coupling in adrenal chromaffin granules, Neuroscience 1: 399.

Bashford, C.L., Chance, B. and Prince, R.C., 1979a, Oxonol dyes as monitors of membrane potential. Their behaviour in photosynthetic bacteria, Biochim.Biophys.Acta. 545:46.

Bashford, C.L., Chance, B., Smith, J.C. and Yoshida, T., 1979b, The behaviour of oxonol dyes in phospholipid dispersions, Biophys.J., 25: 63.

Bashford, C.L. and Pasternak, C.A., 1984, Plasma membrane potential of Lettre cells does not depend on cation gradients but on pumps, J.Membr.Biol., 79:275.

Bashford, C.L. and Pasternak, C.A., 1985a, Generation of plasma membrane potential by the Na^+-pump coupled to proton extrusion, Eur.Biophys.J., in press.

Bashford, C.L. and Pasternak, C.A., 1985b, Plasma membrane potential of neutrophils generated by the Na^+ pump, Biochim.Biophys.Acta, in press.

Bashford, C.L., Alder, G.M., Gray, M.A., Micklem, K.J., Taylor, C.C., Turek, P.J. and Pasternak, C.A., 1985, Oxonol dyes as monitors of membrane potential: The effect of viruses and toxins on the plasma membrane potential of animal cells in monolayer culture and in suspension, J.Cell.Physiol., 123:326.

Galloway, C.J., Dean, G.E., Marsh, M., Rudnick, G. and Mellman, I., 1983, Acidification of macrophages and fibroblast endocytic vesicles in vitro, Proc.Natl.Acad.Sci.USA, 80: 3334.

Harris, E.J. and Pressman, B.C., 1967, Obligate cation exchanges in red cells, Nature 216:918.

Henderson, P.J.F., McGivan, J.D. and Chappell, J.B., 1969, The action of certain antibiotics on mitochondrial, erythrocyte and artificial phospholipid membranes. The role of induced proton permeability, Biochem.J., 111:521.

Hoffman, J.F. and Laris, P.C., 1974, Determination of membrane potentials in human and Amphiuma red blood cells by means of a fluorescence probe, J.Physiol. 239:519.

Hopkins, C.R., 1984, The importance of the endosome in intracellular traffic, Nature 304: 684.

Impraim, C.C., Foster, K.A., Micklem, K.J. and Pasternak, C.A., 1980, Nature of virally mediated changes in membrane permeability to small molecules, Biochem.J., 186:847.

Lew, V.L., Ferreira, H.G. and Moura, T., 1979, The behaviour of transporting epithelial cells. I. Computer analysis of a basic model. Proc.R.Soc.Lond.B., 206:53.

Linnett, P.E. and Beechey, R.B., 1979, Inhibitors of the ATP synthetase system, Meth.Enzymol. 55: 472.

Mehta, S., Bashford, C.L., Knox, P. and Pasternak, C.A., 1985, Chemiluminescence in neutrophils and Lettre cells induced by myxoviruses, Biochem.J. 227: 99.

Segal, A.W., Dorling, J. and Coade, S., 1980, Kinetics of fusion of the cytoplasmic granules with phagocytic vacuoles in human polymorphonuclear leukocytes, J.Cell.Biol., 85:42.

Smith, J.C., Russ, P., Cooperman, B.S. and Chance, B., 1976, Synthesis, structure determination, spectral properties and energy-linked spectral responses of the extrinsic probe oxonol-V in membranes, Biochemistry, 15:5094.

Tatham, P.E.R., Delves, P.J., Shen, L. and Roitt, I.M., 1980, Chemotactic factor-induced membrane potential changes in rabbit neutrophils monitored by the fluorescent dye 3,3'-dipropylthiadicarbocyanine iodide, Biochim.Biophys.Acta, 602:285.

Thomas, R.C., 1972, Electrogenic sodium pump in nerve and muscle cells, Physiol.Rev., 52:563.

Williams, J.A., 1970, Origin of transmembrane potentials in non-excitable cells, J.theor.Biol., 28:287.

Yamoshiro, D.J., Fluss, S.R. and Maxfield, F.R., 1983, Acidification of endocytic vesicles by an ATP-dependent pump, J.Cell Biol., 97: 929.

BACTERIAL MUTANTS RESISTANT TO UNCOUPLERS

Arthur A. Guffanti

Department of Biochemistry
Mount Sinai School of Medicine of the
City University of New York
1 G. Levy Place
New York, N.Y. 10029 USA

INTRODUCTION

Evidence from many laboratories over the last twenty years has firmly established the central role of proton translocation in energy transduction. The tenets of Peter Mitchell's[1] chemiosmotic theory are now almost universally accepted. The energization of solute transport, flagellar motion and oxidative phosphorylation in bacteria have been explained in chemiosmotic terms. The first requisite of the chemiosmotic theory is that the cell membrane must not be highly permeable to protons. Such relative impermeability permits the establishment of a protonmotive force due to the unequal distribution of protons on either side of the membrane. It is the tendency of protons to flow down their electrochemical gradient through specialized channels, such as the F_0 of BF_1F_0 ATPase or proton-solute symporters, that drives ATP synthesis or solute uptake. Among the substantial evidence that proton translocation is essential to energy transduction is the effect of uncouplers of oxidative phosphorylation on membrane proton permeability.[2,3] Acting as weak acids, classic uncouplers such as dinitrophenol and carbonylcyanide m-chlorophenylhydrazone (CCCP) have been shown to dissipate the protonmotive force by permeabilizing membranes to protons. On the other hand, there have also been reports of uncoupler-binding entities within energy transducing membranes.[4-6]

Several years ago, in order to further elucidate the mode of uncoupler action Decker and Lang[7,8] isolated an uncoupler-resistant mutant of Bacillus megaterium. Their studies, and those of others, on bacteria resistant to uncouplers have raised intriguing questions as to the relationship between the protonmotive force and energy transduction in bacteria. With particular emphasis on our own work with uncoupler-resistant mutants of B. megaterium and B. subtilis, I shall review some of the pertinent bioenergetic findings.

Isolation of Uncoupler-Resistant Mutants

Whether it be in strict aerobes such as B. megaterium and B. subtilis, or facultatively anaerobic E. coli, uncoupler-resistant mutants were obtained on agar plates, supplemented with a concentration of CCCP from between 5 to 50 μM, depending upon the particular bacterial species. In my experience with B. subtilis I have found that mutants resistant to 5 μM CCCP could be isolated on plates where malate is the primary carbon source. At concentrations of about 10 μM or higher no mutants could be isolated. Thus, although resistant mutants have been found in the two bacilli and E. coli the resistance is one of degree. For example, the E. coli mutant, designated CM 22, described by Ito et. al.[9] is inhibited by concentrations of CCCP of about 60 μM; whereas, the wild type is inhibited by 20 μM CCCP. This may indicate, as shown in B. megaterium, that there is a concentration of uncoupler that can still totally dissipate the protonmotive force in the mutants, thus rendering them nonviable.

Is Uncoupler-Resistance Due to a Single Mutation?

Uncoupler-resistant mutants invariably demonstrate pleiotropic properties. Both Decker and Lang[7,8] and our laboratory[10] have shown that the respiratory rate and ATPase activity in the B. megaterium uncoupler-resistant C8 strain are different from the wild type (see below). The E. coli mutant of Ito et al.[6] was not only resistant to several uncouplers, but also exhibited the unc phenotype at 42° C. Another E. coli mutant UV6, isolated by Sedgwick et al.[11] was resistant to both CCCP and tri-n-butyltin. In addition, UV6 showed consistently greater proline uptake than the wild type.

Whether the pleiotropic properties are caused by more than one mutation is not absolutely clear, but the evidence points to a single mutational event, at least in some of the reported strains. The frequency of mutation and reversion for B. megaterium C8[10] of one in 10^6 and E. coli CM 22 of one in 10^8 may indicate a single point mutation. Mapping studies, which are underway in our laboratory for B. subtilis, would shed more light on the nature of the mutation(s). Preliminary studies on E. coli CM 22 indicated that the resistant mutant mapped in a region close to the unc operon.[9]

Proton-Solute Symport

Decker and Lang[8] reported that the uptake of glycine and glutamine in B. megaterium C8 was as susceptible to CCCP inhibition as the wild type. They concluded that the uptake of these amino acids was driven by the protonmotive force which was dissipated by CCCP in both the wild type and C8. Sedgwick et al.[11], on the other hand, showed that proline uptake in E. coli UV6 was resistant to concentrations of CCCP that abolished it in the wild type. High concentrations (approximately 50 μM) of CCCP did inhibit proline uptake in UV6, although the ΔpH did not appear to be affected. Date et al.[12] have also isolated an E. coli mutant resistant to the effect of CCCP on proline uptake. They have demonstrated this in both cells and membrane vesicles.

Cytochrome Content and Respiration Rate

The E. coli UV6 strain of Sedgwick et al.[11] appears to be similar in cytochrome content and respiration rate to the wild type. Resistant mutants of B. megaterium[7,10] and B. subtilis do not have significantly different cytochrome content from their relevant wild type strains, but they do show a significant increase in respiration rate on malate, up to two-fold that of the wild type. Ito et al.[9] has also found a two-fold elevated respiration rate in E. coli strain CM 22. Interestingly, the wild type E. coli respiratory rate was stimulated nearly two-fold by the addition of CCCP; whereas, the rate in CM 22 was unchanged. This might indicate that CM 22 and the bacillus mutants are somehow uncoupled from respiratory control.

ATPase Activity

E. coli UV6 described by Sedgwick et al.,[11] exhibited no difference in ATP hydrolytic activity from that of the wild type. In contrast to this result, Decker and Lang's[7] C8 mutant had ATPase activity no higher than 30% of the wild type. Our recently isolated B. subtilis mutants appear to have elevated levels of ATPase activity, anywhere from 50% to 100% higher than the wild type. Ito et al.[9] have speculated that their E. coli CM22 strain carries a mutation in a subunit of the BF_1F_0 ATPase. Indirect evidence for this conclusion comes from fact that CM22 exhibits the unc phenotype in a temperature-dependent fashion. That is, at 42° C, in the absence of CCCP, CM22 grew on glucose, but not on succinate. At the permissive temperature, 37° C, CM22, as well as the wild type, grew on either glucose or succinate. This evidence, along with the fact that CM22 is resistant to tributyltin and sodium azide, putative ATPase inhibitors, may indicate a mutation in the ATPase. Preliminary, mapping studies were also reported to support a possible mutation in the unc operon.

The Effect of Uncouplers on the Protonmotive Force and proton permeability

In whole cells or everted membrane vesicles of E. coli CM22 uncouplers did not appear to transport protons across the membrane barrier as they did in the wild type.[9] The suggestion is that uncouplers act through a membrane component rather than simply as lipophilic proton conductors. At this time it is not clear whether such a conclusion is warranted. The experiments of Ito et al.[9] on the movement of protons in whole cells would seem to presume a ΔpH of the same size for both wild type and CM22, a presumption that is not necessary valid. The best test of uncoupler action on proton permeability would be in an experimental system of the type described by Scholes and Mitchell[13] and Maloney[14]. In such an assay a pH change is imposed by adding acid to the external medium, and subsequent proton influx is followed with a pH electrode. Contrary to the results with CM22, Sedgwick et al.[11] have shown that E. coli UV6 readily takes up protons when CCCP is added. Thus, the uncoupler appears to be working as expected to make the membrane proton permeable. The puzzling result with UV6 is that although CCCP can be shown to move protons inwardly the ΔpH, as measured by benzoic acid uptake, was not affected. The authors suggest that the protons may be equilibrating across the outer membrane in E. coli but are trapped by oligosaccharides in the periplasmic space.

The effect of CCCP on the protomotive force in B. megaterium or B. subtilis resistant mutants appears to conform with the conventional view that the uncoupler works as a lipophilic proton conductor. In whole cells of B. megaterium CCCP has been shown to lower the protonmotive force. In fact the protonmotive force measured depends upon the amount of CCCP added in both the wild type and C8. Recently we have seen in B. subtilis that, at pH 7.4 in the presence of nigericin, where the sole component of the protonmotive force is the ΔΨ (the transmembrane electrical potential), 2 μM CCCP lowers the ΔΨ to about the same extent in the wild type and mutants from −160mV to −70mV. Results with the two bacillus species would therefore indicate that the mutants are not resistant because CCCP fails to dissipate the protonmotive force.

Oxidative Phosphorylation

Unfortunately data have not yet been presented on how E. coli CM22 carries out oxidative phosphorylation in the presence of CCCP. Sedgwick et al.[11] has shown that E. coli UV6 has a lower ΔGp than the wild type because of a very high intracellular phosphate concentration, but the concentrations of ATP and ADP reported are about one tenth of the values usually found in bacteria.

Decker and Lang[7,8] made the striking observation that the C8 mutant of B. megaterium could synthesize ATP in the presence of CCCP concentrations that nearly totally dissipate the protonmotive force. In recent years our group has investigated oxidative phosphorylation in B. megaterium further. Whole cells of B. megaterium can be starved by washing in buffer at pH 7.4, and then reenergized by adding back malate. The subsequent malate-dependent ATP synthesis is inhibited by DCCD, implicating the BF_1F_0. When malate-dependent ATP synthesis in whole cells or membrane vesicles was compared in the wild type and C8, in the presence of increasing concentrations of CCCP, it was evident that C8 was more resistant to the effects of the uncoupler.[10,15] For example, at a CCCP concentration of 2 μM the wild type synthesized 0.5 mM ATP, while C8 synthesized 1.77 mM. In both cases the protonmotive force, consisting only of a $\Delta\Psi$ at pH 7.4, was lowered considerably: from -114 mV to -60 mV in the wild type and from -120 mV to -62mV in C8. When the ratio between the phosphorylation potential (ΔGP) and the protonmotive force was plotted, it was close to 3.0 in uninhibited cells of both strains. However, as the protonmotive force was titrated down with the CCCP, the ratio became increasingly larger and did not remain constant as might be expected if ΔGp and the protonmotive force are truly in equilibrium. The ratio was consistently higher in C8 than in the wild type.

In contrast to the results for ATP synthesis driven by malate, presumably through oxidative phosphorylation, the wild type exhibited a greater ability to synthesize ATP then C8 via an artificially imposed K^+ diffusion potential. This has led us to propose that C8 may be better coupled to a localized electrochemical proton gradient, putatively established by malate respiration, than to a bulk gradient.

Several models for localized proton flow have been presented. Some of the current thoughts on the subject have been summarized by several of the leading investigators in the field.[16] Recently Ferguson[17] has presented an excellent overview of the question of localized versus delocalized proton pathways. He has pointed out, as have van Dam's[18] group, that a deviation from a constant ΔGp to protonmotive force ratio does not necessarily preclude delocalized chemiosmosis. At least some of the deviation from equilibrium could be due to adenylate kinase activity producing ATP. P^5-di (adenosine-5'-) phentaphosphate, as suggested by van Dam's group,[18] may be useful to inhibit adenylate kinase activity. We do know, however, that in B. megaterium the ATP synthesized was accounted for by a stoichiometric disappearance of ADP from the cytoplasmic pool, and no increase in AMP concentration was detected. Very recent results with CCCP-resistant mutants of B. subtilis have also indicated the lack of a constant relationship between the ΔGp and the protonmotive force when whole cells are inhibited with CCCP. Further studies on CCCP-resistant mutants, controlled for possible artifacts such as interference by adenylate kinase, false low values for the proton-motive force etc., should help us better understand exactly how the uncouplers work, and should contribute further to the debate over delocalized versus localized proton pathways.

Acknowledgment

Work in the author's laboratory was supported by research grant GM28454 from the National Institutes of Health.

References

1. P. Mitchell, Coupling of phosphorylation to electron and hydrogen transfer by a chemiosmotic type of mechanism, Nature 191:141 (1961).
2. U. Hopfer, A.L. Lehninger, and T.E. Thompson, Protonic conductance across phospholiped bilayer membranes induced by uncoupling agents for oxidative phosphorylation, Proc. Natl. Acad. Sci. U.S.A. 59; 484 (1968).
3. S.G.A. McLaughlin and J.P. Dilger, Transport of protons across membranes by weak acids, Physiol. Rev. 60:825 (1980).
4. W. G. Hanstein, Uncoupling of oxidative phosphorylation, Biochim. Biophys Acta 456:129 (1976).
5. W.G. Hanstein and Y. Hatefi, Characterization and localization of mito-chondrial uncoupler bindings sites with an uncoupler capable of photo-affinity labeling, J. Biol. Chem. 249:1356 (1974).
6. N.V. Katre and D.F. Wilson, A specific uncoupler-binding protein in Tetrahymena pyriformis and Paracoccus denitrificans, Biochim. Biophys. Acta 593;224 (1980).
7. S.J. Decker and D.R. Lang, Mutants of Bacillus megaterium resistant to uncouplers of oxidative phosphorylation, J. Biol Chem. 252:5936 (1977).
8. S.J. Decker and D.R. Lang, Membrane bioenergetic parameters in uncoupler-resistant mutants of Bacillus megaterium, J. Biol. Chem. 253:6738 (1978).
9. M. Ito, Y. Ohnishi, S. Itoh, and M. Nishimura, Carbonyl cyanide-m-chlorophenyl hydrazone-resistant Escherichia coli mutant that exhibits a temperature-sensitive unc phenotype, J. Bacteriol. 153:310 (1983).
10. A.A. Guffanti, H. Blumenfeld, and T.A. Krulwich, ATP Synthesis by an uncoupler-resistant mutant of Bacillus megaterium, J. Biol. Chem. 256:8416 (1981).
11. E.G. Sedgwick, C. Hou, and P.D. Bragg, Effect of uncouplers on the bioenergetic properties of a carbonyl cyanide m-chlorophenylhydrazone-resistant mutant Escherichia coli UV6, Biochim. Biophys. Acta 767:479 (1984).
12. T. Date, C. Zwizinski, S. Ludmerer, and W. Wickner, Mechanism of membrane assembly: effects of energy poisons on the conversion of soluble M13 coliphage procoat to membrane-bound coat protein, Proc. Natl. Acad. Sci. U.S.A. 77:827 (1980).
13. P. Scholes and P. Mitchell, Acid-base titration across the plasma membrane of Micrococcus denitrificans: factors affecting the effective proton conductance and the respiratory rate, Bioenergetics 1:61 (1970).
14. P.C. Maloney, Membrane H^+ conductance of Streptococcus lactis, J. Bacteriol. 140:197 (1979).
15. A.A. Guffanti, R.T. Fuchs and T.A. Krulwich, Oxidative phosphorylation by isolated membrane vesicles from Bacillus megaterium and its uncoupler-resistant mutant derivative, J. Biol. Chem. 258:35 (1983).
16. H.V. Westerhoff, B.A. Melandri, G. Venturoli, G.F. Azzone, and D.B. Kell, A minimal hypothesis for membrane-linked free-energy transduction, Biochim. Biophys. Acta 768:257 (1984).
17. S. Ferguson, Fully delocalized chemiosmotic or localized proton flow pathways in energy coupling? A scrutiny of experimental evidence, Biochim. Biophys. Acta 811:47 (1985).
18. H. Woelders, W.J. van der Zande, A.A.F. Colen, R.J.A.Wanders, and K. van Dam, The phosphate potential maintained by mitochondria in state 4 is proportional to the proton-motive force, FEBS Letts. 179:278 (1985).

REGULATION OF PHOTOSYNTHETIC ELECTRON TRANSPORT BY PROTEIN PHOSPHORYLATION

Alison Telfer

AFRC Photosynthesis Research Group
Department of Pure and Applied Biology
Imperial College of Science and Technology
London SW7 2BB, U.K.

INTRODUCTION

The complexity of the structure of the inner membranes of the chloroplast, the thylakoids, is described elsewhere in these proceedings by Barber.
The thylakoids are the site of the early reactions of photosynthesis and
contain a number of pigment-protein complexes which function in the conversion of light energy into chemical energy (1). The resultant transfer of
electrons from H_2O to $NADP^+$ requires the cooperation of two reaction centres
acting in series which are known respectively as photosystem two (PS2) and
photosystem one (PS1). Each photosystem has its own light harvesting array
of chlorophyll molecules serving as an antenna to the reaction centre. The
most recent advance in the understanding of thylakoid structure is the
realisation that PS1 and PS2 are inhomogenously distributed within the
plane of the thylakoid membrane (2,3). PS1 and its light harvesting complex (LHC1) are probably restricted to the non-appressed membranes and PS2
and the third major pigment-protein complex, the light harvesting chlorophyll
a/b protein (LHC2), are found preferentially in the appressed regions of the
grana. At neutral pH the thylakoids carry a net negative charge and the
maintenance of the intricate membrane structure requires the presence of a
sufficient level of screening cations. The segregation of the photosystems
within the plane of the membrane is due to the difference in surface charge
density of the various chlorophyll-protein complexes. The appressed regions
of the membranes presumably have a lower net negative charge density (σ)
than the non-appressed regions. This is discussed in detail by Barber elsewhere in these proceedings.

REGULATION OF PHOTOSYNTHETIC ELECTRON TRANSPORT

When light intensity is the factor limiting the rate of photosynthesis
it is clear that efficient electron transport requires equal excitation of
the two photosystems. The action spectra of their activities measured *in
vitro* in isolated thylakoids indicate that the absorption spectra of PS2
and PS1 are not identical (4). *In vivo* when light becomes limiting within
the canopy, i.e. under shade conditions, there is a dramatic change in the
quality of light due to the absorption of the blue and red regions of the
spectrum by the upper portion of the canopy (5). This is obviously a
dynamic situation and the degree of enrichment in far red light will vary
with leaf movement and the passage of sun-flecks during the day. Thus
there is a need for a reversible short-term adaptive mechanism which can

constantly balance the activity of the two photosystems as the quality of
light absorbed varies.

It was Bonaventura and Myers (6) who conclusively demonstrated that
oxygen evolving photosynthetic organisms are able to regulate the supply
of quanta to PS1 and PS2. They showed that on rapidly changing the spectral
quality of the exciting light to induce a sudden imbalance in the relative
excitation of the two photosystems there was a subsequent slow increase in
photosynthetic efficiency. The terms State 1 and State 2 were coined to
describe the adaptive states in which light initially preferentially absorbed by PS1 or PS2, respectively, was found over a period of minutes to be
more evenly balanced between the two photosystems. These experiments were
carried out by measuring changes in modulated oxygen and chlorophyll fluorescence yields in algae. Changes in modulated oxygen yield cannot be
measured directly using an O_2 electrode with the leaves of higher plants
but the expected changes in chlorophyll fluorescence have been observed
using an optical fibre system which measures light emitted from the leaf
surface (7). Recently it has been demonstrated that modulated oxygen
release from leaves can be detected by a photoacoustic technique (8).
Canaani et al (9) using this method showed similar adaptive changes in
modulated oxygen yield induced by light of different spectral quality to
those originally seen in algae.

LIGHT DISTRIBUTION BETWEEN THE PHOTOSYSTEMS

How do changes in light distribution between the two photosystems take
place and how do the models proposed relate to the heterogeneous arrangement of chlorophyll-protein complexes in the plane of the membrane? Two
main mechanisms have been suggested: (i) a change in the degree of 'spillover' or energy transfer from PS2 to PS1 (ii) a change in the size of the
absorption cross-section of each photosystem (see 6,10). The former mechanism requires close interaction between the chlorophyll-protein complexes
of PS2 and PS1 which as mentioned above are normally segregated into the
appressed and non-appressed regions of fully stacked thylakoid membranes.
Unstacking, brought about by suspension in a medium of low cationic strength, leads both to randomization of complexes (11) and a large increase in
'spillover' (12). The latter mechanism involves a direct change in the
relative size of the antenna of the two photosystems and this could be
achieved in stacked membranes by the migration of some LHC2 from the appressed to the non-appressed regions, allowing the direct interaction of the
mobile LHC2 with PS1. The lateral charge displacement model of Barber (2)
predicts that this would occur if sufficient net negative charge was introduced to the surface-exposed portion of LHC2.

The 'spillover' and absorption cross-section models, explaining the
mechanism of changes in light distribution to the two photosystems, can
be distinguished by their effects on chlorophyll fluorescence yield, which
at room temperature only comes from the chlorophylls associated with PS2.
A change in absorption cross-section of PS2 should result in an equal change
in the initial level of fluorescence (F_o) and the maximal level (F_m) seen
when all PS2 traps are closed. On the other hand as energy transfer from
PS2 to PS1 is in competition with primary photochemistry a change in 'spillover' preferentially quenches variable fluorescence thus lowering the $F_m/$
F_o ratio.

Canaani and Malkin (13) using the photoacoustic technique concluded
that State changes in leaves involve changes in absorption cross-section
rather than changes in 'spillover'. We have recently extended this work,
using the modulated fluorescence technique to study State transitions in
leaves, in order to investigate the mechanism involved in these changes
in energy distribution in more detail (Malkin, Telfer and Barber, in

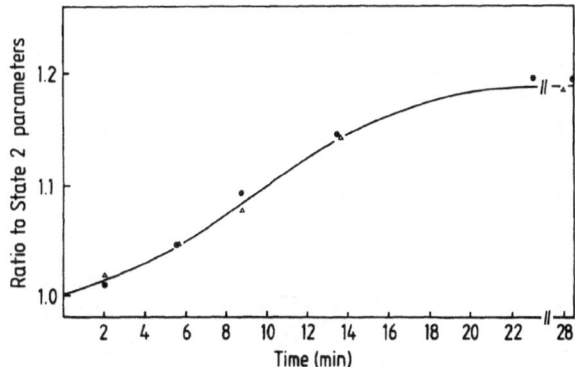

FIG.1 *Showing the relative change in F_m (triangles) and F_0 (circles) of a modulated fluorescence signal (685nm) induced by weak modulated blue light (480nm) during the transition from State 2 to State 1 in a pea leaf.*

preparation). Fig. 1 shows the time course of the relative changes in F_m and F_0 levels of modulated fluorescence from a pea leaf during the transition from State 2 to State 1 (induced by light preferentially absorbed by PS1). The F_0 level was taken as that seen in the presence of excess continuous far red light (light 1) which would be expected to oxidise all PS2 reaction centres. The F_m level was induced by briefly turning off light 1 and adding a high intensity blue green light (light 2) for a few seconds to close all the PS2 traps. Light 2 was then removed and light 1 turned on again to oxidise PS2 and continue the adaptation of the leaf towards State 1.

As can be seen in Fig. 1 F_m and F_0 change equally throughout the time course of the transition from State 2 to State 1 and this is consistent with a change in the absorption cross-section of PS2 accounting totally for the changes in the light distribution which occur in higher plants during State transitions.

REVERSIBLE PROTEIN PHOSPHORYLATION

A possible biochemical mechanism for State changes, observed *in vivo*, has been proposed (14) based on the initial discovery of Bennett (15) that thylakoid membranes undergo reversible protein-phosphorylation with LHC2 being the major protein phosphorylated. It has been demonstrated that a membrane-bound protein kinase is activated when PS2 is over-excited compared to PS1 and that phosphorylation of LHC2 is accompanied by redistribution of excitation energy in favour of PS1.

Relative over-excitation of PS1 oxidises the intersystem electron carriers and deactivates the kinase. Dephosphorylation of LHC2 by a membrane-bound phosphatase allows redistribution of quanta in favour of PS2. The activation-state of the kinase has been correlated with the redox state of the plastoquinone pool (the major intersystem electron carrier) although the mechanism of control is unknown (see Fig. 2).

Telfer and colleagues (16,17) demonstrated that isolated thylakoids will undergo State transitions (as measured by the modulated fluorescence technique) similar to those seen in leaves only if they are provided with the correct conditions to allow reversible phosphorylation of LHC2 to occur. This is shown in Fig. 3 where isolated pea thylakoids supplied with an exogenous electronacceptor adapt towards State 2, in response to light preferentially absorbed by PS2, only in the presence of added ATP. The

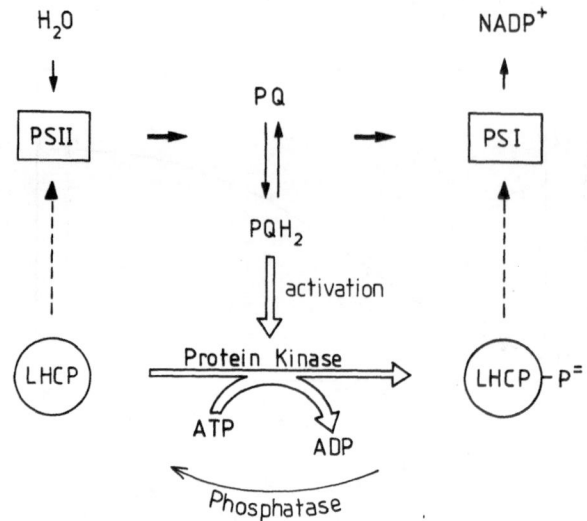

FIG.2. *Model of the mechanism by which reversible phosphorylation of LHC2 is controlled by the redox state of the plastoquinone (PQ) pool.*

ratio of the chlorophyll fluorescence levels without and with light 1 respectively (F-/F+) indicates the extent of over excitation of PS2 relative to PS1. Equal excitation of the two photosystems (or over excitation of PS1, as in the presence of light 1) gives a ratio of 1.00 Before addition of ATP the ratio, in the experiment of Fig. 3, was 1.80 and this was unaffected by 15 min illumination with light 2. However, on addition of ATP the F- level decreased to a new steady state (t$\frac{1}{2}$ approx. 2 min.) and the F-/F+ ratio decreased to 1.41. Continuous illumination with light 1 for approx. 15 min partially reversed this adaptation towards State 2 and increased the F-/F+ ratio to 1.56.

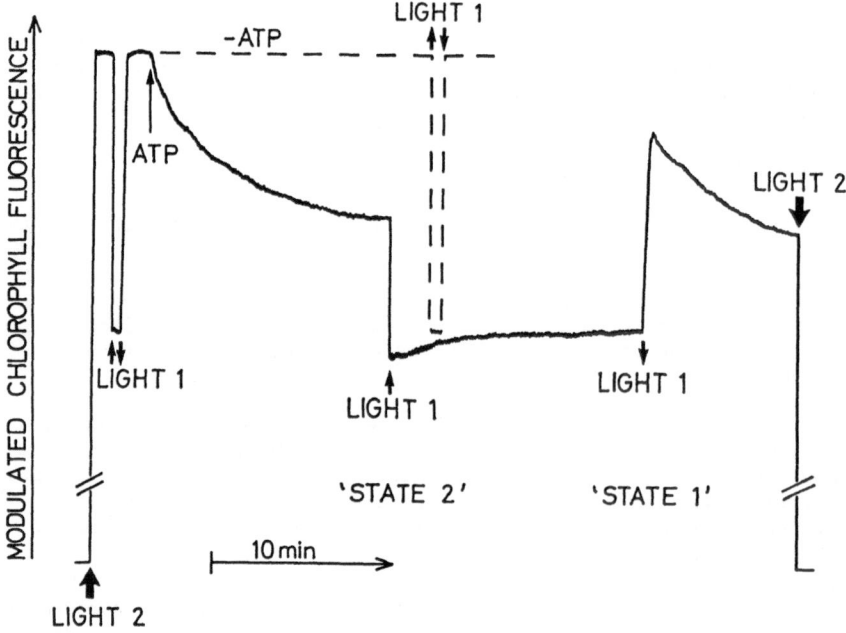

FIG.3. *Demonstration that State 1-State 2 transitions in isolated pea thylakoids require ATP. Light 2 modulated at 80Hz, 480nm, 70μE.m^{-2}.s^{-1}. Light 1 continuous, 713nm, 15μE.m^{-2}.s^{-1}. Modulated chlorophyll fluorescence, 685nm, was detected using a lock-in-amplifier.*

The steady state level of chlorophyll fluorescence in 'State 2' seen with isolated thylakoids reflects the balance between redox-activated protein kinase and phosphatase activity. Addition of NaF, an inhibitor of phosphatase, allowed an almost complete balance to be achieved. The fluorescence decrease induced by light 2 was more extensive giving an F-/F+ ratio of 1.05 (data not shown). The somewhat smaller fluorescence changes seen in isolated thylakoids (Fig.3. & refs.16,17) relative to those seen in intact algae (6) or leaves (7,16) are probably due to the effect of isolation of thylakoids on kinase activity and its activation system.

We confirmed that State transitions in isolated thylakoids, as defined by changes in their chlorophyll fluorescence yield, are accompanied by reversible phosphorylation and dephosphorylation of the polypeptides of LHC2 (Fig. 4). Thylakoids were initially adapted to State 1 and State 2 in the presence of $[^{32}P]$ - ATP and then the time course of phosphorylation of the polypeptides of LHC2 during adaptation to State 2 (and dephosphorylation during adaptation to State 1) was determined in samples subjected to polyacrylamide gel electrophoresis and Cerenkov counting of the excised LHC2 bands.

These data support the hypothesis that *in vivo* regulation of energy distribution is brought about by redox-controlled reversible phosphorylation of LHC2.

A MOBILE POOL OF LHC2

How does reversible phosphorylation of LHC2 control energy distribution between the photosystems? *In vitro* it has been demonstrated that, provided thylakoids are suspended in a medium which induces stacking, there is a 'mobile' pool of LHC2 which after phosphorylation migrates from the appressed to the non-appressed regions of the membrane. Evidence for this is that phosphorylation induces the following:

(i) A relative decrease in the chlorophyll a/b ratio of the non-appressed membranes and an increase in the amount of LHC2 in these membranes (e.g. ref. 18).

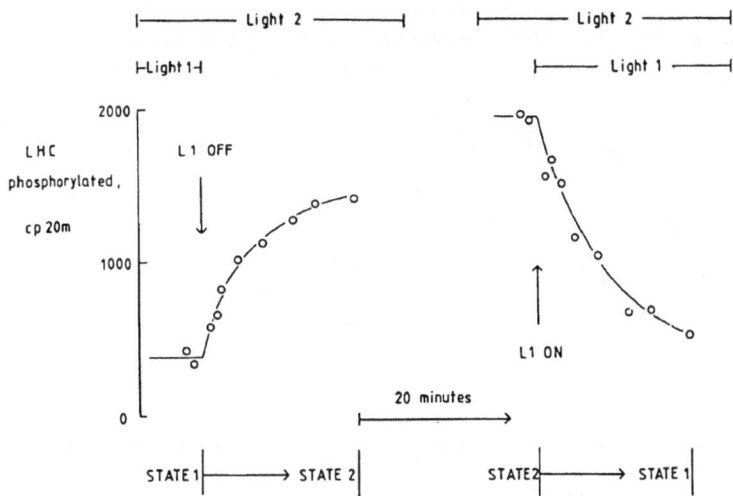

FIG.4. *Phosphorylation and dephosphorylation of the polypeptides of LHC2 during State 1-State 2 transitions in isolated pea thylakoids, adapted from Telfer et al (17).*

(ii) A greater activity of ^{32}P labelled LHC2 in a stromal membrane fraction than in inside-out vesicles, derived respectively from non-appressed and appressed membranes (19).

(iii) A decrease in the degree of stacking, although this is more dramatic at suboptimal concentrations of screening cations (20).

(iv) A reversible change in the relative distribution of freeze-fracture particles attributed to 'free' LHC2 between the appressed and non-appressed regions (21,22).

The consequence of the movement of a proportion of LHC2 between the appressed (PS2 containing) region of the thylakoid and the non-appressed (PS1 containing) area is that it results in a change in absorption cross-section of PS2 and not in the degree of energy transfer from PS2 and PS1. These mechanisms can be distinguished by their effect on chlorophyll fluorescence, the former causing equal lowering of F_m and F_O and the latter preferentially quenching F_m. Fig. 5 and references 23 and 24 clearly demonstrate that the former is the case in isolated thylakoids provided that the level of screening cation is high. As shown in Fig. 1 *in vivo* State transitions, also cause an equal change in F_m and F_O i.e. result in a change in absorption cross-section. One can therefore conclude that the *in vivo* cation level of the stroma is fully saturating for stacking and that it is indeed a 'mobile' pool of LHC2 which brings about the redistribution of energy between the photosystems, associated with State transitions.

EVIDENCE FOR AN ELECTROSTATIC MECHANISM

It was Barber (2,25) who initially proposed that the addition of phosphoryl groups to the surface exposed portion of LHC2 increases its surface charge density and hence by analogy with the electrostatic control of stacking causes randomization of chlorophyll-protein complexes. This hypothesis has been modified to account for the fact that it appears that only LHC2 moves under *in vivo* conditions. It is suggested that the increased charge on LHC2 causes its dissociation from PS2 and migration from the appressed regions which have a low net surface charge density. What evidence is there that an electrostatic mechanism is involved? We have titrated the cation-induced fluorescence increase in both phosphorylated and non-phosphorylated membranes with cations of different valencies. In all cases at high concentrations of cation there is an approximately 20% decrease in F_m which is ascribed to the reduction in the absorption cross-section of PS2 (23,24). However, in addition with each cation tested there is an increase in the concentration requirement ($C_{\frac{1}{2}}$) for the

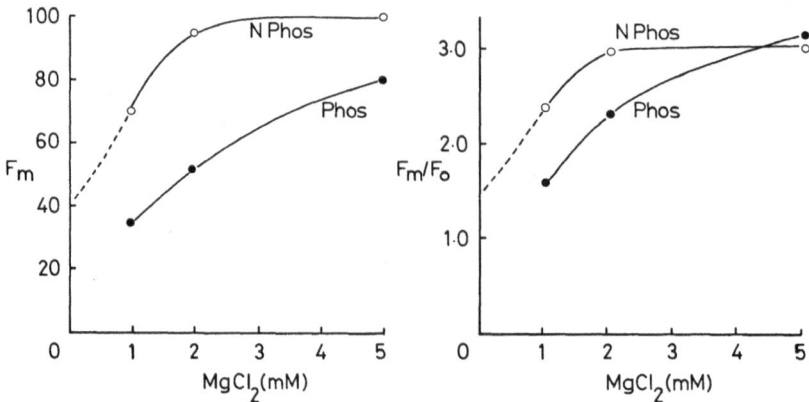

FIG.5. *The effect of Mg^{2+} concentration on the F_m and F_m/F_O ratios in phosphorylated and non-phosphorylated pea thylakoids.*

206

TABLE I. *The effect of LHC2-phosphorylation on the $C_{\frac{1}{2}}$ for cations of different valencies to induce the chlorophyll fluorescence increase*.*

Cation	NPhos	Phos	Phos/NPhos
TEC^{3+}**	$1.5 \times 10^{-5}M$	$2.5 \times 10^{-5}M$	1.67
Mg^{2+}	$1.2 \times 10^{-3}M$	$2.1 \times 10^{-3}M$	1.75
K^+	$7.3 \times 10^{-2}M$	$1.0 \times 10^{-1}M$	1.37

*Data from ref. 26
** TEC^{3+} - tris(ethylenediamine)cobaltic cation.

fluorescence rise (Table I). The implication of this phosphorylation-induced change in $C_{\frac{1}{2}}$ is that there has been an increase in the net charge density of the thylakoid (see Barber, these proceedings).

However, phosphorylation does not result in a significant change in the extent of 9-aminoacridine (9AA) fluorescence quenching by thylakoids or in the $C_{\frac{1}{2}}$ for the release of this quenching by divalent cations (see Fig.6). Chow et al (7) concluded that the net change in surface charge density induced by phosphorylation would be too small to be detected by the 9AA technique. This does not however explain the change in $C_{\frac{1}{2}}$ for the cation-induced chlorophyll fluorescence rise.

Table II presents data on thylakoids which would be expected to show a change in net surface charge density. Plants grown in 'low' light have significantly more stacked membranes and more chlorophyll associated with PS2 compared to PS1 than those grown in 'high' light (27). In Table II adaptation to 'low' light is indicated by the decrease in chlorophyll a/b ratio. This is accompanied by a decrease in the $C_{\frac{1}{2}}$ for the Mg^{2+}-induced chlorophyll fluorescence rise and using the 9AA technique Davies et al (28) found a decrease in the surface charge density (σ). The ratios of the $C_{\frac{1}{2}}$s for the chlorophyll fluorescence rise in 'high' and 'low' light thylakoids are in the order of 1.2 while in non-phosphorylated and phosphorylated membranes (Table I) the ratios were considerably greater (1.37-1.75). However, whereas the 'high' and 'low' light plants showed a con-

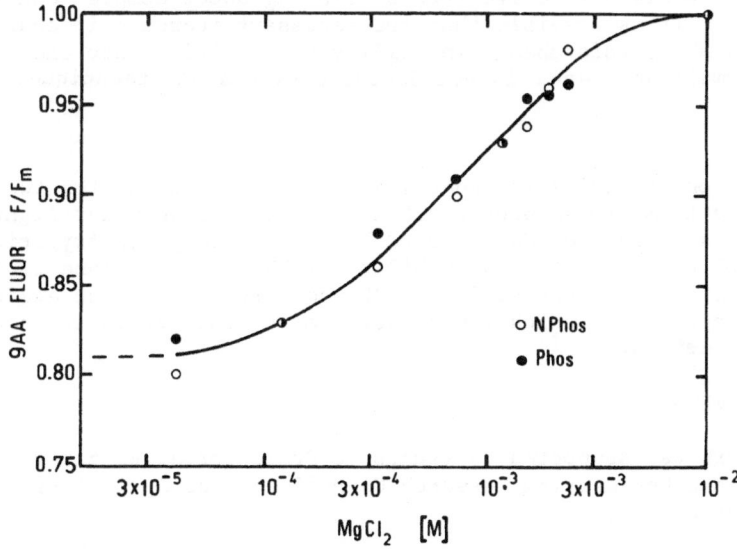

FIG.6. *The absence of effect of protein phosphorylation on the ability of Mg^{2+} to release quenching of 9 aminoacridine (9AA) fluorescence by pea thylakoids.*

TABLE II Comparison of various parameters measured in thylakoids isolated from 'high' and 'low' light-grown plants.

Parameter	'high'	'low'	'high'/'low'
(i) Davies et al (28)-Lettuce			
Chl a/b	4.09	3.40	1.20
$C_{\frac{1}{2}}$ for Mg^{2+} induced chl fluor rise	4.1mM	3.3mM	1.2
σ (C.m^{-2})	−0.0264	−0.0194	1.36
(ii) Telfer - Pea			
Chl a/b	3.15	2.85	1.11
$C_{\frac{1}{2}}$ for Mg^{2+} - induced chl fluor rise	0.76mM	0.62mM	1.23

siderable difference in surface charge density ('high'/'low' = 1.36) no difference was detectable on phosphorylation (7). It is clear from these comparisons that although the effect of phosphorylation must be related to a change in the electrical properties of the membrane it is different from that seen on adaptation to different environmental conditions. In the latter case there has been a change in the ratio of the various chlorophyll-protein complexes which have been predicted to have different surface charge densities (PS1 'high' charge and PS2 'low' charge). The net surface charge density on a chlorophyll basis is thus closely corre-llated to other factors which relate to the proportions of PS2, LHC2 and PS1. The consequence of phosphorylation of LHC2 is presumably different. In this case there may be an increase in the surface charge density of a population of LHC2 above a critical level so that these complexes can no longer partition into the appressed region. It seems that the coulombic repulsive forces cannot be reduced sufficiently by screening cations to allow van der Waals attractive forces to bring about complex-complex interaction. It is possible that the necessary increase in charge density which brings about this change in ability to partition into the appressed regions is small and hence is not dectable by the 9AA technique.

CONCLUSION

Regulation of excitation energy distribution between the two photo-systems so as to maximise photosynthetic electron flow under light limit-ing conditions is controlled *in vivo* by reversible phosphorylation of a 'mobile' pool of LHC2. Lateral mobility of LHC2 in the plane of the thylakoid membrane is controlled by changes in its surface electrical charge characteristics which can be detected by changes in chlorophyll fluorescence levels.

ACKNOWLEDGEMENTS

This work was supported by grants to Professor J. Barber from the Science and Engineering Research Council and the Agricultural and Food Research Council.

REFERENCES

1. R.K. Clayton, "Photosynthesis: Physical Mechanisms and Chemical

Patterns," Cambridge University Press, Cambridge (1980)

2. J. Barber, An explanation for the relationship between salt-induced thylakoid stacking and the chlorophyll fluorescence changes associated with changes in spillover of energy from photosystem 2 to photosystem 1, FEBS Lett., 118: 1-10 (1980)

3. J.M. Anderson, Consequences of spatial separation of photosystem 1 and photosystem 2 in thylakoid membranes of higher plant chloroplasts FEBS Lett., 124: 1-10 (1981)

4. A. Ried, Improved action spectra of light reaction 1 and 2, in:"Proc. 2nd Intnl. Congr. Photosynthesis Res.," G. Forti, M. Avron and A. Melandri, eds., 1: 763-772, Junk, The Hague (1972)

5. M.G. Holmes, Spectral distribution of radiation within plant canopies in: "Plants and the Daylight Spectrum," H. Smith, ed., Acad. Press, London, (1981)

6. C. Bonaventura and J. Myers, Fluorescence and oxygen evolution from *Chlorella pyrenoidosa*, Biochim. Biophys. Acta, 189: 366-383 (1969)

7. W.S. Chow, A. Telfer, D.J. Chapman and J. Barber, State 1-State 2 transition in leaves and its association with ATP-induced chlorophyll fluorescence quenching, Biochim. Biophys. Acta, 638: 60-68 (1981)

8. G. Bults, B.A. Horwitz, S. Malkin and D. Cahen, Photoacoustic measurements of photosynthetic activities in whole leaves: photochemistry and gas exchange, Biochim. Biophys. Acta, 679: 452-465 (1982)

9. O. Canaani, J. Barber, and S. Malkin, Evidence that phosphorylation and dephosphorylation regulate the distribution of excitation energy between the two photosystems of photosynthesis *in vivo*, Proc. Natnl. Acad. Sci. U.S.A., 81: 1614-1618 (1984)

10. W.P. Williams, The two photosystems and their interactions, in: "Topics in Photosynthesis, Vol. 2, Primary Processes of Photosynthesis," J. Barber, ed., Elsevier, Amsterdam (1977)

11. L.A. Staehelin, Reversible particle movements associated with unstacking and restacking of chloroplast membranes *in vitro*, J. Cell Biol. 71: 136-158 (1976)

12. W.L. Butler, Energy distribution in the photochemical apparatus of photosynthesis, Annu. Rev. Plant Physiol. 29: 345-378 (1978)

13. O. Canaani and S. Malkin, Distribution of light excitation in an intact leaf between the two photosystems of photosynthesis: changes in absorption cross-section following State 1-State 2 transitions, Biochim. Biophys. Acta, 766: 513-524 (1984)

14. J.F. Allen, J. Bennett, K.E. Steinback and C.J. Arntzen, Chloroplast protein phosphorylation couples plastoquinone redox state to distribution of excitation energy between photosystems, Nature, 291: 21-25 (1981)

15. J. Bennett, Phosphorylation of chloroplast membrane polypeptides, Nature, 269: 344-346 (1977)

16. A. Telfer and J. Barber, ATP-dependent State 1-State 2 changes in isolated pea thylakoids, FEBS Lett., 129: 161-165 (1981)

17. A. Telfer, J.F. Allen, J. Barber and J. Bennett, Thylakoid protein phosphorylation during State 1-State 2 transitions in osmotically shocked pea chloroplasts, Biochim. Biophys. Acta, 722: 176-181 (1983)

18. D.J. Kyle, T.-Y. Kuang, J.L. Watson and C.J. Arntzen, Movement of a sub-population of the light harvesting complex (LHC_{11}) from grana to stroma lamellae as a consequence of its phosphorylation, Biochim. Biophys. Acta, 765: 89-96 (1984)

19. B. Andersson, H.-E., Akerlund, B. Jergil and C. Larsson, Differential phosphorylation of the light harvesting chlorophyll-protein complex in appressed and non-appressed regions of the thylakoid membrane, FEBS Lett., 149: 181-185 (1982)

20. A. Telfer, M. Hodges, P.A. Millner and J. Barber, The cation-dependence of the degree of protein phosphorylation-induced unstacking of pea thylakoids, Biochim. Biophys. Acta, 766: 554-562 (1984)

21. D.J. Kyle, L.A. Staehelin and C.J. Arntzen, Lateral mobility of the light-harvesting complex in chloroplast membranes controls excitation energy distribution in higher plants, Arch. Biochem. Biophys., 222: 527-541 (1983)

22. D.J. Simpson, Freeze-fracture studies on barley plastid membranes: VII structural changes associated with phosphorylation of the light-harvesting complex, Biochim. Biophys. Acta, 725: 113-120 (1983)

23. P. Horton and M.J. Black, On the nature of the fluorescence decrease due to phosphorylation of chloroplast membrane proteins, Biochim. Biophys. Acta, 680: 22-27 (1982)

24. A. Telfer, M. Hodges and J. Barber, Analysis of chlorophyll fluorescence induction curves in the presence of 3-(3,4-dichlorophenyl)-1,1-dimethylurea as a function of magnesium concentration and NADPH-activated light harvesting chlorophyll a/b protein phosphorylation, Biochim. Biophys. Acta, 724: 167-175 (1983)

25. J. Barber, Influence of surface charges on thylakoid structure and function, Annu. Rev. Plant Physiol., 33: 261-295 (1982)

26. A. Telfer, M. Hodges, P.A. Millner and J. Barber, The effect of cations on the mechanism of regulation of excitation energy distribution by reversible phosphorylation of LHC in pea thylakoids, in: "Proc. 3rd Europ. Bioenerg. Conf. Hannover," 3B: 683-684 (1984)

27. N.K. Boardman, Comparitive photosynthesis of sun and shade plants, Annu. Rev. Plant Physiol., 28: 355-377 (1977)

28. E.C. Davies, W.S. Chow and B.R. Jordan, Thylakoid stacking in relation to photosynthetic light adaptation in lettuce, in: "Meeting on photosynthesis," A.F.R.C., Great Portland St., London (1984)

MODULATION OF THE PSI UNIT CROSS SECTION AS MONITORED BY THE LOW-TEMPERATURE FLUORESCENCE RATIO F685/F730 AT 77°K OF THE PIGMENT-PROTEIN COMPLEX CPIa

G. Akoyunoglou and J.H. Argyroudi-Akoyunoglou

Nuclear Research Center "Demokritos", Athens
Greece

Introduction

For optimization of excitation energy utilization, two types of mechanisms have been proposed to operate in photosynthetic organisms: that of spillover of excitation energy from the antennae of PSII to the reaction center of PSI; and that of modulation of the absorption cross section of the photosynthetic units (1-4). Even though distinctly different, the two mechanisms do not exclude each other. The constant quantum yield of photosynthesis observed at wavelengths between 600 and 680 nm, where only the PSII absorbs, has been explained as the result of transfer of excitation energy from PSII to PSI (5). Similarly, the increase in the Chla fluorescence yield of low-salt chloroplasts and in the 77°K fluorescence ratio F685/F730, upon cation addition, have been considered to monitor the control of spillover by Mg^{++}. (In the absence of cations spillover is enhanced, and the PSI Chl fluoresces more). The suppression of the F685/F730 ratio at 77°K in algae preilluminated with PSII light, has been also attributed to spillover (1,2). On the contrary, the state I/II transitions (high/low fluorescence state) observed in algae by manipulation of the light quality (PSI light vs PSII light) have been attributed to changes in the cross section of the PS units (3). The variation in the cross section of the PS units was calculated to be in the order of 10% (3).

The change in the cross section of the PS units was earlier attributed to energy dependent movement of pigment molecules of PSII away or closer to PSI, or movement of Chla molecules within the PSII (6). Recently, it has been proposed that the change in the optical cross section of PSI is physiologically brought about via the phosphorylation of LHC-II (7). Upon phosphorylation of LHC-II (8), negatively charged phosphate groups are introduced in grana by binding on the threonine residues of LHC-II (9): the grana are forced to unstack, and a "mobile" fraction of LHC-II, visualized as an 8 nm particle (10), moves laterally to stroma lamellae; there, it increases the optical cross section of PSI (II), resulting in increased F730/F685 ratio at 77°K (7).

211

Following the first reports of Izawa and Good on low-
-salt induced unstacking of grana (12), and the work of
Homman (13) and Murata (1,2) showing that addition of cations
to such low-salt unstacked chloroplasts results in the increase of the Chla fluorescence yield, various studies have
shown that grana stacking and Chla fluorescence yield increase are two processes closely correlated (14). The cation requirement for both effects was found to be 2 mM for divalent
and about 50-100 mM for monovalent cations (14). Based on
the findings that (a) similar concentrations of cations were
also required for the dissociation of the organized pigment-
-protein complexes (LHCP oligomers and CPIa) to their constituents, (15) and (b) that the organization and assembly of
the complexes during thylakoid biogenesis is correlated with
changes in the Chla fluorescence yield (16), it was questioned whether the organization of the CPs themselves may be responsible for the cation-induced fluorescence changes.
Indeed, further work showed that parallel to the dissociation of the CPs a drastic increase in the F685/F730 ratio at
77K occurs in the thylakoids or in the isolated complexes
themselves (17,18). Thus, in low-salt conditions, or at high
pH, where the highly organized complexes prevail the F730/
/F685 ratio is high, while in high-salt conditions or low pH,
where the monomeric constituents of the complexes prevail
the F685/F730 ratio is high. Of all complexes the one most
drastically affected was found to be CPIa.

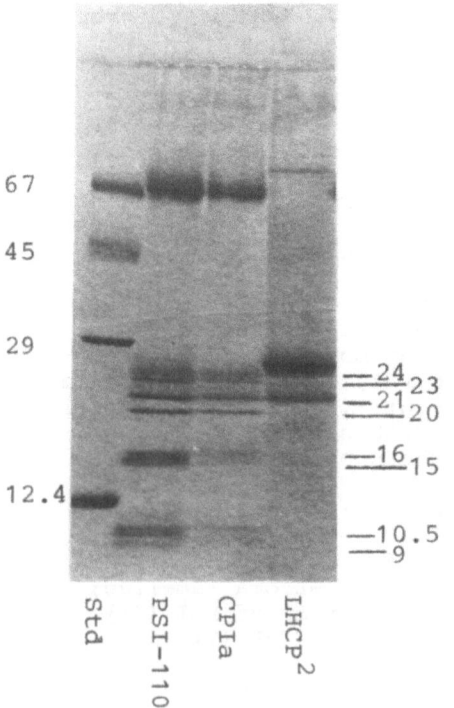

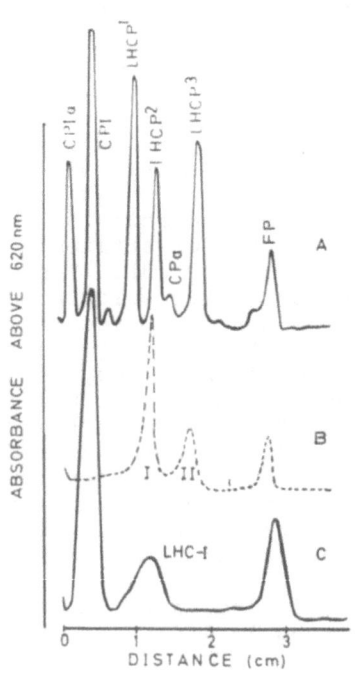

Figure 1. Polypeptide resolution by dissociating Urea-
SDS-PAGE of the isolated CPIa
and LHCP oligomer (dimer)(19)
and of PSI-110 (21,22).

Figure 2. SDS-PAGE profile
of pigment-protein complexes
present in thylakoids (A),
and in CPIa (C) and LHCP[2] (B)
separated from (A).

Table 1. Pigment and Polypeptide composition of CPIa, CPI and LHC-I.

| COMPLEX | Chla/Chlb | b-car | lut | nx | vx | Chl/P700 | Polypeptides (kDa) |
		(moles/100 moles Chl)					
CPIa	4.5	18.5	3.0	-	-	100	67, 24, 23, 21,20, 16,14,11,10,9
CPI	13.0	12.0	-	-	-	30	67
LHC-I	3.4	-	10.2	-	1.3	high	24, 23, 21, 20

CPIa and CPI were isolated by SDS-sucrose density gradient centrifugation (19) and LHC-I as in figure 2C. Pigment analysis by Antonopoulou (23).

CPIa is a pigment-protein complex originating in PSI, isolated either after SDS-sucrose density gradient centrifugation (19) or SDS-PAGE at 4°C (20) from SDS-solubilized thylakoids. This complex is identical to the PSI-110 particles isolated after Triton X-100 extraction of thylakoids (21,22) (fig. 1). It is composed of the CPI-P700 core complex of PSI and the light-harvesting complex LHC-I. It contains Chla, Chlb, 100 Chl per P700 and polypeptides of 68, 24, 23, 21, 20, 16-14, and 11-9 Kd (21,22). The 68 Kd polypeptide is derived from the CPI core, the 24-20 Kd ones from the LHC-I. The LHC-I can be removed from CPIa after partial dissociation of the complex (achieved by addition of 100 mM Na$^+$ in the stacking gel) followed by SDS-PAGE (fig. 2C). It has an electrophoretic mobility equal to that of the LHCP-2 (the dimer form of LHC-II) (21,22; fig. 2A); it can be also separated from the latter complex (fig. 2B). Both CPIa and PSI-110 emit at 730 nm (77K); CPI removed from either complex emits at 722 nm; LHC-I, removed from PSI-110 contains the 24-20 Kd polypeptides and emits at 730 nm, while the LHC-I removed from CPIa contains the 21 Kd polypeptide and emits at 717 nm (21,22). Based on the fact that stroma thylakoid LHCP-2 is enriched in the 21 Kd polypeptide and in long wavelength emitting Chl, it has been proposed that the red emission of the PSI unit originates in the 21 Kd pigment--protein complex of LHC-I (21,22). The pigment composition of the LHC-I removed from CPIa is shown in Table 1, along with that of CPI and CPIa (23).

The cation-induced changes on the 77K fluorescence spectrum of CPIa are shown in fig. 3 along with those observed in chloroplasts suspended in Tricine and in the other complexes isolated. In figure 4 the changes induced in the 77K fluorescence emission spectrum of the isolated LHC-I are also shown. It is clear that the spectra of all complexes originating in PSI are drastically affected by cation- addition or dissociating conditions.

To further test whether these changes are indeed of structural nature, we used glutaraldehyde (GA) to fix the pigment-protein complexes either, in their organized state (low-salt) or in their dissociated state (high-salt), and tried to see how their structure affects the F685/F730 ratio (24). GA fixation was found to prevent the cation-induced increase in the F685/F730 ratio in thylakoids washed and suspended in Tricine (low-salt). For full inhibition of the Mg^{++} effect 10 μmoles GA/mg Chl were required (fig. 5A). Similarly, GA fixation prevented the reversal of the cation effect on the F685/F730 ratio induced by Tricine washing

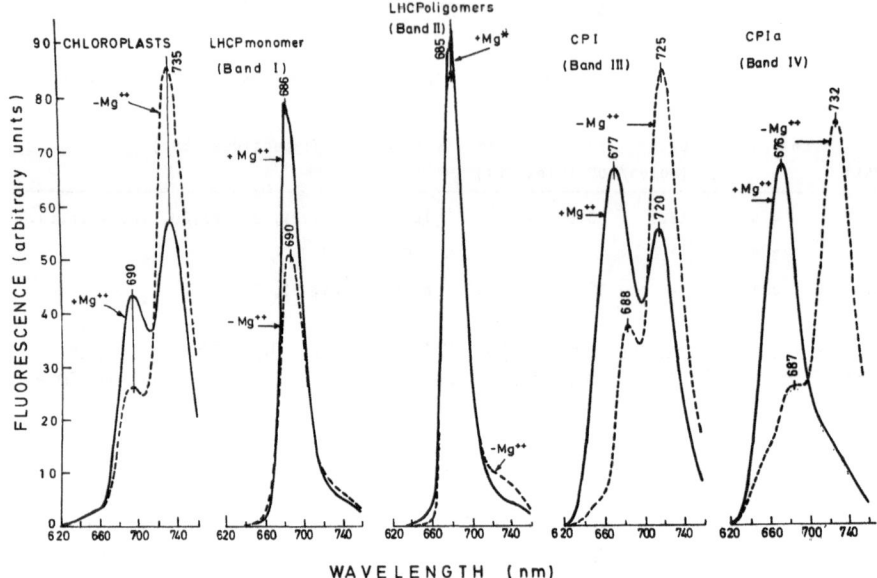

Figure 3. The effect of MgCl$_2$ (10 mM final) on the 77°K fluorescence spectra of pigment-protein complexes, isolated after SDS-sucrose density gradient centrifugation (17,19) from SDS-solubilized thylakoids. The effect of MgCl$_2$ on the 77°K fluorescence spectrum of chloroplasts suspended in 0.05M Tricine-NaOH pH 7.3, is also shown for comparison.

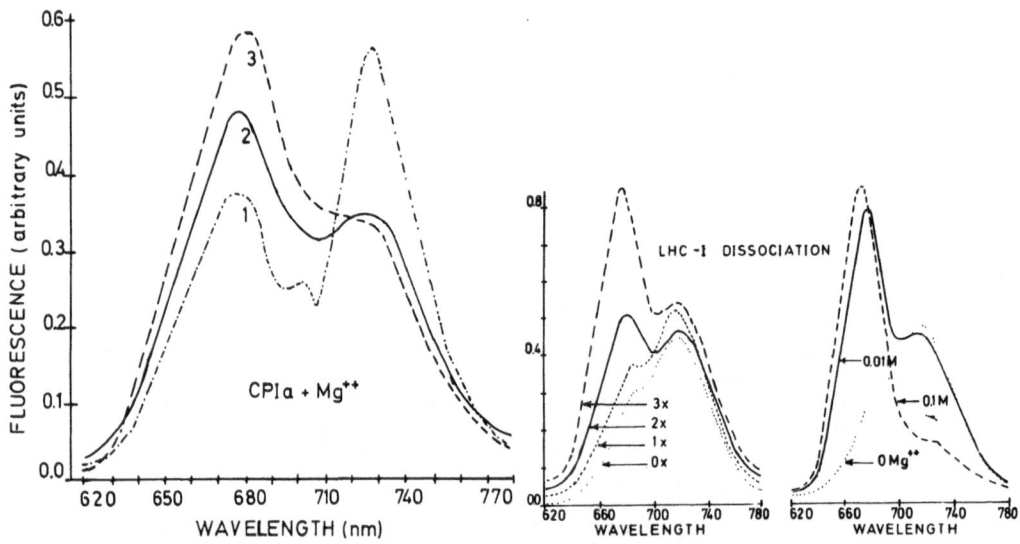

Figure 4. Dependence of the 77°K fluorescence spectra of CPIa and LHC-I on MgCl$_2$ concentration. CPIa was isolated by sucrose-density gradient centrifugation (1):0.09mM Mg^{++}, (2):1mM Mg^{++}, (3):1.6 mM Mg^{++}. LHC-I was separated by SDS-PAGE as in figure 2C from electrophoretically isolated CPIa. Spectra were recorded in situ. Left: the gel slice was freeze-thawed 0 to 3 times ($\overline{0x}$-$\overline{3x}$); right:the gel slice was immersed for 1 min at 25°C in solution of 0, 0.01 or 0.1M MgCl$_2$, and after immersion in liquid N$_2$ the spectra were recorded.

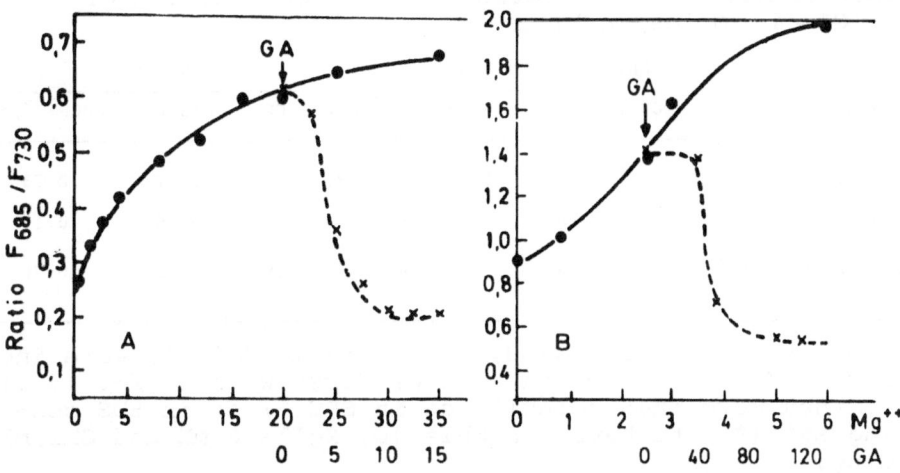

Figure 5. Effect of Mg^{++} on the 77K fluorescence ratio F685/
/F730 of thylakoids (A) washed and suspended in Tricine-NaOH,
pH 7.3; (B) solubilized in Detergent (see text) (24). Incubation mixtures had 1 mg Chl/ml. Incubation with GA was for 45
min prior to the addition of Mg^{++} at 20 mM (A) or 2.5 mM (B).
GA in umoles/mg Chl; Mg^{++} in mM final. Fluorescence measurements at 30 ug Chl/ml.

(Table 2). These findings were more or less predicted, since
GA fixation of thylakoids is expected to prevent any changes
due to cation addition or removal (including large scale
changes of stacking-unstacking). To test whether GA affects
the stabilization of the structure of the complexes prevailing at a certain ionic condition, it was necessary to study
how GA affects the complexes themselves. The study was thus
extended to thylakoids solubilized in detergent (0.1% SDS-
0.2% DOC-0.2% Triton X-100 in 5% sucrose and 0.05 M Tris-
borate, pH 9.5). In this case, Mg^{++} was added with the detergent, to have SDS/Chl= 10, prior to or after GA fixation.
Mg^{++} affected the F685/F730 ratio at lower concentrations than

Table 2. Reversal of the Mg^{++}-induced increase in the F685/
/F730 ratio at 77K by Tricine washing of thylakoids (prior to
or after their fixation with glutaraldehyde at 40 umoles/mg
Chl for 30 min).

| | | THYLAKOID TREATMENT | | |
	CONTROL	$+Mg^{++}$	$+Mg^{++}+$TRICINE	$+Mg^{++}+$GA+TRICINE
RATIO F685/F730	0.5	1.0	0.7	1.1

Chloroplasts in Tricine-NaOH, pH 7.3, (1 mg Chl/ml) were incubated with Mg^{++} at 10 mM and either washed in Tricine and
repelleted, or first incubated with GA, then washed with
Tricine and repelleted. All pellets were diluted in Tricine
to 30 ug Chl/ml.

Table 3. Mg^{++}-Induced dissociation of supramolecular pigment-
-protein complexes as affected by prior GA fixation of thylakoids (GA = 40 µmoles/mgChl).

TREATMENT	Chl DISTRIBUTION (%)						RATIO	
	CPIa	CPI	LHCP[1]	LHCP[2]	LHCP[3]	FP	CPIa/CPI	LHCP[1+2]/[3]
Control (-Mg, -GA)	6	16	13	12	33	20	0.37	0.75
+MgCl$_2$ (*)	0	18	2	6	49	25	0.00	0.17
+GA, then +Mg^{++}(**)	10	19	7	14	30	19	0.52	0.70

Chloroplasts in Tricine-NaOH, pH 7.3 (2 mg Chl/ml) were incubated with MgCl$_2$ for 30 min (*), or first with GA for 15 min
(***); they were then solubilized in SDS(*) or in SDS containing Mg^{++}(**) to have SDS/Chl= 10, Mg^{++}= 6 mM and Chl/ml=
= 650 ug.

those needed in the case of the non-solubilized thylakoids,
since in this case the detergent effect is superimposed (see
fig. 5A, 5B). The cation effect was inhibited in this case
also by prior GA fixation, but higher GA concentrations were
required for full inhibition (80-100 µmoles GA/mg Chl). This
finding could be again explained as the result of insuffici-
ent solubilization of the GA-fixed thylakoids, in which case
cation addition would be of no effect. However, all thylakoid material entered the polyacrylamide gel; furthermore,
upon SDS-PAGE, it became evident that GA fixation prevented
the cation-induced dissociation of the LHCP oligomers and of
CPIa, as shown in Table 3. This suggests that the cation-
-induced increase in the F685/F730 ratio is prevented by
prior GA fixation of thylakoids possibly due to the fixation
of the supramolecular structures of the complexes. These
findings, therefore, support the earlier conclusion that the
changes in the F685/F730 ratio in thylakoids are of structural nature and most probably they reflect structural changes
occuring in the PSI unit and its components.
        Support to this conclusion comes also from the study of
the step-wise assembly in vivo of the CPIa complex, and the
parallel changes in the emission spectrum of the plastids
during development (16,25-27). In the early stages of greening (etiolated plants exposed to intermittent light)no organized CPIa can be detected nor a 730 nm emission (25-27);
only a 714 nm band is evident, gradually shifting to 722 nm.
This indicates the presence of the core complex of PSI, CPI
in these thylakoids (26). After transfer of the plants to
continuous light, where LHC-I is synthesized, CPIa is assembl-
ed and the red emission is further shifted to 730 nm (25,27).
The use of phycocyanin as internal standard made possible the
study of the changes observed in the absolute value of the
emission peaks at 690 and 730 nm during chloroplast development (25). It became evident that both emissions increase
when increased biosynthesis of the complexes occurs, and then
decrease as greening is prolonged. The decrease was followed
by a further red shift to 735 nm, suggesting that this emission reflects the organization of CPIa into supramolecular
structures (CPIa)^n, more efficient in the utilization of the
excitation energy. Figure 6 summarizes these findings, and

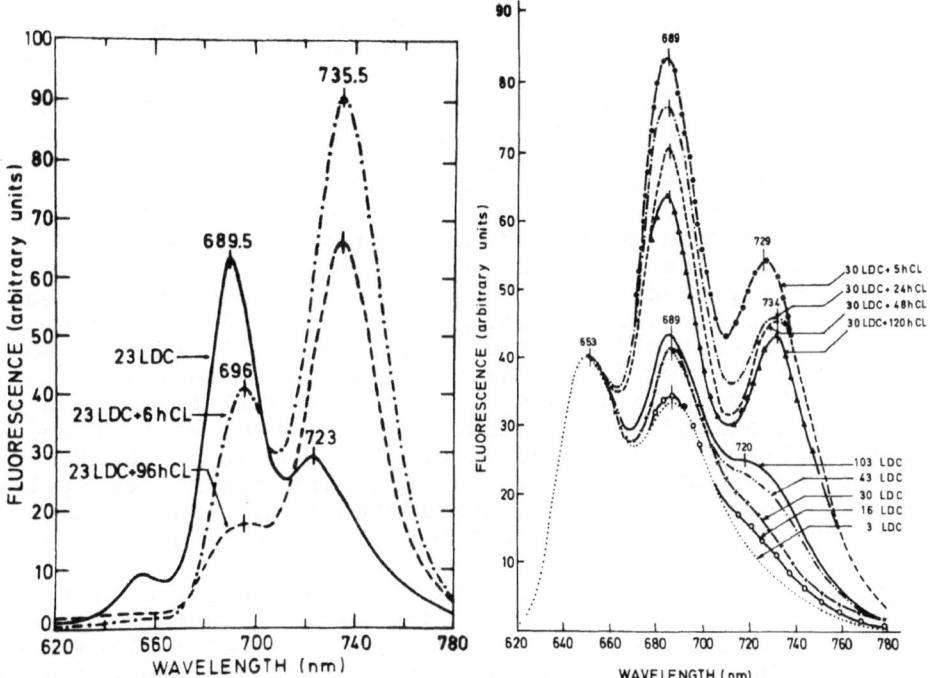

Figure 6. 77K emission spectra of 6-d etiolated bean leaves exposed to light-dark cycles (LDC) and then to continuous light (a); or of plastids isolated from the leaves (b) in the presence of phycocyanin, PC, at PC/Chl= 0.8 (w/w) and at 30ug Chl/ml.

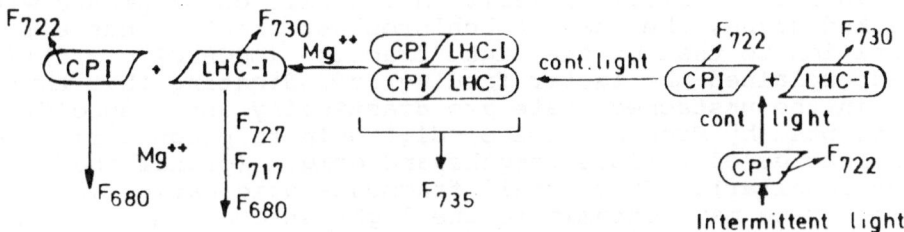

Scheme 1. CPIa dissociation in vitro and physiological assembly in vivo.

scheme 1 shows a model of CPIa dissociation in vitro, and its assembly in vivo, along with the 77K emission spectral changes observed.

The results presented above suggest that high-salt conditions which favor the dissociation of CPIa to its components induce also an increase in the F685/G730 ratio at 77K and a blue shift of the peaks; on the contrary, low-salt conditions which favor the organization of the CPIa supramolecular structure from its components, induce also an increase in the F730/F685 ratio and a red shift of the peaks. Based on these results, the cation-induced changes in the Chla fluorescence yield at room temperature may be explained as monitoring the decrease (high-salt) or increase (low-salt) of the PSI unit cross section. The change in the PSI unit size is thought to

be brought about by the association or dissociation of LHC-I
from the CPI core in the unit. Thus, the chloroplasts, depending on ionic environment may have altered light absorption properties so that the light energy at a given time may
be distributed between the two photosystems in favor of the
one or the other photosystem. This hypothesis is similar to
that proposed by Bonaventura and Myers (3) who attributed
the State I/II transitions in algae induce by manipulation
of the light quality, to changes in the distribution of the
absorbed energy between the two photosystems. This hypothesis requires a reorganization process of the PSI unit components to occur upon cation addition or removal, which involves
short range lateral movement of complexes.

As already discussed above, the cation-induced increase
in the Chl a fluorescence yield and in the F685/F730 ratio
(77K), in low-salt chloroplasts has been attributed to the
control of spillover from PSII to PSI. Thus, in the normal
stacked state very limited spillover from PSII to PSI can
occur, since all of the evidence suggests that most of the
PSII particles are localized in grana and most of the PSI
particles in stroma lamellae (28-31). The control of spillover by cations is therefore considered to be brought about
via the spatial separation of the PSII and PSI upon grana
stacking.

To account for the increased F730/F685 ratio at 77K, observed in chloroplasts suspended in low-salt buffers (grana
unstacking conditions (12), Barber has proposed (32) that upon
unstacking there is a randomization of the PSII and PSI particles throughout the thylakoid. This was based, on one hand
on the electron microscope studies (33) showing randomization
of particles upon low salt unstacking, and on the other hand
on the finding that contrary to the situation observed in
high-salt granal chloroplasts, where the isolated grana are
enriched in LHCP (LHCP/CPI= 4) and the isolated stroma thylakoids in CPI (LHCP/CPI=1) (29), in low-salt chloroplasts with
unstacked grana, the heavy subchloroplast fraction has similar
composition to that of the light fraction (LHCP/CPI= 2.5) (29).
We had explained the latter finding as indicating that thylakoids in the unstacked state are drastically and unspecifical-
ly disrupted by French press or digitonin to produce mixtures
of thylakoids of various lengths and origin (granal and/or
stroma lamellar). Thus, small fragments originating in unstacked grana may contaminate the light stroma lamellar fraction, and larger fragments derived from unstacked grana but
also from stroma lamellar parts may contaminate the heavy
fraction, so that all fractions produced end up with similar
composition and activities (29). Scheme 2A shows the model of
particle randomization via lateral movement of complexes, proposed by Barber (32). According to this, the light subchloroplast fraction of low-salt disrupted chloroplasts is composed
of one type of membranes, depleted in CPI (●) and enriched in
LHC-II (□). According to the alternate hypothesis (scheme
2B), this fraction is composed of a mixture of thylakoid fragments, originating in the CPI-rich stroma lamellae, but also
in the LHC-II-rich unstacked regions of grana. To distin-
guish between the two possibilities we made use of the well
known effect of Mg^{++} on the aggregation of LHC-II rich thylakoids (29,34). In case that the light subchloroplast fraction
was indeed a mixture of stroma and grana lamellar fragments,
addition of Mg^{++} would be expected to induce aggregation of
the LHC-II-rich grana fragments, but not of the CPI-rich

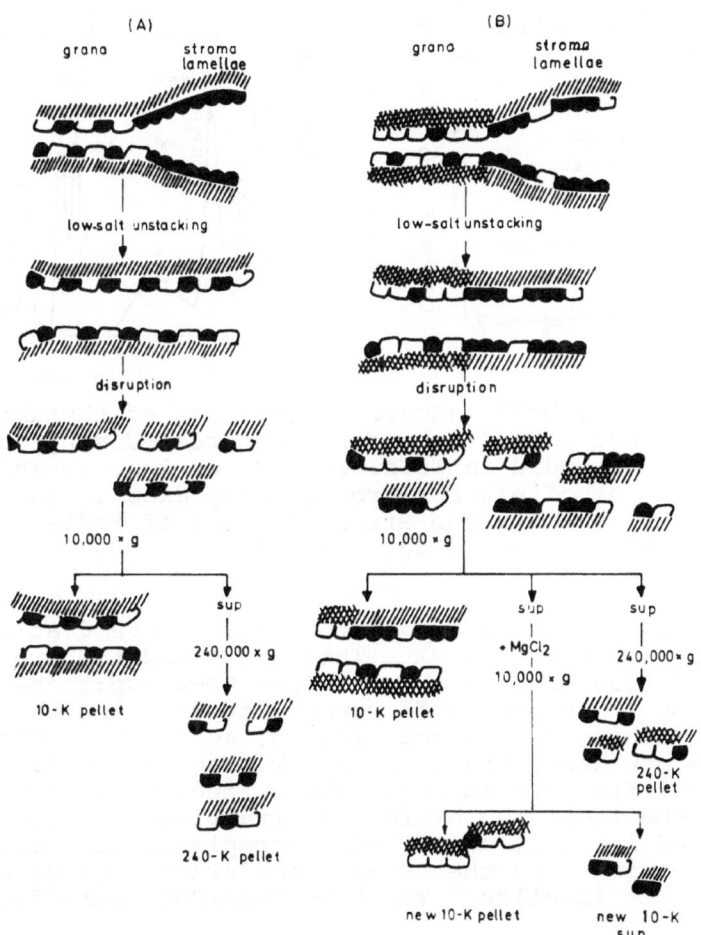

Scheme 2. Simplified models of grana unstacking in chloroplasts suspended in low-salt buffers. (A):model of lateral
protein diffusion proposed by Barber (32). (B):model explaining the results of this work. ⌐ꓶ:the LHC-II complex, ◖ :the
CPI complex, ⫽⫽⫽⫽⫽: the lipid bilayer of the thylakoid. Heavily shaded areas of lipid identify the thylakoid parts coming
from grana.

stroma lamellae. On the contrary, in case of lateral movement
of complexes, all subchloroplast fragments would be expected
to be enriched in LHC-II and to be aggregated by Mg^{++}.

Pea or bean chloroplasts, allowed to unstack for 30 min in
ice-cold Tricine-NaOH, pH 7.3 (400 µg Chl/ml), were disrupted
by French press (31); the subchloroplast fragments were separated by differential centrifugation at 1,000xgx10 min and then
at 10,000xgx30 min. The 10-K sup (Chla/Chlb=3, diluted to 150
µg Chl/ml) was incubated with Mg^{++} for 20 min in ice and then
it was centrifuged at 10,000xgx20 min. The new 10-K pellet was
collected and the new sup was further centrifuged to collect a
new 240-K pellet. The pellets were analyzed for complexes by
SDS-PAGE (20). Fig. 7 (left) shows the dependence of thylakoid
aggregation on Mg^{++} concentration; fig. 7 (right) shows the pigment-protein complex composition of the fractions obtained. As
shown, aggregation of fragments started at 1 mM Mg^{++}, and the
plateau was reached at 4-5 mM.

The Chla/Chlb ratio in the new supernatant increased grad-

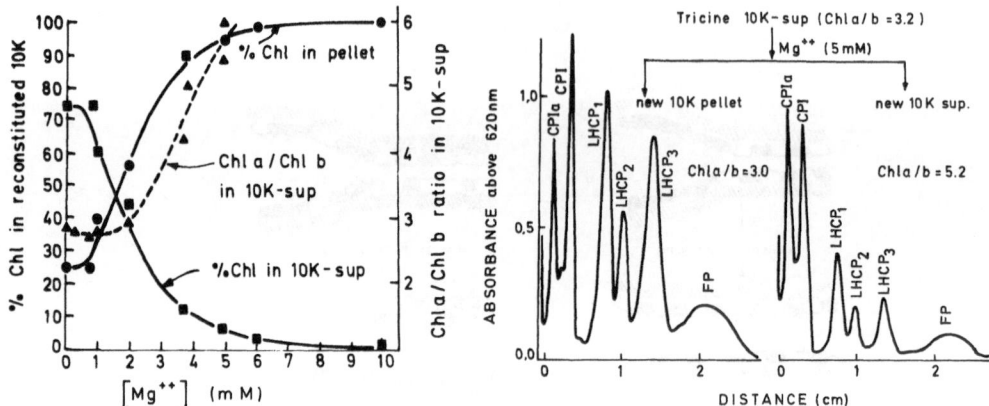

Figure 7. (left): Mg⁺⁺-induced aggregation of thylakoid fragments present in 10K-sup of French press-disrupted pea chloroplasts suspended in Tricine-NaOH, pH 7.3. (right): SDS-PAGE profiles of pigment-protein complexes in the new 10-K pellet and the new sup after addition of $MgCl_2$ at 5 mM to the original 10-K sup (fig. 7 left).

ually from 2.8-3.0 at zero Mg⁺⁺ to values above 6 at Mg⁺⁺ concentrations higher than 4 mM. This indicated the presence of two types of fragments in the original 10-K sup:those rich in Chlb, sedimenting at low Mg⁺⁺ concentrations, and those rich in Chla, remaining in the sup. Indeed, as shown in Fig. 7 (right) the new heavy fraction is enriched in LHC-II, while the new light fraction in CPI. These findings suggest, therefore, that the light subchloroplast fraction produced after disruption of low-salt unstacked chloroplasts is a mixture of thylakoids coming from the LHC-II rich grana regions and the CPI rich stroma lamellae. The high LHCP/CPI ratio found in

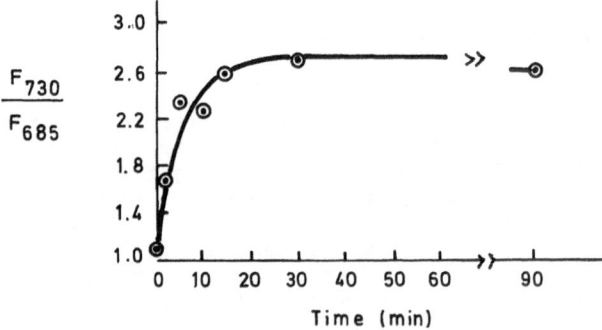

Figure 8. The F730/F685 ratio of pea chloroplasts prepared and washed in O,3M sucrose-0.05M phosphate, pH 7.2-0.01M KCl and then suspended in 50 mM Tricine-NaOH, pH 7.3 at 400 µg Chl/ml (0°C). At various times 75µl aliquots were removed, diluted to 30 µg Chl/ml in ice-cold Tricine-NaOH and a sample was frozen immediately at 77°K. Zero time is that of Tricine addition to chloroplasts.

this fraction, therefore, may not be the result of lateral movement of LHC-II from grana to stroma lamellae during unstacking, but rather the result of contamination of stroma lamellae by grana fragments.

This result may, therefore, suggest that no lateral movement of complexes occurs at least for the first 30 min from suspending the chloroplasts in Tricine at 4°C. A possibility exists, however, that this period of incubation is not adequate for complete grana unstacking, so that appressed grana thylakoids still remain in the 10K-sup. The latter is suggested from the electron microscopy study of Staehelin (33) who showed that for complete randomization of particles the chloroplasts need to remain in Tricine for about 45 min to 1 hour.

It should be noted here, however, that the fluorescence changes observed upon Tricine-induced unstacking in our experiments are over within 30 min from Tricine addition to the isolated chloroplasts (figure 8). Assuming therefore that the fluorescence changes are indeed due to the randomization of particles, randomization had to be completed within the 30 min period incubation of chloroplasts in Tricine. Our findings therefore suggest that randomization may not be required for the fluorescence changes to be observed. If this is correct, then our proposal that the fluorescence changes monitor the modulation of the PSI unit cross section may be used as an alternate hypothesis.

References

1. Murata, N. (1969) Biochim. Biophys. Acta 172, 171-181
2. Murata, N. (1969) Biochim. Biophys. Acta 172, 242-251
3. Bonaventura, C. and Myers, J. (1969) Biochim. Biophys. Acta 189, 366-383
4. Butler, W.L. (1976) Brookhaven Symp. Biol. 28, 338-344
5. Avron, M. (1969) In: An Introduction to Photobiol. Swanson, C., ed., Prentice-Hall, N.J., pp. 143-156
6. Govindjee and Papageorgiou, G. (1971) In: Photophysiol. Vol 6, Giese, A.C., ed., Academic Press, N.Y. pp. 1-46
7. Allen, J.F., Bennett, J., Steinback, K. and Arntzen, C.J. (1981) Nature 291, 25-29
8. Bennett, J. (1977) Nature 269, 344-346
9. Mullet, J.E. (1983) J. Biol. Chem. 258, 9941-9948
10. Kyle, D.J., Staehelin, L.A. and Arntzen, C. (1983) Arch. Biochem. Biophys. 222, 527-541
11. Haworth, P., Kyle, D.J., Horton, P. and Arntzen, C.J. (1982) Photochem. Photobiol. 36, 743-748
12. Izawa, S. and Good, N.E. (1966) Plant Physiol. 41, 544-553
13. Homman, P. (1969) Plant Physiol. 44, 932
14. Argyroudi-Akoyunoglou, J.H. and Akoyunoglou, G. (1977) Arch. Biochem. Biophys. 179, 370-377
15. Argyroudi-Akoyunoglou, J.H. (1980) Photobiochem. Photobiophys. 1, 279-287
16. Castorinis, A., Akoyunoglou, G. and Argyroudi-Akoyunoglou J.H. (1982) Photobiochem. Photobiophys. 4, 283-291
17. Argyroudi-Akoyunoglou, J.H., Castorinis, A. and Akoyunoglou G. (1982) Photobiochem. Photobiophys. 4, 201-210
18. Argyroudi-Akoyunoglou, J.H. and Akoyunoglou, G. (1983) Arch. Biochem. Biophys. 227, 469-477

19. Argyroudi-Akoyunoglou, J.H. and Thomou, H. (1981) FEBS Lett. 135, 177-181
20. Anderson, J.M., Waldron, J.C. and Thorne, S.W. (1978) FEBS Lett. 92, 227-233
21. Argyroudi-Akoyunoglou, J.H. (1984) FEBS Lett. 171, 47-53
22. Kuang, T.Y., Argyroudi-Akoyunoglou, J.H., Nakatani, H., Watson, J. and Arntzen, C.J. (1984) Arch. Biochem. Biophys. 235, 618-627
23. Antonopoulou, P. and Akoyunoglou, G. (in preparation)
24. Papakonstantinou, E., Argyroudi-Akoyunoglou, and Akoyunoglou (in preparation)
25. Argyroudi-Akoyunoglou, J.H., Castorinis, A, and Akoyunoglou G. (1984) Israel J. Bot. 33, 65-82
26. Argyroudi-Akoyunoglou, J.H. and Akoyunoglou G. (1979) FEBS Lett. 104, 78-84
27. Mullet, J.E., Burke, J.J. and Arntzen, C.J. Plant Physiol. 65, 823-827
28. Akoyunoglou, G. and Argyroudi-Akoyunoglou, J.H. (1974) FEBS Lett. 42, 135-140
29. Argyroudi-Akoyunoglou, J.H. (1976) Arch. Biochem. Biophys. 176, 267-274
30. Anderson, J.M. (1982) Photobiochem. Photobiophys. 3, 225-241
31. Sane, P.V. Goodchild, D.J. Park, R.B. (1970) Biochim. Biophys. Acta 216, 162-178
32. Barber, J. (1980) FEBS Lett. 118, 1-10
33. Staehelin, L.A. (1976) J. Cell Biol. 71, 136-158
34. Argyroudi-Akoyunoglou and Tsakiris, S. (1977) Arch. Biochem. Biophys. 184, 307-315.

STUDIES ON ENERGY PARTITIONING BETWEEN THE TWO PHOTOSYSTEMS

R.C. Jennings, F.M. Garlaschi, and G. Zucchelli

Centro di Studio del CNR, Dipartimento di Biologia dell'Università di Milano, Via Celoria 26, Milano, Italy

INTRODUCTION

Of central importance for the photosyntetic process in plants is the efficient utilisation of absorbed quanta. In terms of the commonly accepted Z scheme of photosynthetic electron transport this depends on the efficient reduction of NADP by the two photosystems functioning in series. It is in fact widely held that the quantum efficiency of NADP reduction is high. According to Sun and Sauer (1) it approaches the theoretical maximum of 4 quanta per NADP reduced in the wavelength range 640nm - 678nm with spinach chloroplasts. This would imply a nearly symmetrical quantum distribution between the two photosystems.

With the development of techniques which permit the fractionation and separation of the various chlorophyll-protein complexes with only low levels of chlorophyll solubilisation (2,3)it has become possible in recent years to determine the distribution of light harvesting pigments between the photosystems. Thus Anderson and coworkers (3,4) have demonstrated that in sun plants including spinach, about 60% of the chlorophyll seems to be associated with PSII and about 40% with PSI. In shade plants this imbalance in pigment distribution between the photosystems is even more marked. At first sight it would seem difficult to reconcile this distribution of chlorophyll between the two photosystems with very high quantum efficiencies for non-cyclic electron flow. Several control mechanisms which could conceivably correct and regulate this imbalance may be proposed. Thus Bonaventura and Myers (5) initially suggested that accompanying the state I - state II transitions _in vivo_ the relative optical cross sections of the two photosystems may change. This view has been recently supported (6).The phosphorylation of thylakoid polypeptides and particularly of LHCP is thought to lead to a decreased PSII and an increased PSI cross section (7-13). Evidence exists which suggests that LHCP phosphorylation may in fact provide the mechanistic basis for the state I to state II transition (14-16) in green plants. An imbalance in photon capture by the two photosystems could also be corrected by energy "spillover" i.e. by the direct transfer of energy from the PSII antenna to PSI. In this case there is no change in the cross section of the two photosystems but instead PSII antenna chlorophyll transfers a fraction of its excitons to PSI. In thylakoids "spill-over" may be regulated by the cation concentration (17-19). It has also been suggested that membrane phosphorylation may lead to increased "spill-over" (20). This latter phenomenon may be of particular importance at low cation concentrations (21) due to decreased electrostatic screening.

Apart from these mechanisms, of a biochemical nature, which may be invoked to correct the pigment imbalance between the two photosystems, the physical packaging of the chlorophyll into the chloroplast membranes could in principle alter the absorption characteristics significantly. Thus such factors as the "sieve effect" (22) or selective multiple scattering within the chloroplasts (23) may influence the relative photon absorption flux of the two photosystems. In this case the relative pigment distribution between the two photosystems may not be indicative of their relative light absorption.

In the present paper we present data on each of the above mentioned mechanisms. Using an approach developed in this laboratory (19) we have examined some aspects of the interactions between the chlorophyll-protein complexes which accompany (a) a reduction in the concentration of the electrostatic double layer screening cation, (b) spinach thylakoid phosphorylation and (c) the state I – state II transition in Chlorella. Evidence is also presented with spinach chloroplasts which indicates that photon capture by PSII may be significantly decreased with respect to that of PSI due to the presence of grana by virtue of the "sieve" effect.

MATERIALS AND METHODS

Spinach chloroplasts were prepared from freshly harvested leaves in a medium containing Tricine (30mM, pH8), sucrose 0.4M, NaCl 10mM and unless otherwise indicated $MgCl_2$ 10mM. After filtration through cheese cloth and centrifugation at 1,500 g for 5 minutes they were resuspended for about 2 minutes in the above medium minus sucrose. Following a subsequent centrifugation the chloroplasts were resuspended in either (a) Tricine (30mM, pH8), NaCl 10mM, sucrose 0.1M, NaF 10mM and $MgCl_2$ 10mM for thylakoid phosphorylation experiments or (b) Tricine (30mM, pH8), NaCl 10mM, sucrose 0.2M and unless otherwise indicated $MgCl_2$ 10mM for experiments on light absorption.

Thylakoid phosphorylation was performed at 20°C-22°C in the same medium as that used for resuspension, with the addition of ATP 1mM at a chlorophyll concentration of about 400μM. Kinase activation was achieved in the dark with the NADPH-ferrodoxin system to reduce plastoquinone (24, 9,13) as described previously (13). This chemical kinase activation procedure leads to a quantitavely similar polypeptide labelling pattern using $[\gamma-P^{32}]$ ATP as the more commonly used light activation procedure (Islam,K. and Jennings,R.C. unpublished observation). When required the thylakoids were subsequently suspended in media containing different cation concentrations, as indicated in the text, for at least 10 minutes at 20°C prior to commencing the measurements.

Chloroplast absorption spectra were measured at 22°C in a Jasco Uvidec-510 spectrophotometer with and without opal glass diffusor plates. Absorption spectra were corrected for light scattering by the method of Latimer and Eubanks (25). The chlorophyll concentration was routinely about 3μM.

Chlorophyll fluorescence titration with artificial quenchers of chlorophyll fluorescence (m-dinitrobenzene, DNB and dibromothymo quinone, DBMIB) was performed by successive addition of the same concentration of artificial quencher to the chloroplast suspension (26,19) in the presence of DCMU (25μM). The results were represented in the form of the double quenching plot as previously described (19). F_n is the fluorescence after n additions of artificial quencher; F_{n-1} is the fluorescence after n-1 additions of artificial quencher; F_i is the initial fluorescence value at the maximal (Fm) level prior to artificial quencher addition. The chlorophyll concentration was 3-4μM. The excitation wavelength unless otherwise indicated was 440 nm and the emission wavelength was 682 nm.

Digitonin fractionation of thylakoids was performed essentially as described by Boardman and Anderson (27) in the chloroplast suspension medium (b), described above, minus sucrose. The digitonin concentration was 0.5% and fractionation was for 15 minute at 0°C and a chlorophyll concentration of 350μM. After dilution (100-fold) in the same medium minus digitonin and centrifugation for 15 minutes at 5000 g the heavy pellet was carefully resuspended in a similar medium containing sucrose (0.2M).

Chlorella vulgaris was grown in a medium containing 5mM KNO_3, 0.5mM K_2HPO_4, 0.5mM KH_2PO_4, 2mM $MgSO_4$, 1mM KCl, 0.2% (v/v) Hutner's micronutrients, under continuous fluorescent light at an intensity of 6W/m^2 and at 20°C. The culture flasks were shaken at 150 rpm to provide adequate aeration. Exponentially growing cells (approximately 10^6 cells/ml) were collected by low speed centrifugation and resuspended in fresh culture medium at a chlorophyll concentration of 4μM and dark adapted for 2 hours before being used for experiments.

State I- state II transitions were performed by irradiating dark adapted cells for 10 minutes immediately before the measurements. White light was filtered through B-40 Balzers filters at 708 nm (state I) or 638 nm (state II) and the intensity was adjusted with neutral filters to provide the same absorbed energy at both wavelengths (1.85 W/m^2 or 0.85 W/m^2, depending on the experiment). These low intensities were chosen to avoid High Energy State fluorescence changes, as suggested by Catt et al. (28). After illumination DCMU (50μM) was added and fluorescence titration measurements with DNB were commenced 30-60 seconds later. Each titration consisted of 5 successive addition of DNB and was completed in about 1.5 minutes. During this period the state I/state II condition seemed stable as no fluorescence yield changes were noted.

(a) Cation Concentration Changes

Upon lowering the divalent cation concentration of a chlorophyll suspension below saturating levels (around 2.5mM Mg ions) the fluorescence yield decreases, as was initially observed by Homann (29) and Murata (17). Following Murata's initial suggestion (17) this is thought to be mostly due to an increased "spillover" of energy from PSII to PSI. We have titrated the Fm fluorescence for thylakoids incubated with both a saturating concentration of Mg ions (2.5 mM) and a subsaturating concentration (0.5mM). This latter concentration was chosen as it brought about a fluorescence yield decrease which was similar in extent (around 20%) to that induced by the experimental treatments discussed below in sections (b) and (c). The data are represented in the form of the double quenching plot (fig. 1) following Jennings (19). In the terminology of this type of analysis the ratio of the intercepts of any two extrapolated straight lines to the $F_{n-1} - F_n/F_{n-1}$ axis is known as the quencher interaction index ratio (19, 26). In figure 1 it can be seen that the quencher interaction index ratio for chloroplasts incubated with 0.5mM and 2.5mM MgCl$_2$ (control) equals the ratio of the maximal fluorescence values for these chloroplasts prior to quencher addition (the F_i ratio). These data indicate that lowering the Mg ion concentration leads to an homogeneous type of quenching (19). This means that the endogenous quencher (in this case PSI via the "spillover" mechanism) interacts approximately equally with all PSII units and not just with a fraction of them. There is no _a priori_ reason why this should

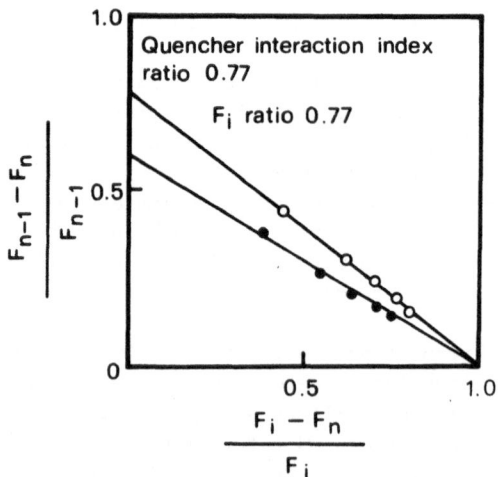

Fig. 1. Double quenching plots for the Fm fluorescence of spinach thylakoids incubated in 2.5mM (o) and 0.5mM (●) MgCl$_2$. The fluorescence quencher used was DBMIB (0.6μM additions). For details see Materials and Methods.

be the case as one could easily imagine that PSI may interact with only a fraction of the PSII units. Analysis of the sigmoidicity of the fluorescence induction curves measured in the presence of DCMU (30,31) indicated that the good PSII-PSII energy transfer was still operative at 0.5mM Mg in the experiments (data are not presented) though this almost disappeared upon total Mg ion removal due to the break up of the LHCP-PSII matrix, as is usually observed. We have also observed (19) that the fluorescence quenching is equal for excitation with 475nm (absorbed predominantly by the LHCP complex) and 435nm (relatively enriched in PSII-complex absorbed wavelengths) upon lowering the Mg ion concentration to 0.5mM. This indicates that down to this concentration in spinach chloroplasts little or no uncoupling of LHCP from the PSII complex occurs (19). Thus we conclude that at 0.5mM $MgCl_2$ the LHCP-PSII matrix was still substantially intact and presumeably located in the remaining partition zones. Thus in order to explain the homogeneous quenching interaction with PSI upon lowering the Mg ion concentration to 0.5mM, it is necessary to suggest that a fraction of the PSI complexes moved towards the LHCP-PSII matrix containing areas of the membranes and was able to establish an energy coupling interaction with them. By virtue of the PSI-PSII energy transfer these PSI complexes then effectively quenched right across the LHCP-PSII matrices, thus giving rise to the homogeneous quenching observed.

We therefore conclude that in the case of relatively small changes in energy distribution between the two photosystems via a cation concentration - "spillover" type mechanism it is the PSI units which would move towards the LHCP-PSII matrices and not the PSII units which become detached from the matrices moving subsequently towards the PSI containing membrane zones (fig. 4).

(b) Spinach Thylakoid Phosphorylation

The phosphorylation of chloroplast membranes leads to the quenching of PSII fluorescence by about 15%-20% (9-13). As mentioned in the Introduction this is thought to be due to the detachment of PSII antenna and to its subsequent transfer to PSI. We have examined this hypothesis using the double quenching analysis briefly described above in section (a) (fig. 2A). When phosphorylated membranes were incubated with a saturating concentration of Mg ions (5mM) the slopes of the double quenching plots for control and phosphorylated thylakoids were equal (i.e. quenching interaction index ratio of 1) though the maximal fluorescence (F_i) ratio was 0.84. As discussed by Jennings (19) and Torti et al. (13) this is indicative of an extreme type of heterogenous quenching. This means that the fluorescence of only a fraction of the fluorescing bed of chlorophyll was quenched and very strongly. We have noticed under these conditions, by analysis of the fluorescence induction curves in the presence of DCMU (30,31) that efficient PSII-PSII energy transfer occurs under these conditions (unpublished data), thus confirming the observation of Horton and Black (7). This indicates that the LHCP-PSII matrices are not disrupted under these conditions after phosphorylation. The double quencher analysis is incompatible with a "spillover" type interaction between PSI and the LHCP-PSII matrices as this would have led to a quencher interaction index ratio significantly less than one (19). The data are

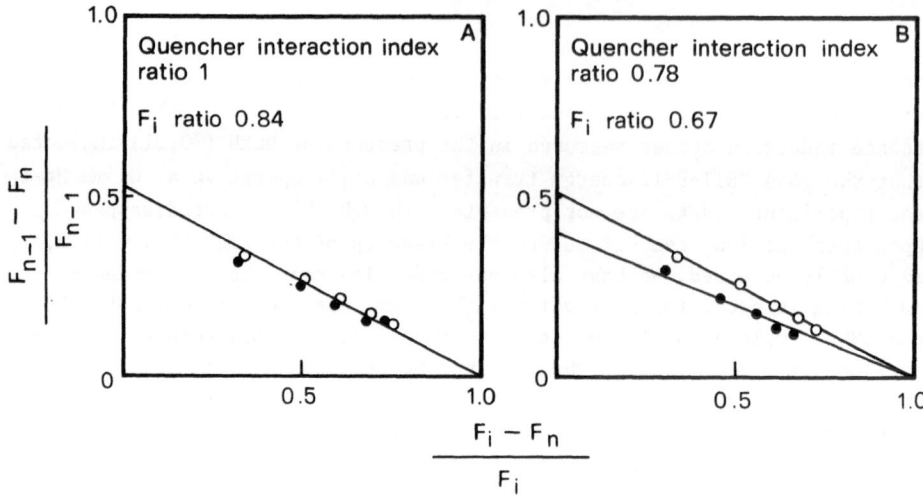

Fig. 2. Double quenching plots for the Fm fluorescence of phosphorylated (●) and non phosphorylated (o) spinach thylakoids. The
fluorescence quencher used was DNB (125μM additions). In (A)
the fluorescence titration was carried in the presence of
5mM $MgCl_2$; in (B) in the presence of 1mM $MgCl_2$. In both cases
phosphorylation was performed in the presence of 10mM $MgCl_2$.

best explained in terms of a part of the PSII antenna becoming detached
from the LHCP-PSII matrices and subsequently moving towards and interacting
strongly with the fluorescence quencher (PSI). That this antenna complex
is in fact LHCP (fig. 4) is strongly supported by the data in table 1
where it can be seen that quenching was significantly greater when the
light used to excite fluorescence was relatively enriched in wavelengths
absorbed principally by LHCP (475nm) than when absorption by the PSII
complex was relatively favoured (435nm).

Horton and Black (21) initially demonstrated that when phosphorylated
thylakoids were incubated with subsaturating concentrations of screening
cations the phosphorylation-induced fluorescence quenching was greatly
increased. We have performed the double quenching analysis also under
this condition of a subsaturating concentration of Mg ions (1mM). The
data (fig. 2B) show that the quencher interaction index ratio is significantly greater than the maximal fluorescence (Fi) ratio, thus indicating
heterogeneous quenching (19), though the degree of heterogeneity is much
less than in the presence of Mg 5mM. Similar data has recently been
reported by Hodges and Barber (11). The fluorescence quenching was also
greater with the 475nm exciting light than with the 435nm light (table 1),
thus demonstrating that LHCP detachment and quenching also occured when
phosphorylated chloroplasts were suspended in the presence of a low Mg
ion concentration. We therefore suggest that at low Mg ion concentrations,
in addition to the movement of LHCP from the LHCP-PSII matrix to PSI
another quenching interaction, probably of the "spillover" kind, also
occurs. This interpretation is in line with the analysis of Haworth et
al. (20) though we emphasise that these two distinct interaction types

Table 1. Fluorescence quenching induced by the phosphorylation of spinach thylakoid membranes and the state I - state II transition in <u>Chlorella</u> cells using two different excitation wavelengths.

Spinach thylakoid phosphorylation		
Mg ion concentration	435nm	475nm
10mM	19.2%	21.6%
1mM	30.8%	32.9%

Chlorella state transition		
	435nm	475nm
	20.3%	21.5%

Spinach thylakoids were phosphorylated as described in the Materials and Methods section and subsequently incubated with either 10mM or 1mM $MgCl_2$. The <u>Chlorella</u> state transitions were carried out at an absorption flux of 1.85 W/cm^2. Fluorescence was measured in the presence of DCMU using 435nm and 475nm as excitation wavelengths. The data are expressed as the percent fluorescence decrease induced by thylakoid phosphorylation or by conversion from the state I to the state II condition. Analysis of the variance of 20 measurements in the case of thylakoid phosphorylation and 56 measurements in the case of the Chlorella state transitions, showed the quenching at the two wavelengths to be significantly different at the 99% confidence level.

only occur when phosphorylated thylakoids are incubated in the presence of subsaturating concentrations of screening cations.

(c) <u>State I - State II Transition in Chlorella</u>

The state I - state II conditions of Bonaventura and Myers (5) reflect a situation in which energy is distributed preferentially towards PSII and PSI respectively. The state changes are achieved after illumination with wavelengths absorbed preferentially by either PSII (state II) or PSI (state I). We have used a similar approach to that already described in the preceding section on isolated thylakoid phosphorylation to analyse the state transition in <u>Chlorella</u>.

In figure 3 is presented the double quenching analysis for the state transitions performed at two different light intensities. We emphasise that at each light intensity the light I and light II absorption fluxes

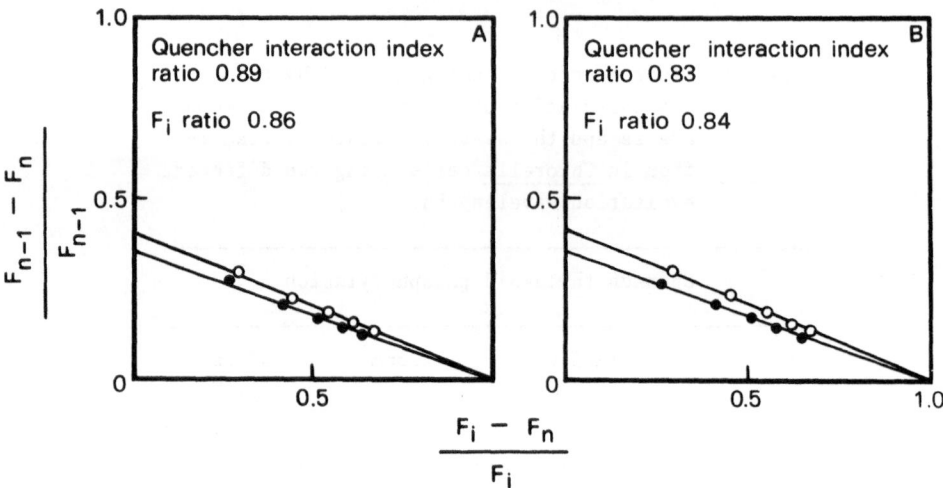

Fig. 3. Double quenching plots for the Fm fluorescence of <u>Chlorella</u>
cells in the state I (o) and state II (●) condition. The
fluorescence quencher used was DNB (125µM additions). In
(A) the absorption flux used to bring about the state
transition was 0.85 W/cm^2 while in (B) it was 1.85 W/cm^2.

were equal. The state II/state I quencher interaction index ratio in both
series of experiments is slightly greater than the maximal fluorescence
(Fi) ratio. This is indicative of a weakly heterogeneous quenching due to
the state I – state II conversion. In fact the situation is not very
different from that described in section (a) for lowering the Mg ion
concentration. Measurement of the state II associated fluorescence quenching
with 475nm and 435nm excitation light (table 1) indicates that this was
slightly greater in the case of the former, relatively enriched in LHCP-
absorbed wavelengths. This difference, though small, is highly significant
statistically based on analysis of the variance (99% confidence level)
of the Fm 475nm/Fm 435nm ratio for 56 separate determinations performed
on 9 different cell batches.

We therefore conclude on the basis of these two types of measurements
that the transition into the state II in Chlorella cells involves both
some LHCP detachment and its subsequent migration towards and association
with the fluorescence quencher (PSI; fig. 4) and another kind of quenching
interaction which seems to be essentially of an homogeneous kind. This is
most likely of the "spillover" type and should therefore represent the
association of some PSI with the LHCP-PSII matrices (fig. 4) as was
described in section (a) above for lowering the Mg ion concentration.

(d) <u>Grana Formation and Light Absorption</u>

In figure 5 are presented absorption spectra (corrected for light
scattering), for chloroplasts resuspended at different Mg ion concentrations and also in the presence of the trivalent cation tris (ethylenediamine)-cobalt (TEC). Clearly treatment with different cations, which is

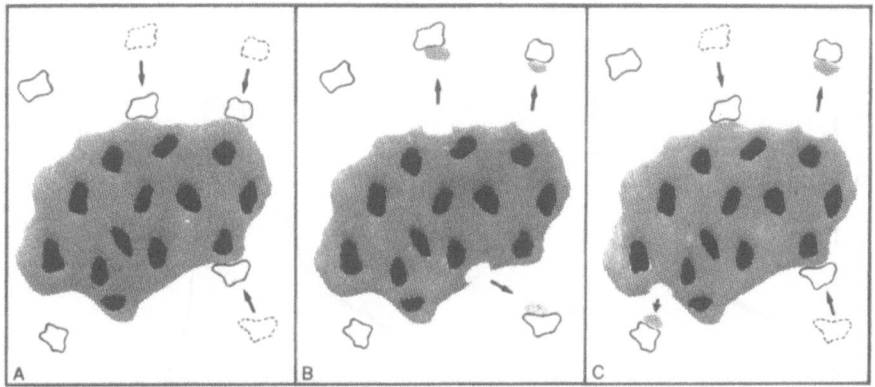

Fig. 4. Schematic representation of the suggested vectorial movements
of the chlorophyll-protein complexes following A) the transfer
of spinach chloroplasts from a saturating concentration of
Mg ions to a subsaturating concentration. B) spinach thylakoid phosphorylation and incubation in the presence of a
saturating concentration of Mg ions. C) the state I - state
II transition in Chlorella cells. These conclusions are
based on the double quenching and 475nm/435nm excitation
wavelength quenching ratio analyses presented in sections
(a) - (c). PSI is assumed to be the LHCP-PSII fluorescence
quencher on the basis of evidence presented in the literature (see Introduction). Hatched areas represent LHCP. Shaded
shapes represent the PSII complex. Open shapes represent the
PSI complex. Arrows indicate the suggested directional
movements.

known to maintain the granal structure, leads to substantial absorption
decreases, with no apparent change in the peak position. A similar effect
is also seen in the blue region of chlorophyll absorption (unpublished
observation). These data are similar to those reported by Henkin and
Sauer (32) for $MgCl_2$ addition to chloroplasts. From the ratio of the absorption spectra (fig. 6) it can be seen that the absorbance decrease
induced by membrane stacking is a positive (non linear) function of the
absorbance at that particular wavelength. In addition, it can be seen in
figure 5 that the half band width is decreased upon membrane unstacking.
These two aspects are consistent with the interpretation that the absorption changes are due to optical flattening phenomena associated with
the sieve effect. Thus, it is concluded that the organisation of thylakoid
membranes into grana leads to decreased absorption of light.

Figure 7 shows how the measured effect of grana formation on light
absorption depends on the concentration of the chloroplast suspension. By
extrapolating to infinitely low chloroplast concentrations, it can be
seen that granal chloroplasts absorb about 17% less light at the red absorption peak than unstacked chloroplasts. Over the spectral range 400nm
- 750nm this corresponds to a decreased light absorption of 10% (unpublished data).

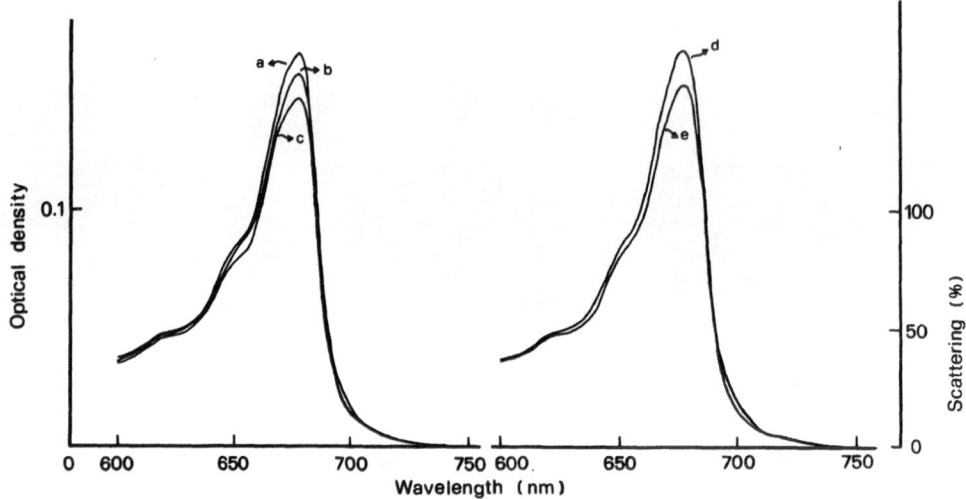

Fig. 5. Absorption spectra of stacked and unstacked spinach chloroplasts.

    a,d – Substantially unstacked chloroplasts incubated with
         $MgCl_2$, 0.1 mM.

      b – Partially stacked chloroplasts incubated with $MgCl_2$,
         1.1mM.

    c,e – Fully stacked chloroplasts incubated with $MgCl_2$ 10mM or
         TEC, 50 μM.

The chloroplasts were diluted to a chlorophyll concentration of
about 3μM from a single concentrated stock of stacked chloroplasts ($MgCl_2$, 10mM). Incubation was for 4 minutes at 20°C
prior to measurement of the absorption spectra, which were subsequently corrected for light scattering (see Materials and
Methods).

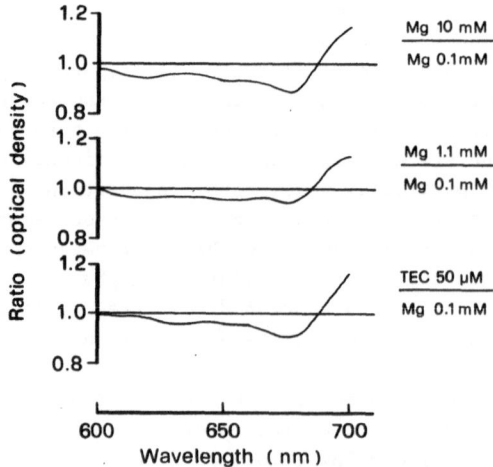

Fig. 6. The ratio of the absorption
spectra presented in Fig.5.

232

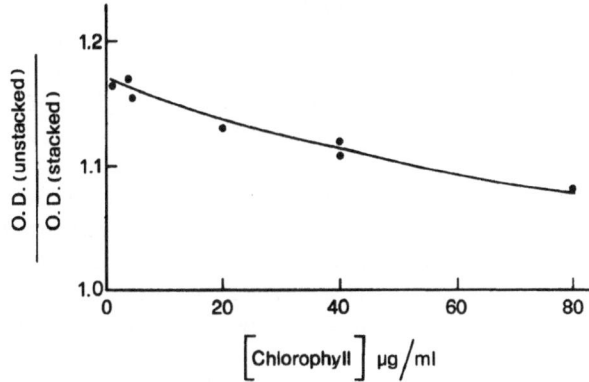

Fig. 7. The effect of membrane stacking on light
absorption at the red absorption peak as
a function of the chlorophyll concentration. Stacked chloroplasts were prepared
and stored in the presence of $MgCl_2$
(5mM), whereas unstacked chloroplasts
were washed and stored in the absence of
$MgCl_2$.

As the sieve effect is essentially caused by a kind of "shading" of
the pigment molecules in zones of high local pigment concentrations, it is
of interest to understand if the effect of grana formation on photon
capture is largely at the level of each single granum or whether it is
substantially caused by the "shading" of one granum by another.

Table 2. Absorption of Stacked and Unstacked Spinach Chloroplasts
and a Grana Preparation.

	Chloroplasts +$MgCl_2$	Chloroplasts −$MgCl_2$	Heavy Digitonin Pellet +$MgCl_2$
Optical Density (677 nm)	0.133	0.155	0.136
Chlorophyll a/b Ratio	2.62	2.62	2.30

The optical density was measured at 677 nm (corrected for light
scattering) of a suspension of stacked granal chloroplasts
($MgCl_2$, 5mM), unstacked chloroplasts ($MgCl_2$, zero) and the
heavy digitonin pellet ($MgCl_2$, 5mM). In all cases, the concentration of chlorophyll a was 2.3 μg/ml. The digitonin pellet,
prepared as described in the Materials section, contained 48%
of the total chlorophyll.

To this end, we prepared grana stacks by a mild digitonin solubilisation procedure followed by a relatively low-speed centrifugation (see Materials and Methods). The data (table 2) show that the detachment of the grana stacks from each other by digitonin did not lead to a significant increase in light absorption. Thus, it is concluded that the bulk of the sieve effect caused by grana formation is due to the packaging of chlorophyll-protein complexes (mostly associated with PSII) into each single granum.

It should be pointed out that the data presented in table 2 in which a granal preparation is shown to absorb to a similar extent at the same chlorophyll concentration as a suspension of granal chloroplasts, lends strong support to the thesis that the "sieve" effect reported here upon cation addition to thylakoids is in fact due to grana formation.

From Figure 6, it can be seen that granal chloroplasts absorb somewhat more light at wavelengths above 685 nm - 690 nm. The origin of this effect is not understood at present.

Thus it is demonstrated that grana formation and the associated packaging of most PSII and its LHCP antenna into the granal membranes leads to a decreased light absorption. This is expected to mainly concern PSII rather than PSI as it is PSII which is predominantly located in the grana. If we accept Anderson's (3,4) figure of about 60% of the total chlorophyll in PSII it may be concluded that the "sieve" effect associated with grana formation decreases PSII photon capture by about 15-20% in a system of randomly oriented chloroplasts. This figure is expected to be modified in the case of the thylakoids being oriented with respect to the angle of incidence of the light. It is not impossible that modification of chloro-plast orientation with respect to the direction of the incident light could provide a mechanism for modulating the relative absorption flux by the two photosystems.

CONCLUSIONS

A number of mechanisms which may influence the partitioning of energy between the two photosystems have been investigated. It is concluded:

Upon reducing the concentration of the electrostatic double layer screening cations from saturating to subsaturating levels, PSI moves towards the LHCP-PSII matrix containing zones of the membranes and interacts approximately equally with the excitons of the whole matrix.

The consequences of the phosphorylation of isolated spinach thylakoids are different, depending on the concentration of the screening cations. At saturating cation concentration some LHCP detachment from the LHCP-PSII matrices occurs, followed by its translocation towards and association with PSI. At subsaturating cation concentration in addition to the above described mechanism, also a direct PSI interaction with the LHCP-PSII matrices takes place.

The state I - state II transition in Chlorella cells involves both

LHCP detachment from the LHCP-PSII matrices and a direct "spillover" type interaction between PSI and the LHCP-PSII matrices.

Grana formation reduces total light absorption in a suspension of randomly oriented spinach chloroplasts by about 10% due to the "sieve" effect. This is thought to result in a decreased absorption by PSII of around 15-20% with respect to the non granal situation.

REFERENCES

1. Sun, S. and Sauer, K. (1971). Biochim. Biophys. Acta 234, 399-414.
2. Anderson, J.M., Waldron, J.C. and Thorne, S.W. (1978). FEBS Letters 92, 227-233.
3. Anderson, J.M. (1980). Biochim. Biophys. Acta 591, 113-126.
4. Chu, Z-X. and Anderson, J.M. (1984). Photobiochem. Photobiophys. 8, 1-10.
5. Bonaventura,C. and Myers, J. (1969). Biochim. Biophys. Acta 189,366-383.
6. Canaani, O. and Malkin,S. (1984). Biochim. Biophys. Acta 766, 513-524.
7. Horton, P. and Black, M.T. (1981). Biochem. Biophys. Acta 635, 53-62.
8. Anderson, B., Akerlund, H-E., Jergil, B. and Larsson, C. (1982). FEBS Letters 149, 181-185.
9. Telfer, A., Allen, J.F., Barber, J. and Bennett, J. (1983). Biochim. Biophys. Acta 722, 176.
10. Horton, P. and Black, M.T. (1983). Biochim. Biophys. Acta 722, 214-218.
11. Hodges, M. and Barber, J. (1984). Biochim. Biophys. Acta 767, 102- 107.
12. Kyle, D.J., Kuang, T-Y., Watson, J.L. and Arntzen, C.J. (1984). Biochim. Biophys. Acta 765, 89-96.
13. Torti, F., Gerola, P.D. and Jennings, R.C. (1984). 767, 321-325.
14. Chow, W.S., Telfer, A., Chapman, D.J. and Barber, J. (1981). 639, 60-68.
15. Saito, K., Williams, W.P., Allen, J.F. and Bennett, J. (1983). Biochim. Biophys. Acta 724, 94-103.
16. Telfer, A., Allen, J.F., Barber, J. and Bennett, J. (1983). Biochim. Biophys. Acta 722, 176-181.
17. Murata, N. (1969). Biochim. Biophys. Acta 189, 171-181.
18. Barber, J. and Chow, W.S. (1979). FEBS Letters 105, 5-10.
19. Jennings, R.C. (1984). Biochim. Biophys. Acta 766, 303-309.
20. Haworth, P., Kyle, D.J. and Arntzen, C.J. (1982). Biochim. Biophys. Acta 680, 343-351.
21. Horton, P. and Black, M.T. (1982). Biochim. Biophys. Acta 680, 22-27.
22. Duysens, L.N.M. (1956). Biochim. Biophys. Acta 19, 1-10.
23. Thorne, S.W., Duniec, J.T. and Lee J.A. (1983). Photobiochem. Photobiophys. 5, 71-78.
24. Bennett, J. (1979). FEBS Letters 103, 342-344.
25. Latimer, P. and Holmes-Eubanks, C.A. (1962). Archiv. Biochem. Biophys. 98, 274-285.
26. Jennings, R.C., Garlaschi, F.M. and Gerola, P.D. (1983). Biochim. Biophys. Acta 722, 144-149.
27. Boardman, N.K. and Anderson, J.M. (1964). Nature 203, 166-167.
28. Catt, M., Saito, K. and Williams, W.P. (1984). Biochim. Biophys. Acta 767, 39-47.
29. Homann, P. (1969). Plant Physiol. 44, 932-936.

30. Joliot, P., Joliot, A. and Kok, B.(1968). Biochim. Biophys. Acta 153, 635-677.
31. Butler, W.L. (1980). Proc. Natl. Acad. Sci. USA 77, 4697-4701.
32. Henkin, B.M. and Sauer, K. (1977). Photochem. Photobiol. 26, 277-286.

IONIC EFFECTS ON THE LATERAL SEGREGATION OF CHLOROPHYLL-PROTEINS DURING RESTACKING OF THYLAKOID MEMBRANES

Cecilia Sundby, Ulla K. Larsson and Bertil Andersson

Department of Biochemistry, University of Lund
P.O.Box 124, S-221 00 Lund, Sweden

INTRODUCTION

The photosynthetic membrane of chloroplasts is differentiated into appressed and non-appressed (stroma exposed) areas. This structural differentiation is accompanied by a functional heterogeneity in the lateral plane of the membrane[1]. The photosystem 1 complex and the ATP synthase are mainly excluded from the appressed membrane regions and localized in the non-appressed regions. Most of the photosystem 2 complexes and its light-harvesting chlorophyll a/b complex (LHC-II) is located in the appressed regions. This lateral segregation of the thylakoid protein complexes has been postulated to involve repulsive electrostatic and attractive van der Waals forces at closely appressed membrane surfaces[2]. Under low salt conditions the thylakoid membranes destack and the membrane components are randomized along the membrane[3]. Readdition of cations reverses this process and restacking accompanied by lateral segregation of the complexes occurs.

The lateral differentiation of the thylakoid membrane has been studied by different methods[1,2,4], which altogether complements each other to give the information known about the stacking phenomenon. One powerful approach has been the analyzes of thylakoid subfractions derived from appressed and non-appressed thylakoid regions[1,5,6]. Inside-out thylakoid vesicles isolated by aqueous polymer two-phase systems derive from the appressed thylakoid regions[6,7]. Thus, the yield and composition of inside-out thylakoid vesicles can be used to monitor the degree of stacking and estimate the composition of the appressed thylakoids at a given condition. In this study we have changed previous methods for isolating inside-out vesicles in a way which allows a direct, detailed and kinetic analysis of the formation of thylakoid grana appressions under different ionic conditions, pH, temperature and phosphorylation state.

EXPERIMENTALS

Preparation of Destacked and Randomized Thylakoids

Spinach leaves were homogenized in 50 mM sodium phosphate buffer pH 7.4/10 mM NaCl/300 mM sucrose. The homogenate was filtered through a nylon mesh and the chloroplasts were spun down at 1000 g for 3 min followed by one wash in the preparation medium. The chloroplasts were broken osmotically by suspension in 50 mM sodium phosphate buffer pH 7.4/10 mM NaCl/50 mM sucrose, and the thylakoids were pelleted at 2000 g for 5 min.

In the case of phosphorylated thylakoids, phosphorylation of intact thylakoids was performed in a medium containing 50 mM Hepes-KOH (pH 8.0)/ 10 mM $MgCl_2$/100 mM sucrose at a concentration of 200 ug chl/ml. Before phosphorylation, NaF was added to a final concentration of 10 mM in order to inhibit phosphatase activity. Thylakoids were phosphorylated by illumination for 10 min (500 uE x m^{-2} x s^{-1}) in the presence of 0.2 mM $(\gamma-^{32}P)$-ATP (300 000 cpm/nmol ATP). Control thylakoids were kept in the dark without addition of ATP. During destacking and restacking of phosphorylated thylakoids 10 mM NaF was included in all media.

For destacking and randomization of the thylakoids, they were washed three times in 10 mM Tricine pH 7.4/100 mM sucrose to remove all residual cations and subsequently left in this medium for at least 90 min at $5^{o}C$. For a stacked control thylakoid sample, 5 mM $MgCl_2$ was included to the Tricine-medium.

Restacking of Thylakoids and Isolation of Inside-Out Thylakoid Vesicles

The original procedure for obtaining inside-out vesicles by Yeda press disruption of thylakoids and phase partitioning[8] was modified to be optimal for a detailed analysis of the restacking process. These changes included:
(i) Only small sample volumes (5 ml) were passed through the Yeda press allowing the fragmentation to be completed within 30 s.
(ii) By operating the Yeda press at 17.5 MPa instead of 10 MPa, fragmentation was achieved in one passage instead of four.
(iii) The concentrations of each of the two phase polymers was lowered to 5.6% from 5.7% in the original procedure. The 5.6% system was shown to give a more pure preparation of inside-out vesicles and therefore only one partition step was needed for isolation.

These changes, involving only one press treatment and one phase partition step allowed a rapid purification of all inside-out vesicles in a high purity still allowing an accurate and sensitive determination of the yield of vesicles at the different restacking conditions.

Restacking was performed under various controlled conditions in 5 ml aliquots, using destacked and randomized thylakoids corresponding to 1000 ug chl/ml. At a certain temperature, a given ion species at a given concentration was added to the sample. The restacking process was stopped after any certain time by rapidly passing the sample through the Yeda press. In cases where different ion concentrations were used for restacking, the samples were made equal with respect to ion concentration after the fragmentation. The samples were diluted 10-fold with a low salt buffer (10 mM sodium phosphate buffer pH 7.4/5 mM NaCl/100 mM sucrose) and centrifuged in SS-34 tubes 40 000 g for 30 min. The upper 80% of the resulting supernatants, containing the so called Y-100 fraction consisting of small right-sided vesicles deriving from the non-appressed thylakoid regions, was carefully removed with a Pasteur pipette and saved for analysis. The pellets (95% of the starting material), containing a mixture of rightsided and inside-out vesicles, were resuspended in 6 ml of the above low salt buffer, yielding concentrations around 1000 ug chl/ml. To remove starch a short low-speed centrifugation was performed 1000 g for 10 min before 5 ml of each sample was put to a phase system with the final composition 5.6% dextran T500/5.6% polyethylene glycol 3350/20 mM sucrose/10 mM sodium phosphate pH 7.4/5 mM NaCl. The phase system was shaken at $3.0^{o}C$ (temperature critical) and the inside-out vesicles were recovered in the lower phase. The upper phase was carefully sucked off. The lower phase and the interface were thoroughly mixed. For estimation of the proportion of thylakoid material partitioning to the lower phase (i.e. the yield of inside-out vesicles), triplet samples were withdrawn from the upper and lower phases for measurement of the absorbance at 680 nm. The lower phase and the interfacial material were collected by centrifugation. To facilitate the sedimentation of the inside-out vesicles, they were quantitatively removed from the very viscous lower phase, by the addition of 5 ml fresh and room-temperated upper phase. In this way a new phase system is created where the vesicles are collected in a small upper phase. This upper phase is removed, diluted with buffer, and the material is spun down at 100 000 g for 30 min.

Biochemical Analyses of Membrane Composition

Chl a/b determinations were made according to Arnon[9]. Chlorophyll-protein composition was estimated from green SDS-polyacrylamide gels after mild SDS-solubilization mainly according to[10]. The amounts of LHC-II apopolypeptides were determined from SDS-polyacrylamide gels.

In experiments using $(\gamma-^{32}P)$-ATP, the gels were sliced, solubilized in H_2O_2/perchloric acid and counted for radioactivity in Aquasol.

RESULTS

1. Formation of Appressed Thylakoid Membranes in the Presence of Mg^{2+} ions

Inside-out vesicles are formed from appressed regions of stacked thylakoids[7]. As shown in Fig. 1 destacked and randomized thylakoids do not form inside-out vesicles upon fragmentation. When 5 mM $MgCl_2$ is added to destacked and randomized thylakoids, the yield of inside-out vesicles increases continuously with the time of incubation prior to disruption. At 5°C, the yield of inside-out vesicles 90 min after addition of Mg^{2+} approaches the value obtained from fractionation of the stacked control thylakoids. Thus, the yield of inside-out vesicles well monitors the restacking process as judged by electron microscopy.

The analyses of the inside-out vesicles do not only give the relative degree of stacking but also show the composition of the appressions forming during the restacking. While the chl a/b ratio of intact thylakoids

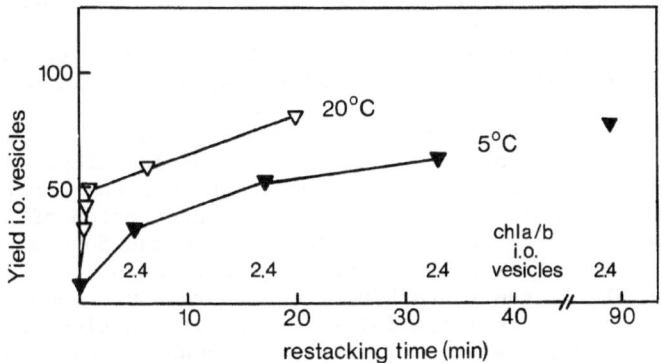

Fig. 1. Restacking with 5 mM $MgCl_2$. The yield of inside-out vesicles is expressed as % of the yield obtained with stacked control thylakoids. The chl a/b ratio for intact thylakoids was 2.9.

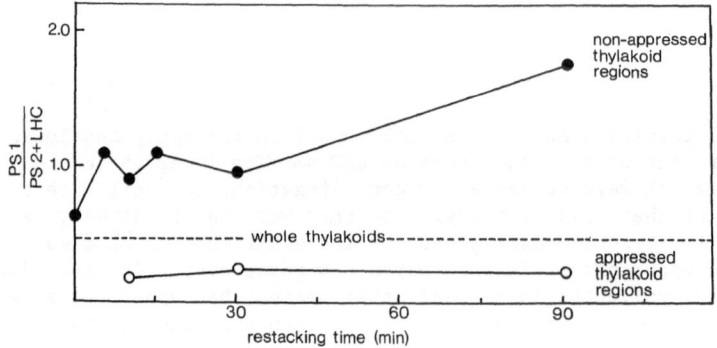

Fig. 2. Restacking with 5 mM MgCl$_2$ at 5°C. The ratio,
PS1/(PS2+LHC), is calculated from the
chlorophyll-protein composition.

is around 2.8, it is 2.3–2.4 for the inside-out vesicles representing the
appressed regions of the stacked control thylakoids (Fig. 1). The chl a/b
ratio of the inside-out vesicles in the restacked samples is low (2.4) and
constant throughout the time course of restacking. This observation sheds
light on whether PS 1 exclusion is a prerequisite for, or a result of, the
formation of appressed membrane regions. Since already the small amounts
of inside-out vesicles formed very shortly after magnesium addition are
PS 1 depleted, we can conclude that exclusion of PS 1 necessarily proceeds
the formation of the tightly appressed membrane regions. Furthermore, in
Fig. 2 it is shown that the low chl a/b values for the inside-out vesicles
is due to the enrichment of PS 2-associated chlorophyll-proteins at the
expense of PS 1-associated chlorophyll-proteins. This is accompanied by a
corresponding increase in PS 1 in the non-appressed parts of the thylakoids,
as judged from the analysis of the Y-100 fraction.

The kinetics of restacking was analysed after transformation of the
restacking curves in Fig. 1 into double-reciprocal plots. In Fig. 3 the
double-reciprocal plot for restacking at 5°C is presented. It can be seen
that restacking at first follows a second-order reaction, describing a
perfect straight line in the double-reciprocal plot. (By integration it
can be shown that the hyberbolic function described by a second-order reaction gives a straight line in a double-reciprocal plot). This indicates
that one may treat restacking as having second-order reaction kinetics,
i.e. that the reaction rate is limited by the probability of collision
between two molecules. The molecule species that should be expected to
collide with each other is LHC-II. Thylakoids loose their stacking capacity if positively charged peptide fragments are removed from LHC-II by
trypsination[11] and Mg-induced aggregation of liposomes with incorporated
LHC-II mimics stacking[12]. In the randomized thylakoid, all LHC-II molecules
can participate in the restacking reaction and therefore be regarded as
the initial concentration of the rate-determining reactant. Since the disorganized thylakoids approximately describes a 3-dimensional and fluid
matrix, the conditions fairly well resembles those assumed for reactions
between molecules in solution. This initial concentration of LHC-II decreases as more and more LHC-II molecules collide and form appressed regions. When some 60–70% of full restacking has taken place (indicated by
an arrow in the plot), the restacking reaction rate proceeds faster than
what should be expected from initial second-order kinetics and the reaction deviates from the straight line in the double-reciprocal plot. At
this stage, the thylakoid membrane system is fairly well organized and
the membranes are aligned at a suitable distance from each other. Thereby
the residual LHC-II molecules, that have not yet reacted, become concentrated in closeness to each other compared with how they were dispersed

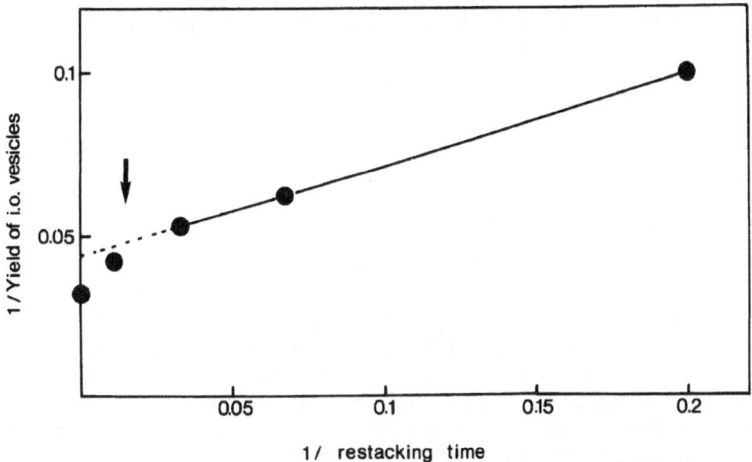

Fig. 3. Double-reciprocal plot of the restacking curve
at 5°C in Fig. 1. Arrow indicates 60-70% of
complete restacking.

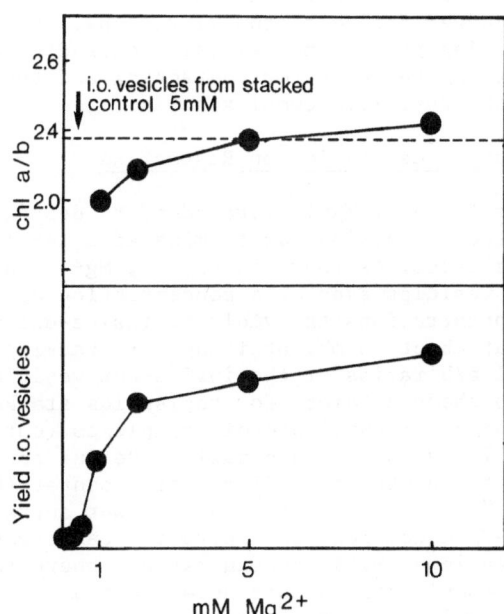

Fig. 4. Restacking for 15 min at 5°C with
increasing concentration of $MgCl_2$.
Bottom: The yield of inside-out
vesicles is expressed as % of the
yield obtained with control thylakoids stacked in 5 mM $MgCl_2$.
Top: Chl a/b ratios of the inside-
out vesicles formed.

in the disorganized randomized membrane system when the reaction started. This concentration effect facilitates the last LHC-II collisions and the reaction rate is accelerated compared to the initial rate. In conclusion, we state that the restacking rate essentially is dependent on the concentration of LHC-II that is available for stacking, and that the reaction rate seems to be facilitated at its final stages by a type of cooperative mechanism when the membranes are close to each other. The double-reciprocal plot for restacking at 20°C described, and deviated from, a straight line in the same way as for 5°C.

2. Temperature Dependence of Stacking

Fig. 1 shows the temperature dependence of the restacking process. Two restacking curves are shown, one for 5°C and one for 20°C which clearly illustrate that membrane adhesion and the lateral redistribution of the chlorophyll-proteins that is induced by Mg^{2+} ions is highly temperature-dependent. At 5°C half-time for full restacking is 10-15 minutes while at 20°C it is less than 1 minute. Thus, restacking proceeds at a rate that is at least 10-fold faster at 20°C than at 5°C. Thus, at physiological temperatures, significant rearrangements in the lateral plane of the membrane can occur within seconds. Interestingly, this is in the same time scale as the state 1-state 2 transitions monitored by fluorescences which are implicated to reflect changes in the energy distribution between the photosystems. Such changes may therefore be explained in terms of a controlled lateral rearrangement of chlorophyll-proteins in the thylakoid membrane. These considerations are in agreement with those made by Briantais et al.[13] after comparison of the time-scale for fluorescence changes and for the lateral segregation as judged from electron microscopy. The temperature dependence of the restacking process may explain contradictory observations in the literature concerning the time for restacking and the accompanying lateral segregation of the thylakoid complexes.

3. Influence of Magnesium Concentration on Restacking

Increasing concentrations of $MgCl_2$ were added to destacked and randomized thylakoids and the restacking was terminated after 15 minutes by passage through the Yeda press. As shown in Fig. 4, Mg^{2+} ions induces the formation of inside-out vesicles even at a concentration as low as 1 mM. With increasing Mg^{2+} concentrations the yield of inside-out vesicles increases but levels out at about 10 mM. Strikingly, restacking with low Mg^{2+} concentrations gives chl a/b ratios of the inside-out vesicles that are significantly lower than those obtained for thylakoids stacked with 5-10 mM $MgCl_2$. Analysis of the chlorophyll-protein complexes (not shown) reveals that these particularly low chl a/b ratios are due to an enrichment of LHC-II compared with the photosystem II reaction center (CP_a). A possible explanation is that the "free LHC-II"[14,15] has a lower surface charge density than the LHC-II-PS 2 complex, and therefore can create membrane appressions even at these lower $MgCl_2$ concentrations where the electrostatic screening is not complete[2]. A more detailed study on the different exclusion on thylakoid complexes from the appressed regions during restacking is presently undertaken.

4. Effect of Hydration Forces on Restacking

Apart from electrostatic forces, repulsive hydration forces can play a role in biological membrane systems[16] and it has been postulated that such hydration forces play a role in the formation of appressed thylakoid membranes[17]. In this study we have tested this possibility by following restacking in the presence of either 5 mM $MgCl_2$ or 5 mM $BaCl_2$. Both these cations have the same capability of electrostatic screening, but the Ba^{2+} ion has a smaller hydration radius compared to the Mg^{2+} ion. Fig. 5 shows

the yield of inside-out vesicles at various times after addition of each
of the two ions to the destacked and randomized thylakoids. No major differences in the formation pattern of appressed thylakoids could be seen,
as deduced from the yield and properties of inside-out vesicles. The chl a/b
ratio is somewhat higher in inside-out vesicles formed in the presence of
Ba^{2+} ions. These results indicate that hydration forces play none or only
a very minor role in the stacking process, compared to the repulsive electrostatic forces.

5. pH Dependence of Restacking

Addition of protons to destacked and randomized thylakoids, i.e.
lowering the pH, will also decrease the repulsive forces. In this study we
have followed the formation of appressions in thylakoids at pH 4.7 and 5.4
(Fig. 6). At pH 4.7, which is quite close to the average isoelectric point
of the outer thylakoid membrane surface[18], inside-out vesicles are formed
instantaneously after addition of protons. However, no time-dependent increase in the yield was seen as was the case after addition of Mg^{2+} ions
(Fig. 1). Moreover, the chl a/b ratio of the inside-out vesicles was not
low, but corresponding to the value for unfractionated thylakoids. Thus,
at pH 4.7 membrane adhesion occurs but no segregation of components in the
lateral plane of the membrane takes place. This is in agreement with electron micrographs of thylakoids suspended in low salt buffers close to the
isoelectric point[7].

By contrast, if protons are added at a lower concentration, corresponding to pH 5.4 (Fig. 6), the formation of appresed thylakoid membranes
resembles very much that obtained with 5 mM $MgCl_2$. The yield of inside-out
vesicles increases with restacking time, and the chl a/b ratios are low
and similar to the ratios obtained for inside-out vesicles formed from

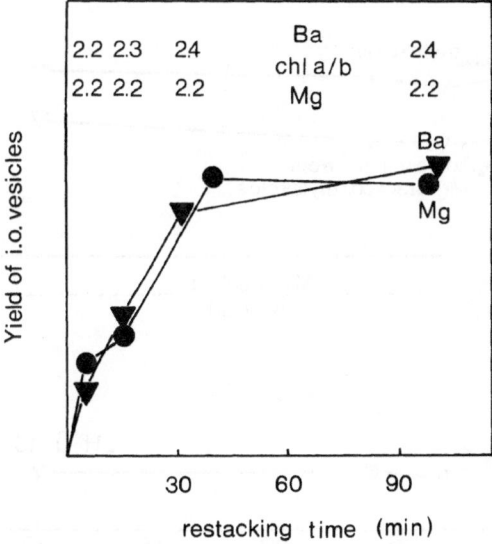

Fig. 5. Restacking with either 5 mM $MgCl_2$
or 5 mM $BaCl_2$ at 5°C. The yield
of inside-out vesicles is expressed
as % of the yield obtained with
control thylakoids stacked in 5 mM
$MgCl_2$. The chl a/b ratios for the
inside-out vesicles formed is shown
on top. For intact thylakoids the
chl a/b ratio was 2.8.

stacked control thylakoids. Thus, lateral segregation of thylakoid components does occur at pH 5.4, but the process is slower. The halftime for
full restacking at 20°C is around 6 min, compared to less than 1 min in the
presence of 5 mM MgCl$_2$. The lateral segregation into stromal and granal
regions occurs at pH 5.4 in the absence of other cations than H[+] has earlier
been shown by electron microscopy[19].

A distinction has been made between "binding ions" that are unable to
evoke fluorescence changes, but cause stacking without lateral protein
segregation and "screening ions" which bring about fluorescence changes
and cause stacking that is accompanied with lateral protein segregation[20].
The former group of ions (e.g. La^{3+}, polylysine, Zn^{2+} and H[+] at pH 4.7)
thus brings about a reduction in the electrostatic repulsive forces by
electrostatic neutralisation and the latter group of ions (e.g. Ca^{2+}, Mg^{2+},
Ba^{2+}) by electrostatic screening.

The results in Fig. 6 demonstrates that even a "binding ion" like H[+]
can bring about some lateral migration, if added in a proper concentration,
so that only a partial and selective neutralisation of membrane components
occurs. At pH 5.4 presumably the PS 2 components and LHC–II are neutralized
while the PS 1 components still carries negative charges. The latter is not
the case at pH 4.7 where presumably also the negative charges of the PS 1
components are neutralised and there is no longer any incitation for PS 1
exclusion from the appressed regions.

6. Stacking and Restacking after Introduction of Negative Surface Charges
   by Protein Phosphorylation

An increase in repulsive forces at constant ionic environments can be
obtained by introduction of additional charges by protein phosphorylation[2].

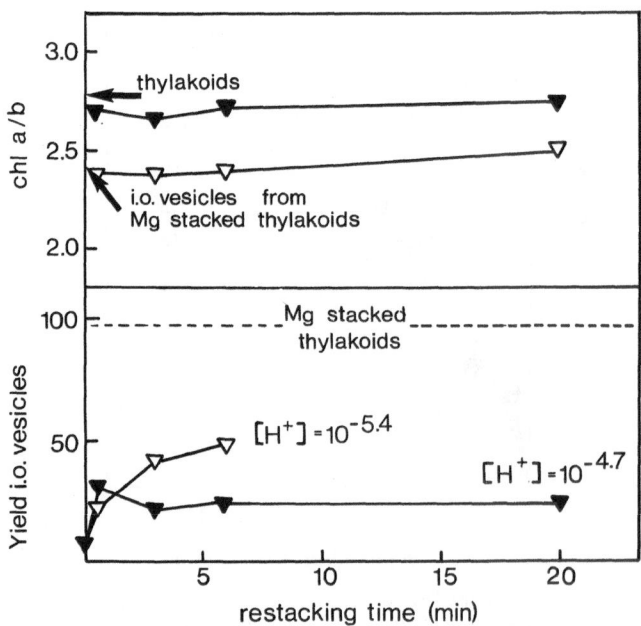

Fig. 6. Restacking at 20°C with H[+] at different concentrations. Bottom: The yield of inside–out
vesicles is expressed as % of the value obtained
with control thylakoids stacked in 5 mM MgCl$_2$.
Top: Chl a/b ratios of the inside–out vesicles
formed.

244

Table 1. Yield of inside-out vesicles (B1) and chl a/b ratios of
appressed (B1) and non-appressed (Y-100) membranes of
unphosphorylated and phosphorylated thylakoids.

		Control	Restacked
Unphosphorylated	Yield of B1[a]	100	93
	Chl a/b B1	2.3	2.3
	Chl a/b Y-100	6.3	5.1
Phosphorylated	Yield of B1[a]	72	70
	Chl a/b B1	2.3	2.3
	Chl a/b Y-100	4.6	4.2

[a]The yield of inside-out vesicles (B1) is expressed as % of the yield
from unphosphorylated stacked control thylakoids. Restacking was performed in 5 mM $MgCl_2$ at 20°C for 30 min.

Heavily phosphorylated polypeptides are those belonging to LHC-II[21].
Table 1 shows the yield of inside-out vesicles from unphosphorylated and
phosphorylated thylakoids. After phosphorylation the yield of inside-out
vesicles decreased by 28% compared to the yield from unphoshporylated control thylakoids, demonstrating a partial destacking induced by phosphorylation. The chl a/b ratios of the inside-out vesicles from unphosphorylated
and phosphorylated thylakoids are equal (Table 1). In contrast, the chl a/b
ratio of the stroma lamellae fraction from phosphorylated thylakoids is
lower than that of unphosphorylated thylakoids, reflecting an increased
proportion of LHC-II in the non-appressed stroma lamellae membranes upon
phosphorylation. This is due to a migration of phosphorylated LHC-II from
the appressed grana to the non-appressed stroma lamellae regions[22,23]. In
a previous study[22] the degree of destacking due to phosphorylation was
estimated from the final yield of inside-out vesicles (B3) according to the
original preparation procedure, which included repeated press treatments
and several phase partition steps. The 30% destacking noted here (Table 1)
is probably a more accurate estimation since the yield is directly measured
and all the inside-out material is recovered. A partial destacking following
phosphorylation has also been reported from electron microscopy[14,24] and
light scattering measurements[24]. It was previously shown[22] that although
most phosphorylated LHC-II migrated to the non-appressed stroma lamellae
regions, significant amounts still remained in the appressed grana regions
even several hours after phosphorylation. Is this residual phosphorylated
LHC-II population restricted in its migration and has not reached the non-
appressed regions prior to fragmentation? Or does it, although phosphorylated, indeed belong to the appressed regions? In the latter case phosphorylated LHC-II must be able to participate in the formation of membrane
appression during restacking. We therefore compared the relative amounts
of appressions in phosphorylated control thylakoids and phosphorylated
thylakoids that were restacked in 5 mM $MgCl_2$ subsequent to destacking/ran-
domization. As seen in Table 1, there is no significant decrease in the
yield of inside-out vesicles isolated from the phosphorylated stacked control thylakoids compared to those isolated from the restacked phosphorylated thylakoids.

Moreover, we determined the specific (^{32}P)-phosphate labelling
(cpm/relative amount of protein) in the apopolypeptide bands of LHC-II in

the inside-out vesicles of control and restacked phosphorylated thylakoids (not shown). The (^{32}P)-labelling of LHC-II in the inside-out vesicles from the destacked/restacked thylakoids showed the same specific LHC-II phosphorylation as those isolated from the control thylakoids. Thus, the yield of inside-out vesicles after restacking of phosphorylated thylakoids as well as their specific labelling demonstrates that there exist phosphorylated LHC-II polypeptides that can be included in thylakoid appressions.

This observation may be interpreted in light of the suggestions[14,15] that there should be two populations of LHC-II. The phosphorylated LHC-II that remains in the appressed grana regions may represent a pool of LHC-II bound to PS 2, whereas the phosphorylated LHC-II migrating to the non-appressed stroma lamellae regions may represent a free pool. We have recently found a different polypeptide composition of this migrating pool of LHC-II compared to the total LHC-II (U.K. Larsson and B. Andersson, unpublished results). LHC-II consists of two major apopolypeptides of 27 and 25 kDa and we have found that the specific incorporation of (^{32}P)-phosphate in the 25 kDa polypeptide is 4-5 times higher compared to the 27 kDa polypeptide. According to[2], proteins carrying more negative phosphate should have a higher probability of being excluded from the appressed thylakoid membranes. To examine if the different (^{32}P)-phosphate incorporation was accompanied by a corresponding difference in lateral migration behaviour, the relative amounts of the 27 and 25 kDa polypeptides were compared in unphosphorylated and phosphorylated appressed inside-out and non-appressed stroma lamellae vesicles. The ratio between the amount of the 27 and 25 kDa polypeptides in unphosphorylated inside-out and stroma lamellae thylakoids was 4:1 and this ratio remained roughly constant in the appressed membranes after phosphorylation. In contrast, in the phosphorylated stroma lamellae membranes the ratio between the 27 and 25 kDa polypeptides decreased to 3:1. This lower ratio implies that the 25 kDa polypeptide is quite abundant in the LHC-II population that is migrating to the non-appressed regions upon phosphorylation. Since the amount of LHC-II in the stroma lamellae membranes nearly doubles after phosphorylation[22], the ratio between the 27 and 25 kDa polypeptides should be close to 2:1 in the migrating pool of LHC-II compared to 4:1 for total LHC-II.

DISCUSSION

The differentiation of the continuous thylakoid membrane into appressed and non-appressed regions has been studied by various methods, such as electron microscopy, light scattering or subfractionation. The method described in this work is based on the possibility to analyse subfractions obtained representative of appressed and non-appressed membrane regions during the stacking process and has certain advantages compared to the previously used methods.

The ultrastructural techniques includes thin section electron microscopy and freeze-fracture electron microscopy. Thin section electron microscopy gives a crude quantitative estimation of the degree of stacking but does not give information about the tightness of the appressions or the composition of the different membrane regions. The latter restriction does not apply to freeze-fracture electron microscopy. By this method[3,13] you can follow the differences in number and size of particles during the lateral movement of components in the thylakoid membrane during destacking and restacking. However, at present the correlation of certain freeze-fracture particles with certain protein complexes is not quite clear. For detailed kinetic studies on grana formations the electron microscopy methods have the disadvantage that the time for complete fixation of the membranes after addition of glutaraldehyde can not be accurately controlled.

Light scattering studies are fast, easy and convenient. They also seem to give a reliable estimation of the degree of stacking and appear particularly suited for kinetic analyses. However, the interpretation of the light scattering analyses are obscured by the sensitivity of the method which may pick up other changes than the formation of appressed thylakoids, such as membrane shrinkage and protein conformational changes. For example addition of 30 mM KCl causes changes in 540 nm light-scattering[25] although no lateral segregation is seen by freeze-fracture[3] at this level of monovalent ions.

Prior to this study subfractionation of the thylakoid membrane has been used to study the stacking process. The methods have included French press[3] or digitonin treatment[26,27] of thylakoids under different conditions. These methods give a crude estimate of the amount and properties of the grana stacks rather than the relative amount of appressed membrane as is the case with the present methods. Moreover, they strongly rely on the assumption that only appressed thylakoids can form aggregates large enough to sediment as a heavy fraction during centrifugation. This is not the case at least under conditions with low degree of stacking[28]. Further on, with detergent based subfractionation there is always the risk that components are selectively solubilized from the membrane. Moreover, when stacking is to be studied under various ion or temperature conditions detergent fractionation is of limited use due to the fact that the detergent interaction with biological membranes is strongly dependent on both temperature and ionic environment[29].

Our present method combines several of the advantages of the methods discussed above within suffering from many of the disadvantages.
(i) It is quantitative since it allows an estimation of appressed regions through the yield of inside-out vesicles. Note however, that the yield of inside-out vesicles is only proportional to and not a direct measurement of the degree of membrane appressions. The latter would require an analysis by counter-current distribution that would at least require some 60 partition steps.
(ii) It is qualitative since the composition of the inside-out vesicles can be analysed thereby giving an estimation of the appressed regions. For this purpose it is crucial that no detergents have been present.
(iii) It is a simple and direct method requiring only one fragmentation and one phase partition step. This allows the method to be used for kinetical studies on the stacking process.

ACKNOWLEDGEMENT

We thank Prof. Gösta Pettersson for constructive criticism concerning the kinetical analyses and Ms Ingun Sundén-Cullberg for skilful technical assistance. This work has been supported by the Alice and Knut Wallenberg foundation, the Lennander foundation and the Swedish Natural Science Research Council.

REFERENCES

1. J.M. Anderson and B. Andersson, The architecture of photosynthetic membranes: lateral and transverse organization, Trends Biochem. Sci. 7:288 (1982)
2. J. Barber, Influence of surface charges on thylakoid structure and function, Annu. Rev. Plant Physiol. 33:261 (1982)
3. L.A. Staehelin, Reversible particle movements associated with unstacking and restacking of chloroplast membranes in vitro, J. Cell. Biol. 71:136 (1976)

4. L.A. Staehelin and C.J. Arntzen, Regulation of chloroplast membrane function: protein phosphorylation changes the spatial organization of membrane components, J. Cell Biol. 97:1327 (1983).

5. P.V. Sane, D.J. Goodchild and R.B. Park, Characterization of chloroplast photosystem 1 and 2 separated by a non-detergent method, Biochim. Biophys. Acta 216:162 (1970).

6. B. Andersson, C. Sundby, H.-E. Åkerlund and P.-Å. Albertsson, Inside-out thylakoid vesicles: important tools for the characterization of the photosynthetic membrane, Physiol. Plant. in press (1985).

7. B. Andersson, C. Sundby and P.-Å. Albertsson, A mechanism for the formation of inside-out membrane vesicles, Biochim. Biophys. Acta 599:391 (1980).

8. B. Andersson and H.-E. Åkerlund, Inside-out membrane vesicles isolated from spinach thylakoids, Biochim. Biophys. Acta 503:462 (1978).

9. D.J. Arnon, Copper enzymes in isolated chloroplasts. Polyphenol oxidase in Beta vulgaris, Plant Physiol. 24:1 (1949).

10. J.M. Anderson, J.C. Waldron and S.W. Thorne, Chlorophyll-protein complexes of spinach and barley thylakoids, FEBS Lett. 92:227 (1978).

11. J.E. Mullet and C.J. Arntzen, Simulation of grana stacking in a model membrane system, Biochim. Biophys. Acta 589:100 (1980).

12. I.J. Ryrie, Freeze-fracture analysis of membrane appression and protein segregation in model membranes containing the chlorophyll-protein complexes from chloroplasts, Eur. J. Biochem. 137:203 (1983).

13. J.M. Briantais, C. Vernotte, J. Olive and F.-A. Wollman, Kinetics of cation-induced changes of photosystem II fluorescence and of lateral distribution of the two photosytems in the thylakoid membranes of pea chloroplasts, Biochim. Biophys. Acta 766:1 (1984).

14. D.J. Kyle, L.A. Staehelin and C.J. Arntzen, Lateral mobility of the light-harvesting complex in chloroplast membranes controls excitation energy distribution in higher plants, Arch. Biochem. Biophys. 222:527 (1983).

15. L.A. Staehelin and M. DeWit, Correlation of structure and function of chloroplast membranes at the supramolecular level, J. Cell Biochem. 24:261 (1984).

16. P. Rand, Interacting phospholipid bilayers: Measured forces and induced structural changes, Annu. Rev. Biophys. Bioeng. 10:277 (1981).

17. J.T. Duniec, J.N. Israelachvilli, B.W. Ninham, R.M. Pashley and S.W. Thorne, An ion-exchange model for thylakoid stacking in chloroplasts, FEBS Lett. 129:193 (1981).

18. H.-E. Åkerlund, B. Andersson, A. Persson and P.-Å. Albertsson, Isoelectric points of spinach thylakoid membrane surfaces as determined by cross partition, Biochim. Biophys. Acta 552:238 (1979).

19. P.D. Gerola, R.C. Jennings, G. Forti and F.M. Garlaschi, Influence of protons on thylakoid membrane stacking, Plant Sci. Lett. 16:249 (1979).

20. J. Barber, An explanation for the relationship between salt-induced thylakoid stacking and the chlorophyll fluorescence changes associated with changes in spillover of energy from photosystem 1 to photosystem 11, FEBS Lett. 118:1 (1980).

21. J. Bennet, Phosphorylation of chloroplast membrane polypeptides, Nature 269:344 (1977).

22. U.K. Larsson, B. Jergil and B. Andersson, Changes in the lateral distribution of the light-harvesting chlorophyll-a/b-protein complex induced by its phosphorylation, Eur. J. Biochem. 136:25 (1983).

23. D.J. Kyle, T.-Y. Kuang, J.L. Watson and C.J. Arntzen, Movement of a sub-population of the light-harvesting complex (LHC-II) from grana to stroma lamellae as a consequence of its phosphorylation, Biochim. Biophys. Acta 765:89 (1984).

24. A. Telfer, M. Hodges, P.A. Millner and J. Barber, The cation-dependence of the degree of protein phosphorylation-induced unstacking of pea thylakoids, Biochim. Biophsy. Acta 766:554 (1984).

25. F.-A. Wollman and B.A. Diner, Cation-control of fluorescence emission, light scatter and membrane stacking in pigment mutants of Chlamydomonas reinhardi, Arch. Biochem. Biophys. 201:646 (1980).
26. J.H. Argyroudi-Akoyunoglou, Effect of cations on the reconstitution of heavy subchloroplast fractions (grana) in disorganized low-salt agranal chloroplasts, Arch. Biochem. Biophys. 176:267 (1976).
27. W.S. Chow and J. Barber, Further studies of the relationship between cation-induced chlorophyll fluorescence and thylakoid membrane stacking changes, Biochim. Biophys. Acta 593:149 (1980).
28. H.-E- Åkerlund, B. Andersson and P.-Å. Albertsson, Isolation of photosystem II enriched membrane vesicles from spinach chloroplasts by phase partition, Biochim. Biophys. Acta 449:525 (1976).
29. C. Tanford, "The hydrophobic effect", Wiley-Interscience, New York (1973).

LIPID STRUCTURES AND LIPID-PROTEIN INTERACTIONS IN THYLAKOID MEMBRANES

Kleoniki Gounaris

AFRC Photosynthesis Research Group
Department of Pure and Applied Biology
Imperial College of Science and Technology
London   SW7 2BB, U.K.

INTRODUCTION

Biological membranes are essentially lipoprotein structures and, in general, are described in terms of the fluid mosaic model. In this current view of biomembrane structure, the lipid component forms a closed, stable bilayer while in a fluid liquid-crystalline condition. It thereby provides a regulated and controlled internal environment as well as a matrix for the lateral diffusion of membrane proteins. Such a model implies that any enzyme residing in the bilayer must have lipid associated with the protein surface. Most biomembranes contain a large number of integral proteins and clearly the fraction of lipid involved at protein/lipid interfaces will be higher as the protein to lipid ratio of the membrane as a whole increases. Lipid-protein associations occur in all membrane systems and at the molecular level any interactions, whether hydrophobic, steric or electrostatic, must be governed by common principles.

The physiological properties of biomembranes are considered to be primarily determined by the protein components, which in turn are related to the properties of the lipid matrix, such as its fluidity characteristics. Lipid heterogeneity and diversity has thus been rationalised in terms of fluidity on the basis that intrinsic protein function may be modulated by the local fluidity of the bilayer. These suggestions have been supported by the observations that the physical state of the lipids has marked effects on the enzymatic activity of membrane proteins such that certain integral proteins require a liquid-crystalline environment for function, which is inhibited in the presence of gel-state lipids (1,2). In recent years it has been demonstrated that proteins may also affect the state of lipid component, this being expressed on both the polar region and the hydrophobic chains of the lipids (3). In addition, it has been recognised that a number of membrane lipids exhibit structural polymorphism thus adopting structures other than a bilayer in aqueous environments. Such observations have indicated that the organisation of lipids and proteins in membranes is interdependent and that the exact nature of lipid conformation and dynamics are of great importance when considering intermolecular interactions and membrane organisation (4).

The thylakoid membrane of higher plant chloroplasts will be considered in this presentation. This energy transducing membrane system consists of mainly five high molecular weight intrinsic protein complexes. It has a high protein to lipid ratio and is characterised by an unusual lipid composition. Since knowledge of how lipids associate in aqueous solutions can provide insight into the behaviour and organisation of lipids in membranes, the self-assembly of thylakoid lipids in aqueous systems will be first described. Lipid requirements of membrane protein complexes and their interaction will then be reported.

THE LIPIDS OF THE THYLAKOID MEMBRANES: COMPOSITION AND CONFORMATION

The lipid matrix of the thylakoid membranes consists mainly of polar lipids, most of the neutral lipids being bound to the protein component of the membrane. The thylakoids are distinct from other biomembranes in terms of their polar lipid composition in that glycolipids are the major constituent. Phospholipids account for less than 20% of the polar lipid content. The glycolipids of this membrane system are, in order of abundance, monogalactosyldiacylglycerol (MGDG), digalactosyldiacylglycerol (DGDG) and sulphoquinovosyldiacylglycerol (SQDG). The major phospholipid is phosphatidylglycerol (PG) while phosphatidylcholine (PC) is a minor component. The two galactolipids, MGDG and DGDG, are electroneutral, account for two-thirds of the total polar lipid content and occur in a ratio of 2:1 (mono:-digalactosyldiacylglycerol). The glucolipid SQDG is strongly acidic and together with phosphatidylglycerol are the only negatively charged lipids of the thylakoid membrane accounting for a total of about 20%. The major fatty acid in thylakoid lipids in linolenic (18:3) acid, usually comprising 70-80% of the total. SQDG and PG tend to be more saturated than the galactolipids but in general the average number of double bonds per lipid molecule in the thylakoid membrane is in the region of 4.5-5.0.

Despite the disparity in the nature of the headgroups and fatty acyl composition, DGDG, SQDG and PG form liposomes on hydration. These structures, as examined by freeze-fracture electron microscopy, are of similar bilayer configurations even though there is a variation in the size of the liposomes formed (Fig. 1b). Bilayers in liposomes are the characteristic arrangements adopted by most lipids extracted from membrane sources. Monogalactosyldiacylglycerol, however, which accounts for about 50% of the lipid content of the thylakoid membrane, does not adopt a bilayer configuration (Fig. 1a). Instead, the molecules are arranged in an hexagonal type II configuration, which consists of lipid cylinders enclosing a water channel. This type of arrangement was first described in (5) and it has since been found that a number of lipids of biological origin adopt this structure on hydration.

It is now clear that nearly every biomembrane contains substantial amounts of lipid species which, when purified and dispersed in water at physiological temperatures, do not orient in bilayers but prefer a different type of organisation. The thylakoid membranes are unusual in the sense that they contain large amounts (about half the total) of lipid that would adopt such non-bilayer configurations. In order to account for the polymorphic phase behaviour of membrane lipids a proposal was made which postulates that the preference of a lipid species for a given structure reflects the dynamic molecular shape assumed by the individual components (4,6). Thus geometric expressions have been derived relating the hydrocarbon-water interfacial area, the hydrocarbon chain volume and the hydrocarbon chain length to the aggregated structure. Other examples of hexagonal type II forming lipids are unsaturated phosphatidylethanolamines (7) and monoglucodiacylglycerols (8). Cardiolipin (9) and phosphatidic acid (10) also convert isothermally from the bilayer to the hexagonal type II phase by the addition of Ca^{2+} and Mg^{2+} ions respectively. This type of isothermal polymorphic phase transition has been shown to depend on the concentration of the cations (11).

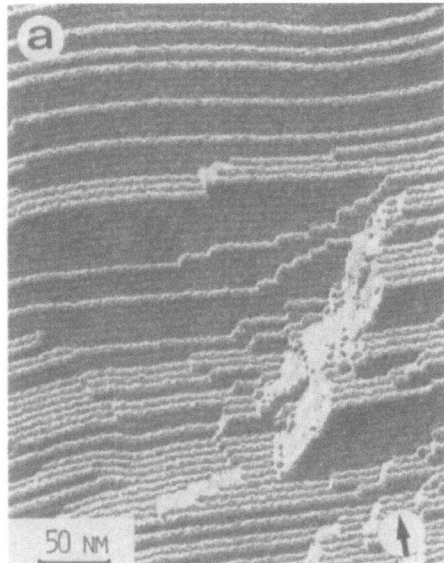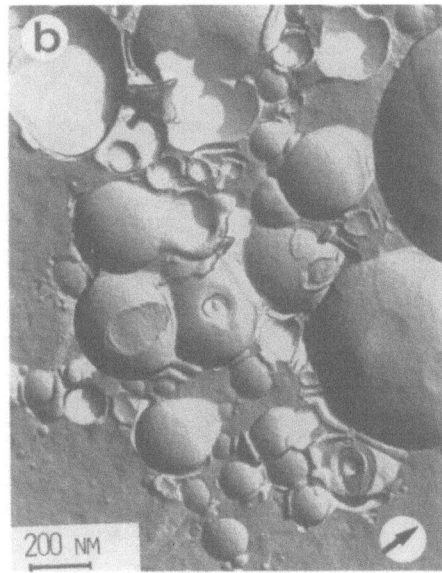

Fig. 1 Typical electronmicrographs obtained from freeze-fracture replicas of aqueous dispersions of (a) monogalactosyldiacylglycerol and (b) the rest of the thylakoid membrane's polar lipids.

The two galactolipids, MGDG and DGDG, dominate the lipid composition of the thylakoid membranes and despite their similarities in chemical structures their properties are quite different. Typical electromicrographs of replicas obtained from mixtures of these galactolipids in a 2:1 molar ratio (MGDG:DGDG) are shown in Fig. 2. A consistent feature of the replicas is the presence of small freeze-fracture particles. This type of molecular organisation has been termed the "lipidic particle" and was demonstrated to correspond in galactolipid mixtures to inverted lipid micelles sandwiched within a bilayer (12). Particles of this type are commonly observed in lipid mixtures originating from biomembranes other than the thylakoids and were first reported in aqueous dispersions of cardiolipin:phosphatidyl-choline mixtures containing Ca^{2+} (13). The requirement for the formation of such organisations is that, in the lipid mixture, at least one component is present which on isolation and hydration can adopt the hexagonal type II configuration. In the case of the galactolipid mixtures this role is fulfilled by MGDG. Different molecular arrangements observed in freeze-fracture replicas of mixed galactolipids (Fig. 2b) are believed to be related and interconvertible under appropriate conditions. In addition it has been shown that factors that interfere with lipid-lipid, lipid-water or water-water interactions interfere with the structural arrangements of the galactolipid mixtures (14).

The structures adopted by the total polar lipid extract of thylakoid membranes are distinctly different (Fig. 3). Small unilamellar liposomes are formed when this lipid mixture is dispersed in water and there is no evidence of inverted lipid micelles of the type that dominate the galactolipid mixture dispersions. The difference in the chemical composition of the two types of mixtures, is the presence of the two anionic lipids, namely SQDG and PG in the total lipid extract. The absence of non-bilayer structures in the latter lipid dispersions can thus be correlated with the

Fig. 2 a,b. Freeze-fracture electronmicrographs of aqueous dispersions of
mixtures of monogalactosyl and digalactosyldiacylglycerol (2:1
molar ratio respectively).

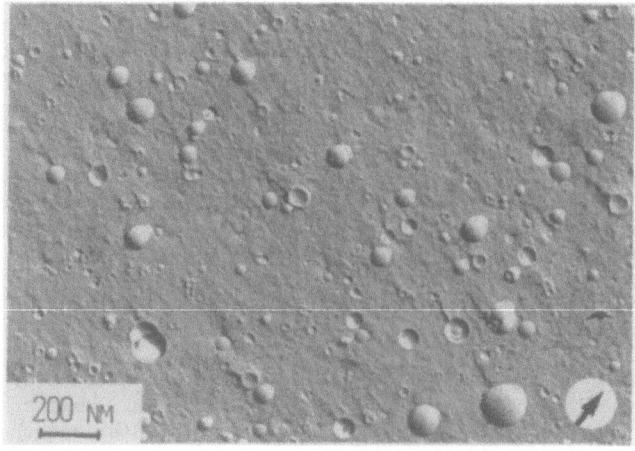

Fig. 3. Electronmicrograph of freeze-fracture relica obtained from aqueous
dispersions of total polar lipid extract in water.

presence of the two anionic lipids in such extracts. The structures observed were found to be highly sensitive to the addition of cations. Addition of different concentrations of inorganic salts to the total polar lipid dispersions led to flocculation. High concentrations of divalent or trivalent salts resulted in the formation of large visible aggregates. The addition of cations resulted in two distinct effects. At low concentrations (< 2mM) the small unilamellar liposomes appeared to fuse to form larger vesicles containing a few inverted lipid micelles. This low concentration effect is triggered with increasing efficiency by monovalent, divalent and trivalent cations. Valence dependencies of this type has been reported (15) for processes involving the electrostatic shielding of negative charged surfaces in aqueous systems. Whether in this case, electrostatic shielding of the negative charges on the acidic lipids or direct binding of the cations occurs is not established.

The aggregation of the lipids induced by higher concentrations of polyvalent but not monovalent cations appears to reflect a different process (Fig. 4). On increasing the Mg^{2+} concentration to greater than 10 mM, much more extensive aggregation took place. Large lipid aggregates containing extensive arrays of inverted micelles are formed (Fig. 4). Structures of this type are not observed on the addition of higher concentrations of monovalent cations but substitution of Mg^{2+} by Ca^{2+} led to a similar effect. The effect of different cations on the total polar lipid extract of thylakoids has been examined in detail (16).

Even in the presence of cations the formation of non-bilayer structures in aqueous dispersions of total polar lipid extracts of thylakoid membranes is influenced by another main factor namely the degree of unsaturation of the fatty acyl chains of the lipids, particularly those of monogalactosyldiacylglycerol. Thus, when this latter lipid is subjected to catalytic hydrogenation of the fatty acyl chains double bonds is incapable of adopting non-bilayer configurations. Typical electronmicrographs of freeze-fracture replicas obtained from aqueous dispersions of MGDG at different levels of saturation are shown in Fig. 5. The native lipid, with an average number of bonds per lipid molecules of about 5.5 shows a characteristic hexagonal type II configuration (Fig. 5a). Samples with an average number of double bonds per lipid molecule below about 4.5 showed bilayer structures (Fig. 5c). It should be noted that the bilayers formed by saturated MGDG are in the form of open sheets and there is no evidence of the formation of closed structures typical of conventional liposomes. In Fig. 5b the direct juxtaposition of bilayer and non-bilayer structures emphasises the dynamic equilibrium existing between the two types of phases. The polymorphic behaviour of MGDG in response to fatty acyl saturation and other factors influencing the structures has been studied in some detail (17).

Despite the fact that non-bilayer configurations are readily formed when thylakoid membrane lipids are dispersed in aqueous media of the type employed in functional studies of the membrane, equivalent structures are not observed in thylakoids. It appears, however, that any treatment which would tend to destabilise lipid-protein interactions in thylakoid membranes or indeed other membrane systems, is likely to induce phase separations of non-bilayer lipids (18,19,20). The absence of these structures in the native membranes implies that other membrane components are capable of suppressing the tendency of the non-bilayer lipids to adopt such configurations. This in turn could suggest that non-bilayer lipids are involved in certain types of interactions which prevent them from the above type of arrangement. In such a case, MGDG must play an important role in the macro molecular organisation of the thylakoid membrane. We have examined the interactions of the thylakoid lipids with the CF_0-CF_1 ATP synthase complex and some of the findings are reported below.

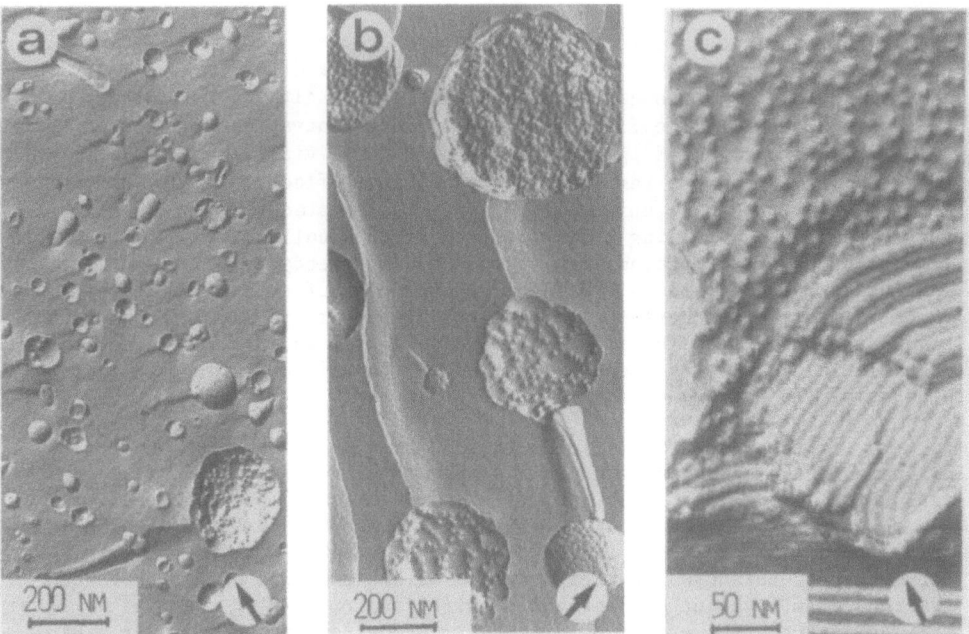

Fig. 4. Electromicrographs of freeze-fracture replicas obtained from dispersions of total polar lipid extract in the presence of $MgCl_2$ (a) 1 mM $MgCl_2$; (b) 7.5 mM $MgCl_2$ and (c) 30 mM $MgCl_2$.

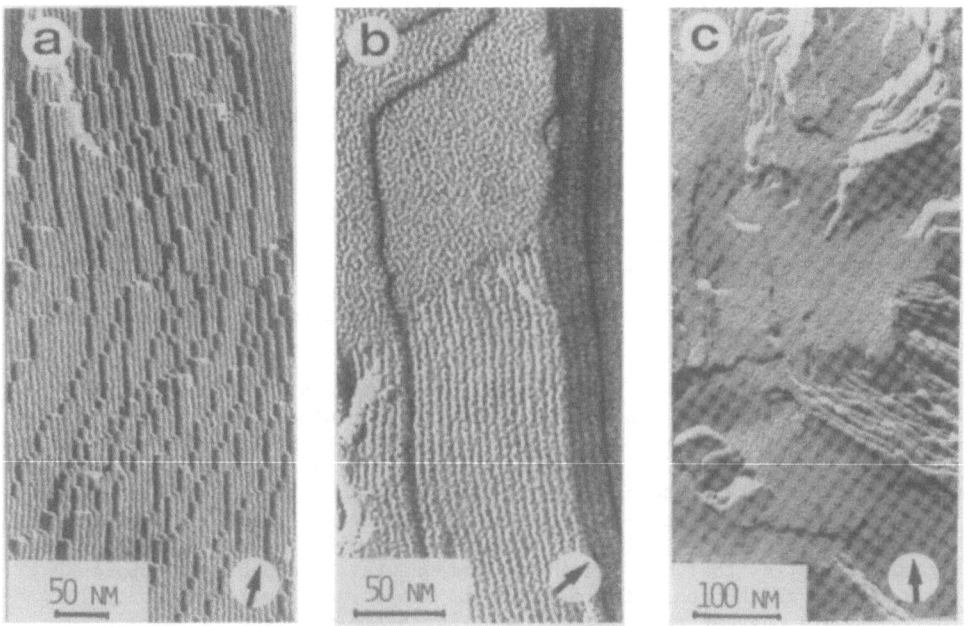

Fig. 5. Electronmicrographs of freeze-fracture replicas obtained from monogalactosyldiacylglycerol aqueous dispersions. The average number of double bonds per lipid molecule are about (a) 5.5; (b) 5.1 and (c) 3.9.

THE EFFECT OF CHLOROPLAST LIPIDS ON THE CATALYTIC PROPERTIES OF CF_0–CF_1 ATP SYNTHASE

It has been long known that the ability of purified CF_0–CF_1 ATP synthase to catalyse P_i–ATP exchange depends on the formation of tight vesicles, impermeable to protons (21). It had also been shown that the presence of soybean phospholipids during reconstitution of the enzyme stimulates both the P_i-ATPase exchange and the Mg-ATPase activity of CF_0–CF_1 (22). This could suggest that phospholipids are required for the activation of the enzyme. In order to examine whether thylakoid membrane lipids have a specific effect on CF_0–CF_1 we have compared the effects of these lipids with that of soybean phospholipids on both the activation of Mg-ATPase and of P_i–ATP exchange (Fig. 6). It was demonstrated that CF_0–CF_1 proteoliposomes prepared from thylakoid lipids have a much higher Mg-ATPase activity than the soybean phospholipid proteoliposomes. In addition, much lower concentrations of thylakoid lipids were required in order to achieve an optimal reconstitution. The P_i-ATP exchange rate however was lower with thylakoid lipids. We have previously demonstrated (22) that the permeability of thylakoid lipids to protons is higher than that of soybean phospholipids and indeed this can account for the lower rate of P_i-ATP exchange in thylakoid lipids CF_0–CF_1 proteoliposomes.

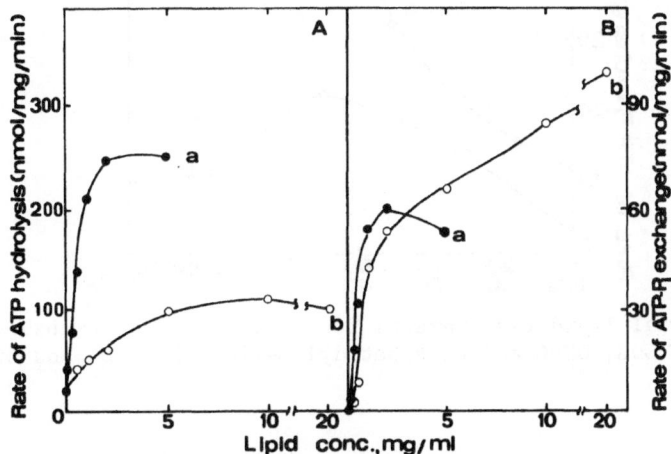

Fig. 6. The effect of thylakoid lipids on ATP hydrolysis (A) and ATP-P_i exchange (B) of CF_0–CF_1 ATP synthase. The optimal concentration of thylakoid lipids (a) for maximal reconstitution is compared with that of soybean phospholipids (b).

In order to examine how the effect of the thylakoid lipids is expressed in the kinetics of ATP hydrolysis and P_i-ATP exchange, the ATP concentration dependence of both reactions in soybean and thylakoid lipid proteoliposomes has been compared. It was observed that thylakoid lipids decrease the K_m (ATP) and increase the V_{max} when compared to soybean proteoliposomes (see ref. 22). This result suggests that interactions of specific thylakoid lipids with the enzyme activate it by increasing its turnover and its affinity for ATP.

To investigate any specific effect of thylakoid lipid classes different lipids and lipid mixtures were examined in their capability of activating the enzyme. It was observed that in the absence of MGDG, there was a sharp decrease in both P_i-ATP exchange and Mg-ATPase activities. Re-addition of MGDG to the lipid mixtures restored both activities. The two galactolipids, MGDG and DGDG, constitute about 75% of the total thylakoid lipid content. We thus examined the effect of these two lipids at varying concentrations on the activity of the enzyme (Fig. 7). It was found that optimal rates are observed at a ratio of MGDG to DGDG of about 70:30 which is slightly above the naturally occurring ratio in the thylakoids. Above this ratio we observed a decline in the rates. This probably reflects phase separation of MGDG and concomitant appearance of the hexagonal type II phase, a situation which would decrease the effectiveness of reconstitution. MGDG could not be substituted for by any other lipid. In addition, hydrogenated MGDG, in which the fatty acid chains have been fully reduced, was found to be ineffective in the activation of P_i-ATP exchange. It therefore appears that this lipid class is required for the activation of the enzyme and furthermore the activation also depends on the high degree of unsaturation of its fatty acyl chains.

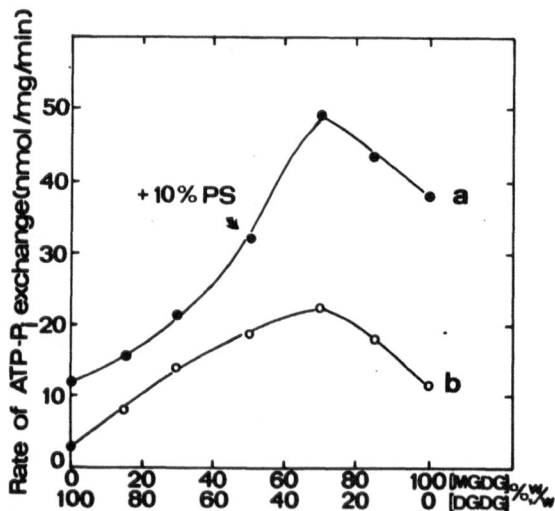

Fig. 7. Optimal lipid requirements for maximal ATP-P_i exchange rates. (a) MGDG, DGDG and phosphatidyl serine, (b) MGDG and DGDG.

An important observation was that the rate of P_i-ATP exchange was highly stimulated when 10% of a charged lipid was introduced to the galactolipid mixture. As shown in Fig. 7 addition of phosphatidyl serine enhanced the rates significantly. The same effect was observed with the charged thylakoid lipids, SQDG and PG, pointing to an additional requirement by the enzyme of negatively charged lipids.

In order to examine whether the non-polar part of the lipid molecules are of importance in the reconstitution of the enzyme we carried out partial catalytic hydrogenation on total thylakoid lipid extract (Fig. 8a). Saturation of the lipids' double bonds resulted in a fall of the enzymatic activity. A much sharper decrease was observed (Fig. 8b) when the fatty acid residues of MGDG only in the lipid mixture were saturated.

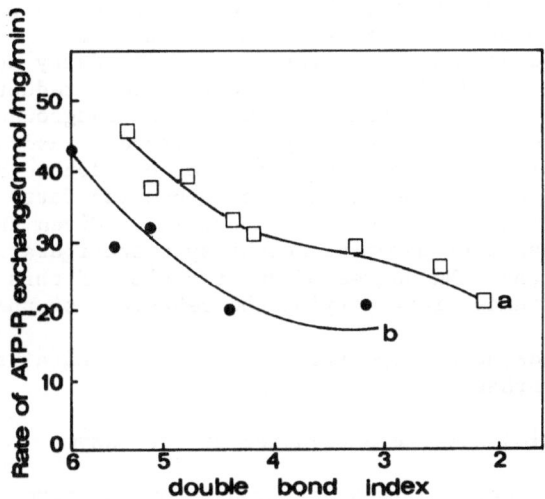

Fig. 8. The dependency of the P_i-ATP exchange rates on the degree of unsaturation of the fatty acid chains. (a) Total thylakoid lipids, (b) MGDG, DGDG and phosphatidylserine (4.5:4.5:1). In (b) only MGDG was catalytically saturated.

SPECIFIC LIPID ASSOCIATIONS OF THE CF_0-CF_1 ATP SYNTHASE

The purification of the CF_0-CF_1 from thylakoid membranes involves an extensive delipidation of the enzyme. We investigated the nature of this isolated protein complex after purification in order to examine whether it specifically associates with lipids which are essential for its activity. We observed that 0.2% of the total lipids of the membrane co-purify with the enzyme. As shown in Table 1, residual lipid associated with the CF_0-CF_1 ATP synthase is composed mainly of the SQDG (23). Prolonged dialysis in the absence of presence of different detergents, or additional purification steps on detergent containing sucrose gradients, resulted in a small decrease of bound sulpholipid and a parallel decrease in ATPase activity. This data suggests a strong association between SQDG and CF_0-CF_1.

TABLE 1 Analysis of lipid composition of spinach CF_0-CF_1 preparations

Preparations	Lipid/Protein(%)	Relative Amount of Lipid			(mol%)
		MGDG	DGDG	SQ	PG
Thylakoid membranes	50	50	25	8	13.5
Crude CF_0-CF_1	8	40.5	13	28	17
Purified CF_0-CF_1	1	–	10(0.5)*	90(5)*	–

*Calculated values of mol lipid/mol CF_0-CF_1 by assuming a MW of 5×10^5 for CF_0-CF_1.

CONCLUSIONS

In this report we showed that the lipids of the thylakoid membrane may play other roles beyond the mere provision of a matrix in the system. The results presented here showed that both the polar and hydrophobic parts of the lipid molecules are of importance when considering a functional membrane organisation. They also point to the conclusion that the protein complexes may be involved in specific lipid interactions rather than simply satisfying a lipid solvation requirement.

In summary we reported that:

(a) The conformation adopted by thylakoid lipids in aqueous environments depend on the presence of MGDG in the lipid mixture, the presence of cations and the degree of unsaturation of the fatty acid chains. The prime function of added cations appears to be a reduction of elecrostatic repulsion between neighbouring lipid headgroups which tends to favour the formation of non-bilayer structures. Hydrogenation of the fatty acyl chains decreases the number of <u>cis</u> double bonds thus allowing increased interactions between neighbouring fatty acid chains and the stabilisation of bilayer configurations. Given that MGDG is the only lipid capable of adopting non-bilayer configurations it is hardly surprising that the degree of unsaturation of this lipid class is of great importance, in typifying the behaviour of the mixture.

(b) Thylakoid lipids have a specific effect on the catalytic activity of CF_O-CF_1 ATP synthase

(c) MGDG is essential for the activation of the enzyme

(d) The lipids are required in a highly unsaturated form for maximal reconstitution

(e) Acidic lipids are required for the optimal reconstitution of CF_O-CF_1

(f) There is a strong association of SQDG with the CF_O-CF_1 ATP synthase enzyme.

We further suggest that MGDG by virtue of its ability to form structures other than bilayer could make the packaging of the protein complex in the membrane possible through interactions other than those normally envisaged in the fluid mosaic model. In addition we believe that MGDG is required in its highly unsaturated form not simply to maintain the appropriate fluidity conditions but also to retain its capability of non-bilayer configurations.

In conclusion, it is interesting to consider that cardiolipin, the major acidic lipid in mitochondria inner membranes, and capable of adopting non-bilayer configurations, is required for the catalytic activation of cytochrome oxidase (24) and F_O-F_1 ATP synthase (25).

ACKNOWLEDGEMENTS

The work on the CF_O-CF_1 synthase was carried out in collaboration with Dr. Uri Pick, Department of Biochemistry, the Weizmann Institute, Israel. Financial support for the work has come via an Agriculture and Food Research Grant to Professor J. Barber.

REFERENCES

1. Warren, G.B., Houslay, M.D., Metcalfe, J.C. and Birdsall, N.J.M. (1975) Cholesterol is excluded from the phospholipid annulus surrounding an active calcium transport protein. Nature 255, 684-687

2. Cronan, J.E. and Gelmann, E.P. (1975) Physical properties of membrane

lipids: Biological relevance and regulation. Bacteriol. Rev. 39, 232-256

3. Taraschi, T.F., de Kruijff, B., Verkleij,A.J. and van Echteld, C.J.A. (1982) Effect of glycophorin on lipid polymorphism. A ^{31}P-NMR study. Biochim. Biophys. Acta 685, 153-161

4. Israelachvili, J.N., Marcelja, S. and Horn, R.G. (1980) Physical principles of membrane organisation, Q. Rev. Biophys. 13, 121-200

5. Luzzati, V. and Husson, F. (1962) The structure of the liquid-crystalline phases of lipid-water systems. J. Cell Biol. 12, 207-219

6. Cullis. P.R. and de Kruijff, B. (1978) The polymorphic phase behaviour of phosphatidylethanolamines of natural and synthetic origin. A ^{31}P-NMR study Biochim. Biophys. Acta 513, 31-42

7. Cullis, P.R., Verkleij, A.J. and Vernergaert, P.H.J.Th. (1978) Polymorphic phase behaviour of cardiolipin as detected by ^{31}P-NMR and freeze-fracture techniques. Effects of calcium, dibucaine and chlorpromazine. Biochim. Biophys. Acta 513, 11-20

8. Wieslander, A. Ulmins, J., Lindblom, G. and Fontel, K. (1978) Water binding and phase structures for different Acholeplasma Laidlawii membrane lipids studied by deuteron nuclear magnetic resonance and X-ray diffraction. Biochim. Biophys. Acta 512, 241-253

9. Deamer, D.W., Leonard, R., Tardieu, A. and Branton, D. (1970) Lamellar and hexagonal lipid phases visualised by freeze-etching. Biochim. Biophys. Acta 219, 47-60

10. Papahadjopoulos, D., Vail, W.J., Pangborn, W.A. and Poste, G. (1976) Studies on membrane fusion. Induction of fusion in pure phospholipid membranes by calcium ions and other divalent metals. Biochim. Biophys. Acta 448, 265-283

11. Vail, W.J. and Stollery, J.G. (1979) Phase changes of cardiolipin vesicles mediated by divalent cations. Biochim. Biophys. Acta 551, 74-78

12. Sen, A., Williams, W.P., Brain, A.P.R., Dickens, M.J. and Quinn, P.J. (1981) Formation of inverted micelles in dispersions of mixed galactolipids. Nature 293, 488-490

13. Verkleij, A.J., Mombers, C., Leunissen-Bijrelt, L. and Ververgaert, P.J.J.Th. (1979) Lipidic intramembranous particles. Nature, 279, 162-163

14. Sen, A., Brain, A.P.R., Quinn, P.J. and Williams, W.P. (1982) Formation of inverted lipid micelles in aqueous dispersions of mixed sn-3-galactosyldiacylglycerols induced by heat and ethylene glycol. Biochim. Biophys. Acta 686, 215-224

15. Barber, J. (1980) Membrane surface charges and potentials in relation to photosynthesis. Biochim. Biophys. Acta 594, 253-308

16. Gounaris, K., Sen, A., Brain, A.P.R., Quinn, P.J. and Williams, W.P. (1983) The formation of non-bilayer structures in total polar lipid extracts of chloroplast membranes. Biochim. Biophys. Acta 728, 129-139

17. Gounaris, K., Mannock, D.A., Sen, A., Brain, A.P.R., Williams, W.P. and Quinn, P.J. (1983) Polyunsaturated fatty acyl residues of galactolipids are involved in the control of bilayer/non-bilayer lipid transitions in higher plant chloroplasts. Biochim. Biophys. 732, 229-242

18. Gounaris, K., Brain, A.P.R., Quinn, P.J. and Williams, W.P. (1984) Structural reorganisation of chloroplast thylakoid membranes in response to heat-stress. Biochim. Biophys. Acta 766, 198-208

19. Crow, L.M. and Crowe, J.H. (1982) Hydration-dependent hexagonal phase lipid in a biological membrane. Arch. Biochem. Biophys. 217, 582-587

20. van Venetie, R. and Verkleij, A.J. (1982) Possible role of non-bilayer lipids in the structure of mitochondria. A freeze-fracture electron microscopy study. Biochim. Biophys. Acta 692, 397-405

21. Winget, G.D., Kanner, N. and Racker, E. (1977) Formation of ATP by the adenosine triphosphatase complex from spinach chloroplasts reconstituted together with bacteriorhodopsin. Biochim. Biophys. Acta 460, 490-499

22. Pick, U., Gounaris, K., Admon, A. and Barber, J. (1984) Activation of the CF_O-CF_1 ATP synthease from spinach chloroplasts by chloroplast lipids. Biochim. Biophys. Acta 765, 12-20

23. Pick, U., Gounaris, K., Weiss, M. and Barber, J. (1985) Tightly bound sulpholipids in chloroplast CF_O-CF_1. Biochim. Biophys. Acta 808, 415-420

24. Robinson, N.C. (1982) Specificity and binding affinity of phospholipids to the high-affinity cardiolipin sites of beef heart cytochrome c oxidase. Biochemistry. 21, 184-188

25. Kagawa, Y., Kandrach, A. and Racker, E. (1973) Partial resolution of the enzymes catalyzing oxidative phosphorylation. Specificity of phospholipids required for energy transfer reactions. J. Biol. Chem. 248, 676-684

ACTION OF BICARBONATE ON PHOTOSYNTHETIC ELECTRON TRANSPORT IN THE PRESENCE

OR ABSENCE OF INHIBITORY ANIONS

Julian J. Eaton-Rye, Danny J. Blubaugh, and Govindjee

Department of Plant Biology
289 Morrill Hall, 505 S. Goodwin Ave.
Urbana, IL 61801 (U.S.A.)

INTRODUCTION

Bicarbonate (or CO_2) was shown by Warburg and Krippahl [1] to stimulate electron transport during the Hill reaction. This phenomenon has been referred to as the bicarbonate (HCO_3^-) effect. The electron transport chain can be dissected into a number of clearly defined partial reactions through the addition of specific inhibitors and electron donors and acceptors. By applying this approach (see $\underline{e.g.}$, [2,3]) the HCO_3^- effect has been shown to be associated with the acceptor side of Photosystem II (PS II):

$$H_2O \longrightarrow OEC \longrightarrow Z \longrightarrow P680 \longrightarrow Pheo \longrightarrow Q_A \longrightarrow Q_B \longrightarrow PQ \qquad (1)$$

The electron donor side of PS II contains the oxygen-evolving complex (OEC), and a bound plastoquinol "Z" that supplies electrons to the reaction center chlorophyll (Chl) $\underline{a}$ P680. The electron acceptor side contains a pheophytin (Pheo) molecule and two plastoquinones, Q_A and Q_B. Electrons are transferred from $Pheo^-$ to Q_A, which can only be reduced to the semiquinone form. After two such events, Q_B is reduced to plastoquinol. The plastoquinol, Q_BH_2, is then able to exchange with the plastoquinone (PQ) pool to provide a second Q_B molecule for subsequent reduction. This two-electron transfer step at the Q_B level is known as the two-electron gate (see $\underline{e.g.}$, [4]). Stoichiometrically, two PQH_2 molecules are formed and two water molecules oxidized for each O_2 evolved. The resultant PQH_2 molecules are oxidized at the plastoquinol-plastocyanin oxidoreductase or cytochrome b_6/f complex. The oxidation of H_2O and PQH_2 supply protons to the internal thylakoid space for chemiosmotic coupling.

The oxygen-evolving mechanism has been described by a kinetic model which recognizes 5 separate oxidation states or S-states [5]:

$$S_0 \Longrightarrow S_0{'} \longrightarrow S_1 \Longrightarrow S_1{'} \longrightarrow S_2 \Longrightarrow S_2{'} \longrightarrow S_3 \Longrightarrow S_3{'} \longrightarrow S_4 \underset{O_2}{\longrightarrow} S_0 \qquad (2)$$

A single photoact advances an S-state from S_n to $S_n{'}$ while the transition from $S_n{'}$ to S_{n+1} represents a recovery reaction before a center is able to utilize a second photon. Molecular O_2 is released during the S_4 to S_0 transition. For a recent review of PS II the reader is referred to [6] and of the OEC to [7].

The Bicarbonate Effect

Although it has been suggested [8,9] that HCO_3^- plays a role on the water oxidation side of PS II, as yet no firm evidence is available to show a significant, large, and direct effect of HCO_3^- before P680. On the other hand, a role of HCO_3^- on the PQ reduction side of PS II, at least in the presence of HCO_2^-, has been firmly established (see reviews [10-13]).

Fig. 1 shows the scheme of electron flow from Q_A^- to the PQ pool via the two electron acceptor Q_B; also shown are the steps that are affected by HCO_2^-/HCO_3^-: the major effect of HCO_3^-, at least in the presence of HCO_2^-, is to facilitate the passage of electrons through the two-electron gate. The specific effect may involve mediation in the protonation reactions. The principal arguments to support the action of HCO_3^-, shown in Fig. 1, are: (1) the reoxidation of Q_A^-, as measured by the Chl a fluorescence yield decay [15,16] or by the absorbance change at 320 nm [17,18], is stimulated ten- to twenty-fold by HCO_3^- in HCO_3^--depleted membranes; (2) in PS II particles, a light- and chemically-induced EPR signal (g = 1.82), attributed to the Q_A^--Fe^{2+} complex, is reversibly increased in amplitude by a factor of about 10 by HCO_3^--depletion [19]; (3) The Chl a fluorescence yield [20] and thermoluminescence [21] after a series of single flashes of light suggest a dramatic slowing down of electron flow after the third and subsequent flashes following HCO_3^--depletion, which is totally reversed upon HCO_3^--addition; this suggests that there is a drastic HCO_3^- effect on the exchange of Q_B^{2-} with PQ at the Q_B-apoprotein and (4) HCO_3^--depletion causes a several-fold change in the affinity of the binding of ^{14}C-atrazine, which binds in the Q_B-apoprotein region [22; cf. 23]. In contrast, attempts to observe a HCO_3^- effect on (a) the OEC to Z^+ electron transfer kinetics, EPR signal II_{vf} [15]; (b) Z to P680$^+$ electron flow by the kinetics of the fast Chl a fluorescence rise [15]; (c) the H_2O to Q_A reaction with ferricyanide as an electron acceptor in trypsin-treated thylakoids [24]; and (d) the kinetics of O_2 evolution after the third flash [12], have failed thus far.

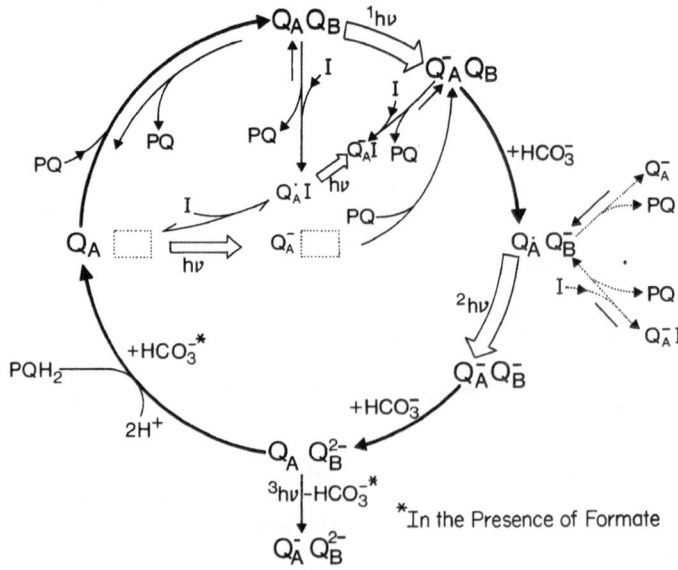

Fig. 1.   A Scheme for the Effect of Bicarbonate in the Presence of Formate on the Electron Transfer on the Acceptor Side of Photosystem II. In this scheme, I stands for Inhibitor/herbicide, Q_A for the first bound-quinone electron acceptor, Q_B for the second bound-quinone electron acceptor, PQ for the plastoquinone pool, ^{n}hv for the nth flash. (The scheme is self-explanatory, and is modified after the scheme presented earlier [13,14].)

Recently it has been reported that HCO_3^- is able to reverse the inhibition of PS II by a number of monovalent anions [25]. In addition to reversing the inhibition by formate, HCO_3^- was also effective when NO_2^-, NO_3^-, F^- and acetate were used as inhibitory anions. This work confirmed and extended the earlier findings of Good [26] by including NO_2^- and supports the notion that HCO_3^- is binding to a more general anion binding site [27]. All of these anions exhibit competitive inhibition of HCO_3^- binding to the anion binding site [25,27]. The binding constant of HCO_3^- has been determined and is approximately 0.08 mM [27]. Although it has been suggested [27-29] that some electron transport proceeds when the anion binding site is empty, only HCO_3^- facilitates electron transfer through the two-electron gate.

THE ACTIVE SPECIES

To determine whether CO_2 or HCO_3^- is the species required, advantage was taken of the pH dependence of the ratio $[CO_2]/[HCO_3^-]$ at equilibrium. This ratio increases with lower pH, because an increase in $[H^+]$ drives the following reaction toward the left:

$$CO_2 + H_2O \underset{K_1}{\overset{}{\rightleftharpoons}} H_2CO_3 \underset{K_2}{\overset{}{\rightleftharpoons}} H^+ + HCO_3^- \underset{K_3}{\overset{}{\rightleftharpoons}} 2H^+ + CO_3^{2-} \qquad (3)$$

K_1, K_2, and K_3 are known equilibrium constants ($K_1 = (1.4 \pm 0.2) \times 10^{-3}$; $K_2 = (3.2 \pm 0.4) \times 10^{-4}M$; $K_3 = 4.70 \times 10^{-11}M$ [28]), from which the relative ratios of reactants to products can be calculated. When $[H^+]$ and the total concentration of all carbonate species are known, then $[CO_2]$ and $[HCO_3^-]$ at equilibrium can be calculated:

$$[HCO_3^-]_{eq} = \frac{[HCO_3^-]_i}{\dfrac{[H^+]}{K_1 \cdot K_2} + \dfrac{[H^+]}{K_2} + 1 + \dfrac{K_3}{[H^+]}} \qquad (4)$$

$$[CO_2]_{eq} = \frac{[H^+]}{K_1 \cdot K_2}[HCO_3^-]_{eq} \qquad (5)$$

The subscripts "eq" and "i" refer to equilibrium and initial conditions, respectively.

Spinach thylakoids were depleted of CO_2 to inhibit the Hill reaction, following a procedure similar to [31]. The percentage of control activity restored to these thylakoids was then measured as a function of the total $[HCO_3^-]$ that was added, and the experiment repeated at a variety of pHs. Figures 2 and 3 show the same data plotted against the equilibrium $[HCO_3^-]$ and against the equilibrium $[CO_2]$, respectively. When plotted against $[HCO_3^-]_{eq}$ there is no apparent pH dependence; even though the $[CO_2]/[HCO_3^-]$ ratio varies four-fold over the pH range, each curve falls on top of the others. This means that $[CO_2]_{eq}$ has no apparent effect on the degree of restoration. On the other hand, when plotted against $[CO_2]_{eq}$, the curves become steeper with increasing pH. From equation (5) it is obvious that the ratio $[HCO_3^-]_{eq}/[CO_2]_{eq}$ is constant at any given pH, but changes proportionately with any change in $[H^+]$. Therefore, at any given $[CO_2]_{eq}$, the $[HCO_3^-]_{eq}$ is greater at higher pH. Thus, the pH dependence of the curves in Fig. 3 and the lack of such dependence in Fig. 2 is evidence that HCO_3^-, not CO_2, is responsible for restoring the Hill reaction. The inset to Fig. 3 shows the effect of $[HCO_3^-]_{eq}$ on the Hill activity, with $[CO_2]_{eq}$ held constant. The $[HCO_3^-]_{eq}$ was calculated from equation (5), and the percent

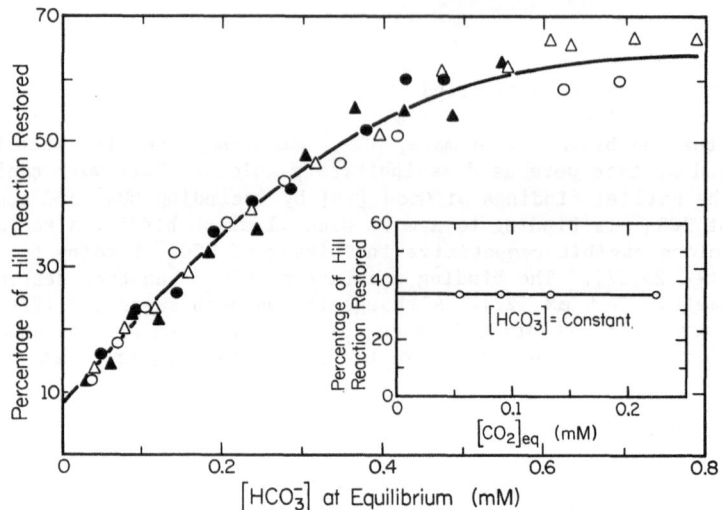

Fig. 2. The Percentage of Control Activity Restored to CO_2-depleted Spinach Thylakoids as a Function of the Equilibrium $[HCO_3^-]$. The Hill activity was measured by following the reduction of 2,6-dichlorophenolindophenol (DCPIP) as the decrease in absorbance at 600 nm. Illumination began 3 min after the addition of $NaHCO_3$. Symbols: closed circles, pH 6.30; closed triangles, pH 6.53; open circles, pH 6.7; open triangles, pH 6.91. Inset: The effect of $[CO_2]_{eq}$ on the Hill activity, with $[HCO_3^-]$ held constant at 0.2 mM, taken from the curves shown in the main portion of the figure (Blubaugh and Govindjee, 1984, unpublished)

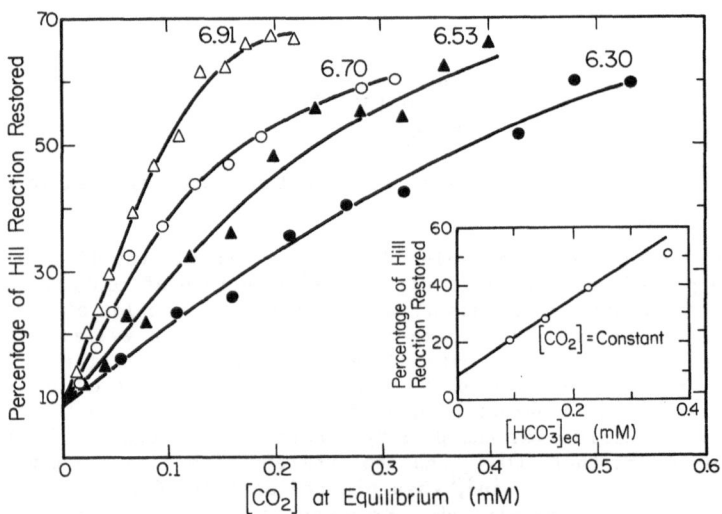

Fig. 3. The Percentage of Control Activity Restored to CO_2-depleted Spinach Thylakoids as a Function of the Equilibrium $[CO_2]$. The symbols and protocol are the same as in Fig. 2. Inset: The effect of $[HCO_3^-]_{eq}$ on the Hill activity, with $[CO_2]_{eq}$ held constant at 0.1 mM, taken from the curves in the main portion of the figure (Blubaugh and Govindjee, 1984, unpublished).

control activity was taken from the curves of Fig. 3 at the point on each curve where $[CO_2]_{eq}$ = 0.1 mM. Following the same method, the inset to Fig. 2 shows the effect of $[CO_2]_{eq}$ on the Hill activity with $[HCO_3^-]_{eq}$ held constant at 0.2 mM. It is clear that the effect of HCO_3^- is independent of $[CO_2]_{eq}$.

In the analysis thus far, no consideration has been given to the possibility of H_2CO_3 or CO_3^{2-} as the active species, primarily because both exist at extremely low concentrations in the system described here. However, both species can be ruled out from the data presented here. The $[H_2CO_3]_{eq}/[CO_2]_{eq}$ ratio is equal to K_1 (1.4×10^{-3}) and is independent of pH. Thus, a plot of activity versus $[H_2CO_3]_{eq}$ would qualitatively appear identical to the data in Fig. 3. Since $[H_2CO_3]_{eq}$ is directly proportional to $[CO_2]_{eq}$ at all pHs, identical arguments can be made for H_2CO_3 as were made for CO_2. Thus, H_2CO_3 cannot be the required species. CO_3^{2-} exists at only 10^{-3}-10^{-4} times the concentration of HCO_3^- at the pHs used here. The ratio of $[CO_3^{2-}]_{eq}$ to $[HCO_3^-]_{eq}$ is equal to $K_3/[H^+]$. Since this ratio is inversely proportional to $[H^+]$, the curves in Fig. 2 would be expected to show a pH dependence if CO_3^{2-} were involved in restoration of the Hill activity. As was the case with CO_2, the lack of such pH dependence in Fig. 2 suggests that CO_3^{2-} is not involved. Only if the system was not at equilibrium at the time of each measurement could the curves in Fig. 2 show no apparent pH dependence if CO_3^{2-} was involved. However, from a time course of the restoration after addition of a half-saturating $[HCO_3^-]$, we have concluded that equilibrium is reached in 2-2.5 min (data not shown). Since the measurements in Figs. 2 and 3 were made 3 min after HCO_3^- addition, CO_3^{2-} can be ruled out as having any involvement.

The possibility that CO_2 (with perhaps some contribution from H_2CO_3) is required for diffusion to the active site is not disputed by the data presented here. These measurements were made under equilibrium conditions and do not reflect the kinetics of bicarbonate binding. Some experiments, done under non-equilibrium conditions, indicate that CO_2 is required. For example, when HCO_3^- is added to HCO_3^--depleted thylakoids, a lag is observed before the Hill reaction is stimulated, whereas CO_2 stimulates the Hill reaction sooner [32,33]. Presumably the initial $[CO_2]$ determines how quickly bicarbonate is able to reach the binding site. Similarly, from the effect of pH on the rate of HCO_3^- binding, it was suggested that CO_2 is required [34]. While the equilibrium $[HCO_3^-]$ in the vicinity of the binding site is shown here to be a critical factor, the $[CO_2]$ could be an important kinetic consideration if the binding site is buried beneath a hydrophobic domain.

THE ANION BINDING SITE

Recently the possibility has been raised [27-29] that the stimulatory HCO_3^- effect is a simple reversal of the inhibitory HCO_2^- effect [26]. Snel et al. [28] and Snel and van Rensen [29] entertain the possibility that both the HCO_2^- and the HCO_3^- effects may be of physiological significance, since formate is produced in peroxisomes. The binding constant for HCO_3^-, calculated by these authors, is about 0.08 mM, whereas the $[CO_2]$ in photosynthesizing chloroplasts is estimated to be only 0.005 mM [35]. Thus, one may argue that under normal conditions, all binding sites are empty, and there may not be any real role for HCO_3^- in vivo. However, since it is $[HCO_3^-]$ and not $[CO_2]$ which determines the degree of functioning, there is no good reason to assume that the sites are empty. For example, if, in the vicinity of the binding site, the pH is 8, then the $[HCO_3^-]$ in equilibrium with 0.005 mM CO_2 is 0.22 mM, well above the binding constant. The location of the HCO_3^- binding site is, therefore, an important question to answer.

It is also possible that the binding constant has been overestimated, because of the salt concentration at which it was determined. This possibility is discussed later. The key question of whether HCO_3^- is required for electron flow in a normal system needs to be tested. Experiments must be done with totally HCO_3^--depleted samples, with the HCO_3^- concentration measured, since a tightly bound HCO_3^- may have been present in many experiments. Furthermore, experiments should be done to see if the HCO_3^- effect can be obtained without the use of HCO_2^-; this is a definite way to test if HCO_3^- stimulation is a reversal of the HCO_2^- effect.

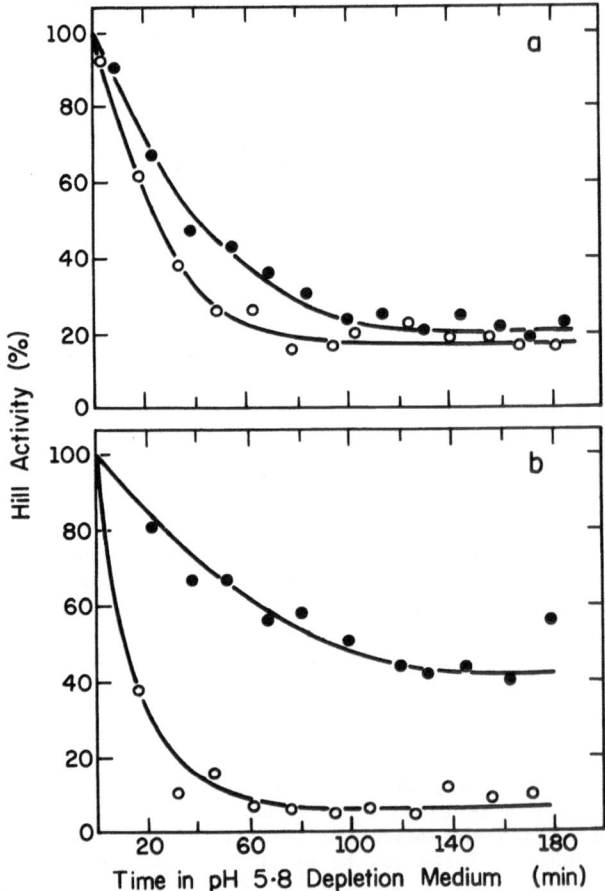

Fig. 4. The Percentage of Hill Activity Remaining after Incubation in a CO_2-free HCO_3^--depletion Medium at pH 5.8. The depletion medium contained 300 mM sorbitol, 10 mM NaCl, 5 mM $MgCl_2$ and 10 mM sodium phosphate (pH 5.8). The Hill activity was assayed in a reaction medium containing 300 mM sorbitol, 10 mM NaCl, 5 mM $MgCl_2$ and 25 mM sodium phosphate (pH 6.5). 25 mM HCO_2^- was also present in both the depletion and reaction media in (b). The spinach thylakoids used in this experiment had been stored in liquid N_2. The electron acceptor was methyl viologen. For other experimental details, see [3]. In (a) 100% = 896 uequiv./mg Chl/hr and in (b) 100% = 584 µequiv./mg Chl/hr (Eaton-Rye and Govindjee, 1985, unpublished)

The Bicarbonate Effect in the Absence of Formate

We have already shown [3] the existence of a stimulatory effect of HCO_3^- on Q_A^- to Q_B electron flow, when HCO_2^- was omitted from both the depletion medium and the reaction medium. This determination was made by following the decay of the Chl $\underline{a}$ variable fluorescence. As was observed in formate-containing membranes [16,20], maximal inhibition occurred after the third actinic flash. However, in the absence of formate, the extent of this inhibition was decreased ($t_{1/2}$ for HCO_3^--depleted in the presence of HCO_2^-, ~ 15 ms; for HCO_3^--depleted in the absence of HCO_2^-, ~ 2.5 ms; for HCO_3^--resupplied and control, ~ 0.35 ms.). In the formate-free case we found that after approximately a 60 min depletion the half-time of Q_A^- reoxidation did continue to increase but the effect became increasingly irreversible. Fig. 4 demonstrates a similar observation in the Hill reaction. Thus it appears, under these conditions, that if the anion binding site is empty, irreversible

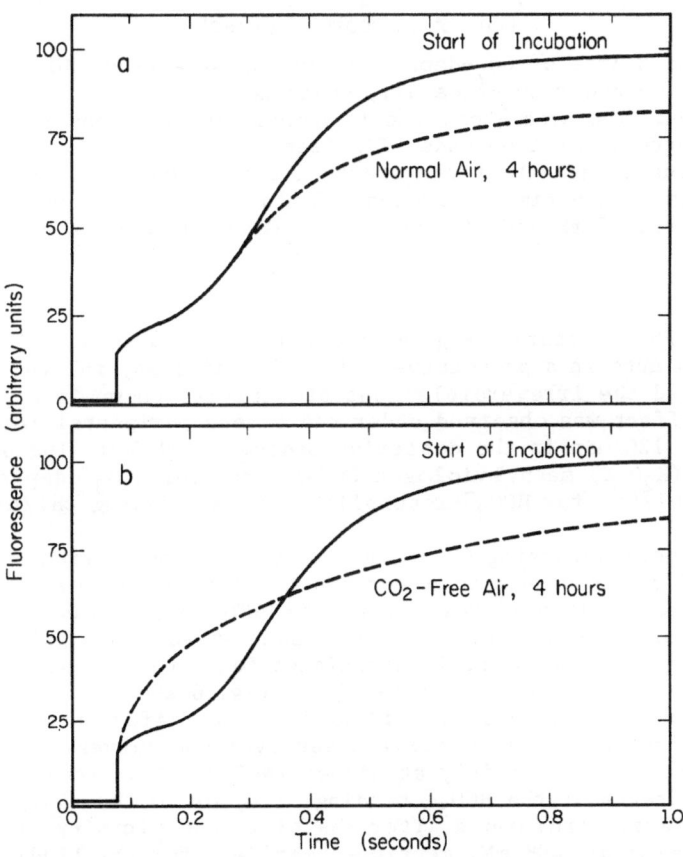

Fig. 5. The Effect of CO_2-depletion, in the Absence of HCO_2^-, on the Variable Chl $\underline{a}$ Fluorescence Transient. (a) The fluorescence transient at the start of incubation and after 4 hr of incubation under a stream of normal air. (b) The fluorescence transient at the start of incubation and after 4 hr of incubation under a stream of CO_2-free air. The area above the curve for the CO_2-depleted sample is approximately 60% of that for the non-CO_2-depleted sample. Spinach thylakoids were suspended in 50 mM phosphate, pH 6.4, 15 mM NaCl, and 5 mM $MgCl_2$ (Blubaugh and Govindjee, 1985, unpublished).

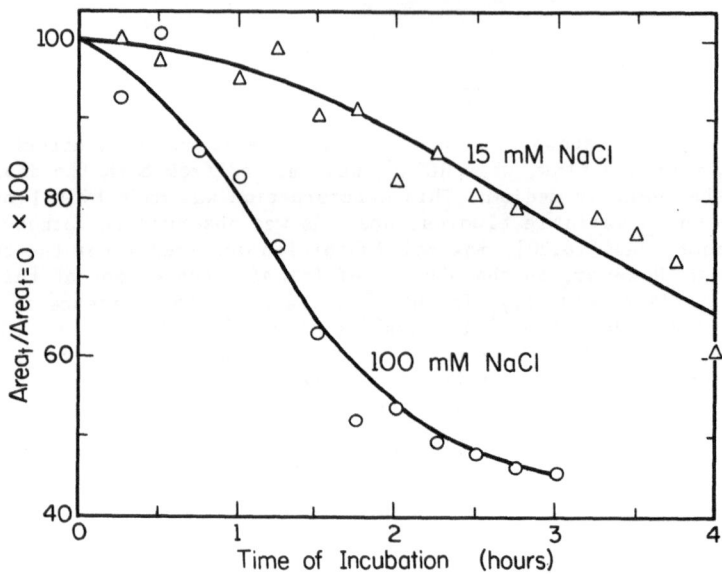

Fig. 6. The Time Course of CO_2-depletion in the Absence of HCO_2^-, by Incubation under CO_2-free air, under Low and High Salt Concentrations. The area above the fluorescence transient is plotted as a function of time under CO_2-free air. Area $_{t = 0}$ is the area above the transient at the start of the incubation, and Area$_t$ is the area at the time indicated. Symbols: circles, 100 mM NaCl; triangles, 15 mM NaCl (Blubaugh and Govindjee, 1985, unpublished).

inactivation of PS II occurs. Fig. 4b shows that the presence of 25 mM formate actually acts in a protective role. In addition, raising the pH to 6.5 also prevented the irreversible loss of activity shown in Fig. 4a. A two-fold HCO_3^- effect was observed under the same experimental conditions as in Fig. 4a after 120 min in the depletion medium at pH 6.5. The electron transport rates (H_2O to methylviologen (MV)) were: for HCO_3^--depleted, 162 µequiv./mg Chl/hr; for HCO_3^--resupplied, 368 µequiv./mg Chl/hr.

Similarly, merely passing CO_2-free air over a formate-free sample of thylakoids at pH 6.4 is enough to alter the kinetics of the variable Chl _a_ fluorescence (Fig. 5b) in a manner similar to HCO_3^--depletion. Passing normal air over an identical sample had no such effect (Fig. 5a). The level of maximum fluorescence was equally diminished for both samples; this effect was probably due to aging of the thylakoids. Fig. 6 shows the time course of this effect for spinach thylakoids suspended at two different salt concentrations, determined from the relative areas over the curves. It is apparent that HCO_3^- is removed more readily at higher [NaCl]. This could be due to ionic strength effects on the HCO_3^- binding site, or to the charge density at the membrane surface, which would alter the pH in the vicinity of the membrane. The value of 0.08 mM, mentioned earlier, for the binding constant of HCO_3^- had been determined at a salt concentration of 200 mM [25]. Since at lower [NaCl] the HCO_3^- appears to be held more tightly, the binding constant may be much lower at physiological ionic strength.

Formate has been shown to be competitive with the HCO_3^- binding site [16,27,36]. This conclusion can now be extended to a number of anions which inhibit the Hill reaction and which also increase the dissociation constant for HCO_3^-. The most effective anion that has been tested is nitrite (NO_2^-) [25]. We have repeated the experiment shown in Fig. 4b, but we replaced the

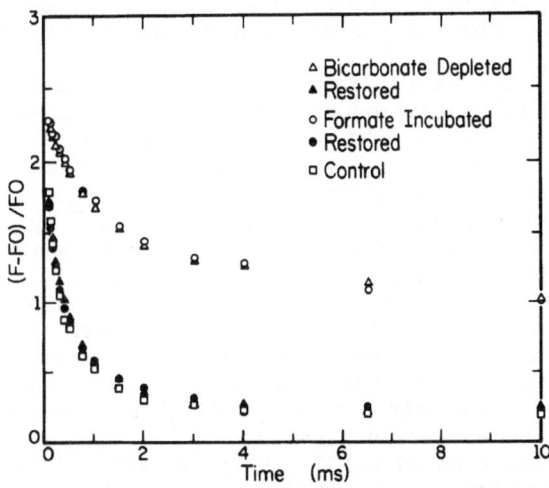

Fig. 7. The Decay of the Variable Chl a Fluorescence after Three Actinic
Flashes Spaced at 1 s. FO is the Chl a fluorescencae yield from
the measuring flash when all Q_A was oxidized and F was the yield
at the indicated time after the actinic flash. The reaction
medium (see Fig. 4) was supplemented with 25 mM HCO_2^-, 0.1 mM
methyl viologen and 0.1 mM gramicidin. Pea thylakoids were used
and the half-times were determined as in [16] (see text). The
figure is redrawn from [16].

25 mM HCO_2^- with 25 mM NO_2^-. Our results indicate that nitrite behaves in a
similar way to formate and prevents the irreversible loss of activity shown
in Fig. 4a. A 7-fold HCO_3^- effect was observed in the presence of 25 mM NO_2^-
after 30 min in the depletion medium at pH 5.8. The electron transport rates
(H_2O to MV) were: for HCO_3^--depleted, 75 μequiv./mg Chl/hr; for HCO_3^--
resupplied, 539 μequiv./mg/ Chl/hr.

The Relationship Between the Anion Binding Site and the Acceptor Side Bicarbonate Effect

Data are presented in Fig. 7 from an earlier study [16], in which the
effect of HCO_3^--depletion in the presence of HCO_2^- and the effect of HCO_2^-
incubation in the presence of atmospheric CO_2 were compared. The decay of
the variable Chl a fluorescence after one or three actinic flashes was nearly
identical with either HCO_2^--incubated or HCO_3^--depleted samples. With both
preparations, the addition of 10 mM HCO_3^- restored the decays to the pattern
seen with control thylakoids. Twenty-five mM HCO_2^- was present in these
experiments. A similar observation has been made for the Hill reaction [37],
but in this instance 100 mM HCO_2^- was required to observe a HCO_3^--reversible
inhibition in HCO_2^--incubated samples [cf. 16]. The results in Fig. 7
suggest that HCO_2^- and HCO_3^- compete for a site on the reducing side of
PS II. It is evident that if HCO_2^- is present at a low HCO_3^- concentration,
forward electron flow is slowed; if HCO_3^- is present in excess, or at
physiological concentrations in the absence of formate, the normal flow is
restored. It was also reported in [16] that during HCO_2^--incubation the
transition from a fast to a slow decay of the variable Chl a fluorescence was
accelerated under any of the following conditions: a) the concentration of

HCO_2^- was increased; b) the pH was lowered; or c) when the CO_2 concentration was lowered. These observations all support the hypothesis that HCO_2^- and HCO_3^- compete for a site on the reducing side of PS II, and they further support the conclusion that HCO_3^-, not CO_2, is the active species.

In a recent study of the Chl a fluorescence transient, Blubaugh and Govindjee [38] found that addition of excess HCO_3^- to non-depleted thylakoids resulted in an enhanced rate of Chl a fluorescence rise in the presence of 3-(3,4-dichlorophenyl)-1,1-dimethylurea (DCMU). This result indicated that addition of excess HCO_3^- to these thylakoids either inhibits a back-reaction with the donor side or prevents some reoxidation of Q_A^- by auxilliary acceptors of PS II (see e.g., [39]). It has also been shown that addition of excess HCO_3^- to non-depleted thylakoids inhibits the Hill reaction in the presence of silicomolybdate (SiMo) [40-42], which is thought to accept electrons at the level of Q_A. This result and the work of Blubaugh and Govindjee [38] suggest that two HCO_3^- binding sites exist. We therefore asked if the HCO_3^- binding site that gave rise to a dissociation constant of 0.08 mM for HCO_3^- was the specific site for HCO_3^- that facilitates electron transfer through the two-electron gate. Since the most effective competitor of HCO_3^- binding so far studied is NO_2^- [25], we studied the effect of NO_2^- incubation on Q_A^- reoxidation. As in the presence of HCO_2^-, we observed a maximum inhibition after the third actinic flash (Fig. 8). The half-time for the NO_2^--incubated sample was ~ 15 ms, which is nearly identical to the result obtained from the data of Fig. 7. A similar result was obtained for NO_2^--incubated samples in the presence of 10 mM hydroxylamine (HA) (data not shown), as was earlier shown with HCO_2^- in the presence of HA [3]. Therefore, our data strongly support the hypothesis that the anion binding site is the site where HCO_3^- binds and facilitates electron transfer through the two-electron gate.

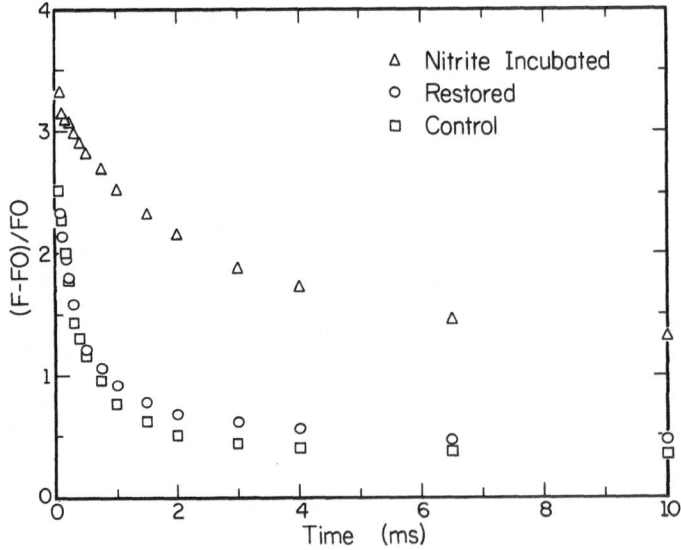

Fig. 8. The Decay of the Variable Chl a Fluorescence after Three Actinic
        Flashes Spaced at 1 s. Spinach thylakoids were used. Other
        conditions were as in Fig. 7 except that 25 mM HCO_2^- was replaced
        by 25 mM NO_2^- (Eaton-Rye and Govindjee, 1985, unpublished).

THE ACTION OF BICARBONATE AT THE LEVEL OF THE TWO-ELECTRON GATE

The operation of the two-electron gate of PS II, neglecting the precise nature of the protolytic steps, is summarized in equations 6-10:

$$Q_A Q_B \rightleftharpoons Q_A^- Q_B \longleftrightarrow Q_A Q_B^- \tag{6}$$

$$Q_A Q_B^- \rightleftharpoons Q_A^- Q_B^- \longleftrightarrow Q_A Q_B^{2-} \tag{7}$$

$$Q_A Q_B^{2-} \longleftrightarrow Q_A + PQH_2 \tag{8}$$

$$Q_A + PQ \longleftrightarrow Q_A Q_B \tag{9}$$

$$Q_A Q_B \rightleftharpoons Q_A^- Q_B \longleftrightarrow Q_A Q_B^- \tag{10}$$

Following a short, single-turnover actinic flash given to dark-adapted thylakoids, an electron is placed on Q_A. This electron is then transferred to Q_B with a half-time of ~ 0.15 ms (see $\underline{e.g.}$, [43]; Eq. 6). The equilibrium established for the sharing of an electron between Q_A and Q_B greatly favors centers in the state $Q_A Q_B^-$ [43]; these centers are open and able to undergo a second photoact (Eq. 7). The second electron is transferred to Q_B^- with a half-time of ~ 0.3 ms (see $\underline{e.g.}$, [43]). The resulting plastoquinol is then able to exchange with a plastoquinone from the PQ pool at the Q_B binding site (Eqs. 8 and 9). If a dark-time of sufficient duration is given before a third actinic flash, the kinetics of electron transfer from Q_A^- to Q_B in equation 10 will resemble those obtained after the first flash (Eq. 6) The minimum dark-time between flash 2 and flash 3 where the kinetics of Q_A^- reoxidation resemble those observed after the first actinic flash is defined here as the turnover time of the two-electron gate. This measurement has recently been made by H.H. Robinson and A.R. Crofts (personal communication) and found to have a half-time of < 2.5 ms in control thylakoids.

We have varied the dark-time between the second and third actinic flashes in HCO_3^--depleted (with HCO_2^- present) and HCO_2^--incubated thylakoids. The dark-times were: 30 ms, 50 ms, 100 ms, 250 ms, 500 ms, and 1 s. In each case maximal inhibition of Q_A^- reoxidation occurred after the third flash and exhibited a $t_{1/2}$ of ~ 15 ms (data not shown). These findings suggest that even after 1 s, centers that have been depleted of HCO_3^- or inhibited by HCO_2^- are unable to exchange with the PQ pool at their Q_B binding site. In each case the effect is fully reversible and control kinetics are completely restored upon the addition of HCO_3^-.

The rate of Q_A^- reoxidation after the third actinic flash in HCO_3^--depleted samples is identical to all subsequent actinic flashes (Fig. 9, also see [16,20]). The same result is obtained with both HCO_2^-- and NO_2^--incubation (data not shown). The decay kinetics after the second flash are intermediate between those following flash 1 and 3 (see [16]).

A possible explanation for these observations is that the binding of inhibitory anions to PS II may alter the association constant, K_0, for Q_B. From equation (9):

$$K_0 = [Q_A \cdot Q_B]/([Q_A] + [PQ_{(pool)}]) \tag{11}$$

Although there is no direct measure of the value for K_0, a number of methods for estimating a value are available [44]. One method is to analyze the decay kinetics of Q_A^- by monitoring the variable Chl $\underline{a}$ fluorescence after a single flash. Biphasic kinetics are observed for this decay; 60-70% of centers undergo oxidation by a first-order process with a half-time of ~ 0.15 ms and the remainder by slower processes of indeterminate order [44]. If it is assumed that the centers exhibiting first-order kinetics represent centers

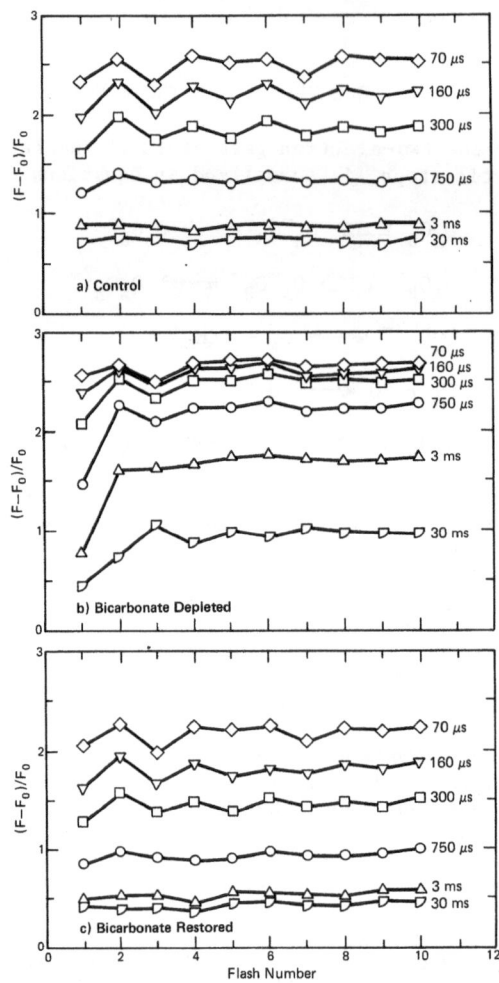

Fig. 9. The Effect of HCO_3^--depletion, in the Presence of 25 mM HCO_2^-, on the Fluorescence Flash Pattern. The times indicated are when the measuring flash was fired. Spinach thylakoids were used; other details are as in Fig. 7 (Eaton-Rye and Govindjee, 1984, unpublished)

in the state Q_AQ_B before the flash, a value of 500 M^{-1} for K_0 can be calculated [44]. We have analyzed our earlier data for HCO_3^--depleted and HCO_2^--incubated thylakoids [16] and found that K_0 is reduced to 200 M^{-1} in these samples. A second effect is also evident from this analysis. The half-time of the fast phase is increased approximately 3-fold (i.e., from ~ 0.2 ms to ~ 0.6 ms) in these samples [cf. 16]. The mechanism of this second effect cannot be explained from the available data.

In addition to the slowing and reduction of the fast phase of Q_A^- reoxidation, a shift in the equilibrium for the sharing of an electron between Q_A and Q_B (see Eq. 6) has been reported in thylakoids that have been HCO_3^--depleted in the presence of formate [45]. A two-fold shift in this equilibrium towards Q_A^- was observed by comparing the rates of the back-reaction with S_2 both in the presence and absence of DCMU. In the absence of DCMU, the back-reaction from Q_B^- to S_2 was unaffected by HCO_3^--depletion, but in the presence of DCMU the back-reaction of Q_A^- with S_2 was inhibited two-fold [45].

The equilibrium for the sharing of an electron between Q_A and Q_B is pH dependent [46]. It has therefore been suggested that the presence of a proton in association with the Q_B site stabilizes the electron on $Q_B^-(H^+)$. A possible explanation for the observed shift in this equilibrium may be that HCO_3^--depletion inhibits protonation at the Q_B site. In addition, the fraction of centers decaying through the rapid first-order process has been shown to be proportional to the fraction of centers in which $Q_B^-(H^+)$ is present before the second flash [46]. Therefore the inhibition on the $Q_B^-(H^+)$ protonation suggested above may also account for the inhibited kinetics of Q_A^- reoxidation observed after the second flash in HCO_3^--depleted membranes [16,20]. By analogy, the maximal inhibition observed after the third flash may result from Q_B^{2-} not becoming protonated and therefore not able to exchange with the PQ pool. This interpretation suggests that the rate-limiting step introduced by HCO_3^--depletion and/or anion inhibition is the rate of protonation of Q_B^{2-}.

Recently the ability of carbonic anhydrase inhibitors to inhibit PS II has been reported [25,42]. This approach to the bicarbonate problem has also led to the suggestion that HCO_3^- functions as a proton donor/acceptor at the anion binding site [25,47]. These workers have also reported an inhibitory effect of HCO_3^- on PS II at pH 8.0 [34,42]. However, the reoxidation of Q_A^- was unaffected by the addition of HCO_3^- in these samples [42]. On the other hand, a slowing of the $S_2' \longrightarrow S_3$ transition was shown, but this phenomenon appears to be distinct from the HCO_3^- effect under consideration here. Addition of 10 mM HCO_3^- to non-depleted thylakoid membranes at pH 6.0 and 7.0 did not inhibit the Hill reaction supported by $K_3Fe(CN)_6$ or MV [48].

The inhibition of the back-reaction of Q_A^- with S_2 in the presence of DCMU [45] presents other interesting possibilities arising as a consequence of HCO_3^--depletion. The apparent increase in the stability of Q_A^- is not reflected in a shift in the mid-point potential of the Q_A/Q_A^- couple [49]. Therefore the observed inhibition of the back-reaction between Q_A^- and S_2 in HCO_3^--depleted thylakoids may indicate an additional effect of HCO_3^--depletion and/or anion incubation on the donor side of PS II . However, this result may also be explained by HCO_3^--depletion altering the redox potential of the Fe^{3+}/Fe^{2+} couple associated with PS II. This interpretation may also explain the increased EPR signal (g = 1.82) attributed to Q_A^--Fe^{2+} observed in PS II particles [19].

CONCLUSIONS

Our findings here support the following conclusions: (1) the active species of the bicarbonate effect is HCO_3^-, not CO_2 or CO_3^{2-}; (2) NO_2^- (as is the case with HCO_2^-) inhibits the electron acceptor side of PS II; this inhibition is also reversed by HCO_3^-; (3) both HCO_2^- and NO_2^- protect PS II against irreversible damage at pH 5.8; and (4) in the complete absence of inhibitory anions, HCO_3^--depletion still inhibits the electron acceptor side of PS II. Furthermore, we suggest that the unique ability of HCO_3^- to stimulate electron transport may arise from its ability to participate in the protolytic reactions of the two-electron gate.

Acknowledgements

We are thankful to the National Science Foundation (PCM 83-06061) for financial support. D.B. was supported by a service award from the National Institute of Health (PHS 5-T32GM7283).

REFERENCES

1. O. Warburg and G. Krippahl, Notwendigkeit der Kohlensäure für die Chinon-und Ferricyanid-Reaktionen in grünen Grana, Z. Naturforsch. 15B:367 (1960).

2. R. Khanna, Govindjee and T. Wydrzynski, Site of Bicarbonate Effect in Hill Reaction: Evidence from the Use of Artificial Electron Acceptors and Donors, Biochim. Biophys. Acta 462:208 (1977).

3. J. J. Eaton-Rye and Govindjee, A Study of the Specific Effect of Bicarbonate on Photosynthetic Electron Transport in the Presence of Methyl Viologen, Photobiochem. Photobiophys. 8:279 (1984).

4. B. R. Velthuys, Electron-dependent Competition Between Plastoquinone and Inhibitors for Binding to Photosystem II, FEBS Lett. 126:277 (1981).

5. B. Kok, B. Forbush and M. McGloin, Cooperation of Charges in Photosynthetic O_2 Evolution-1: A Linear Four Step Mechanism, Photochem. Photobiol. 11:457 (1970).

6. H. J. van Gorkom, Electron Transfer in Photosystem II, Photosynth. Res. 6:97 (1985).

7. Govindjee, T. Kambara and W. Coleman, The Electron Donor Side of Photosystem II: The Oxygen Evolving Complex, Photobiochem. Photobiol. 40: in the press (1985).

8. H. Metzner, K. Fischer and O. Bazlen, Isotope Ratios in Photosynthetic Oxygen, Biochim. Biophys. Acta 548:287 (1979).

9. A. Stemler, Inhibition of Photosystem II by Formate: Possible Evidence for a Direct Role of Bicarbonate in Photosynthetic Oxygen Evolution, Biochim. Biophys. Acta 593:103 (1980).

10. Govindjee and J.J.S. van Rensen, Bicarbonate Effects on the Electron Flow in Isolated Broken Chloroplasts, Biochim. Biophys. Acta 505:183 (1978).

11. W. F. J. Vermaas and Govindjee, Unique Role(s) of Carbon Dioxide and Bicarbonate in the Photosynthetic Electron Transport System, Proc. Indian Natl. Sci. Acad. B47:581 (1981).

12. W. F. J. Vermaas and Govindjee, Bicarbonate or Carbon Dioxide as a Requirement for Efficient Electron Transport on the Acceptor Side of Photosystem II, in: "Photosynthesis Vol. II: Development, Carbon Metabolism and Plant Productivity," Govindjee, ed., Academic Press, New York (1983).

13. Govindjee, I. C. Baianu, C. Critchley and H. S. Gutowsky, Comments on the Possible Roles of Bicarbonate and Chloride Ions in Photosystem II, in: "The Oxygen Evolving System of Photosynthesis," Y. Inoue, A. R. Crofts, Govindjee, N. Murata, G. Renger, and K. Satoh, eds., Academic Press, Tokyo (1983).

14. Govindjee, The Oxygen Evolving System of Photosynthesis, in: "Advances in Photosynthesis Research Vol. I," C. Sybesma, ed., Martinus Nijhoff/Dr. W. Junk Publishers, The Hague (1984).

15. P. Jursinic, J. Warden and Govindjee, A Major Site of Bicarbonate Effect in System II Reaction: Evidence from ESR Signal II_{vf}, Fast Fluorescence Yield Changes and Delayed Light Emission, Biochim. Biophys. Acta 440:322 (1976).

16. H. H. Robinson, J. J. Eaton-Rye, J. J. S. van Rensen, and Govindjee, The Effects of Bicarbonate Depletion and Formate Incubation on the Kinetics of Oxidation-Reduction Reactions of the Photosystem II Quinone Acceptor Complex, Z. Naturforsch. 39c:382 (1984).

17. U. Siggel, R. Khanna, G. Renger and Govindjee, Investigation of the Absorption Changes of the Plastoquinone System in Broken Chloroplasts: The Effect of Bicarbonate Depletion, Biochim. Biophys. Acta 462:196 (1977).

18. J. Farineau and P. Mathis, Effects of Bicarbonate on Electron Transfer Between Plastoquinones in Photosystem II, in: "The Oxygen Evolving System of Photosynthesis," Y. Inoue, A. R. Crofts, Govindjee, N. Murata, G. Renger, and K. Satoh, eds., Academic Press, Tokyo (1983).

19. W. F. J. Vermaas and A. W. Rutherford, EPR Measurements on the Effects of Bicarbonate and Triazine Resistance on the Acceptor Side of Photosystem II, FEBS Lett. 175:243 (1984).

20. Govindjee, M. P. J. Pulles, R. Govindjee, H. J. van Gorkom and L. N. M. Duysens, Inhibition of the Reoxidation of the Secondary Electron Acceptor of Photosystem II by Bicarbonate Depletion, Biochim. Biophys. Acta 449:602 (1976).

21. Govindjee, H. Y. Nakatani, A. W. Rutherford and Y. Inoue, Evidence from Thermoluminescence for Bicarbonate Action on the Recombination Reactions Involving the Secondary Quinone Electron Acceptor of Photosystem II, Biochim. Biophys. Acta 766:416 (1984).

22. R. Khanna, K. Pfister, A. Keresztes, J. J. S. van Rensen and Govindjee, Evidence for a Close Spatial Location of the Binding Sites for CO_2 and for Photosystem II Inhibitors, Biochim. Biophys. Acta 634:105 (1981).

23. W. F. J. Vermaas, J. J. S. van Rensen, and Govindjee, The Interaction Between Bicarbonate and the Herbicide Ioxynil in the Thylakoid Membrane and the Effects of Amino Acid Modification on Bicarbonate Action, Biochim. Biophys. Acta 681:242 (1982).

24. J. J. S. van Rensen and W. F. J. Vermaas, Action of Bicarbonate and Photosystem II Inhibiting Herbicides on Electron Transport in Pea Grana and in Thylakoids of a Blue-Green Alga, Physiol. Plant. 51:106 (1981).

25. A. Stemler and J. B. Murphy, Bicarbonate: Reversible and Irreversible Inhibition of Photosystem II by Monovalent Anions, Plant Physiol. 77:974 (1985).

26. N. E. Good, Carbon Dioxide and the Hill Reaction, Plant Physiol. 38:298 (1963).

27. A. Stemler and J. B. Murphy, Determination of the Binding Constant of $H^{14}CO_3^-$ to the Photosystem II Complex in Maize Chloroplasts: Effects of Inhibitors and Light, Photochem. Photobiol. 38:701 (1983).

28. J. F. H. Snel, A. Groote-Schaarsberg and J. J. S. van Rensen, Mechanism and Physiological Role of Bicarbonate Action on Electron Flow: A Possible Link Between Photorespiration and Photosynthetic Electron Flow, in: "Advances in Photosynthesis Research Vol. I.", C. Sybesma, ed., Martinus Nijhoff/Dr. W. Junk Publishers, The Hague (1984).

29. J. F. H. Snel and J. J. S. van Rensen, Reevaluation of the Role of Bicarbonate and Formate in the Regulation of Photosynthetic Electron Flow in Broken Chloroplasts, Plant Physiol. 75:146 (1984).

30. W. Knoche, Chemical Reactions of CO_2 in Water, in: "Biophysics and Physiology of Carbon Dioxide," C. Bauer, G. Gros and H. Bartels, eds., Springer Verlag, New York (1980).

31. W. F. J. Vermaas and J. J. S. van Rensen, Mechanism of Bicarbonate Action on Photosynthetic Electron Transport in Broken Chloroplasts, Biochim. Biophys. Acta 636:168 (1981).

32. G. Sarojini and Govindjee, On the Active Species in Bicarbonate Stimulation of Hill Reaction in Thylakoid Membranes, Biochim. Biophys. Acta 634:340 (1981).

33. G. Sarojini and Govindjee, Is CO_2 An Active Species in Stimulating the Hill Reaction in Thylakoid Membranes?, in: "Photosynthesis II. Electron Transport and Photophosphorylation," G. Akoyunoglou, ed., Balaban International Science Services, Philadelphia, Pa. (1981).

34. A. Stemler, Forms of Dissolved Carbon Dioxide Required for Photosystem II Activity in Chloroplast Membranes, Plant Physiol. 65:1160 (1980).

35. J. D. Hesketh, J. T. Wooley and D. B. Peters, Predicting Photosynthesis, in: "Photosynthesis Vol. II.: Development, Carbon Metabolism and Plant Productivity," Govindjee, ed., Academic Press, New York (1983).

36. J. F. H. Snel and J. J. S. van Rensen, Kinetics of the Reactivation of the Hill Reaction in CO_2-Depleted Chloroplasts by Addition of Bicarbonate in the Absence and in the Presence of Herbicides, Physiol. Plant. 57:422 (1983).

37. J. F. H. Snel, D. Naber and J. J. S. van Rensen, Formate as an Inhibitor of Photosynthetic Electron Flow, Z. Naturforsch. 39c:386 (1984).

38. D. J. Blubaugh and Govindjee, Comparison of Bicarbonate Effects on the Variable Chlorophyll $\underline{a}$ Fluorescence of CO_2-Depleted and Non-CO_2-Depleted Thylakoids in the Presence of Diuron, Z. Naturforsch. 39c:378 (1984).

39. W. F. J. Vermaas and Govindjee, The Acceptor Side of Photosystem II in Photosynthesis, Photochem. Photobiol. 34:775 (1981).

40. R. Barr and F. L. Crane, Control of Photosynthesis by CO_2: Evidence for a Bicarbonate-Inhibited Redox Feedback in Photosystem II, Proc. Indiana Acad. Sci. 85:120 (1976).

41. F. L. Crane and R. Barr, Stimulation of Photosynthesis by Carbonyl Compounds and Chelators, Biochem. Biophys. Res. Comm. 74:1362 (1977).

42. A. Stemler and P. Jursinic, The Effects of Carbonic Anhydrase Inhibitors Formate, Bicarbonate, Acetazolamide, and Imidazole on Photosystem II in Maize Chloroplasts, Archiv. Biochem. Biophys. 221:227 (1983).

43. H. H. Robinson and A. R. Crofts, Kinetics of the Oxidation-Reduction Reactions of the Photosystem II Quinone Acceptor Complex, and the Pathway for Deactivation, FEBS Lett. 153:221 (1983).

44. A. R. Crofts, H. H. Robinson and M. Snozzi, Reactions of Quinones at Catalytic Sites: A Diffusional Role in H-Transfer, in: "Advances in Photosynthesis Research Vol. I.," C. Sybesma, ed., Martinus Nijhoff/Dr. W. Junk Publishers, The Hague (1984).

45. W. F. J. Vermaas, G. Renger and G. Dohnt, The Reduction of the Oxygen-Evolving System in Chloroplasts by Thylakoid Components, Biochim. Biophys. Acta 764:194 (1984).

46. H. H. Robinson and A. R. Crofts, Kinetics of Proton Uptake and the Oxidation-Reduction Reactions of the Quinone Acceptor Complex of PS II from Pea Chloroplasts, in: "Advances in Photosynthesis Research Vol. I.," C. Sybesma, ed., Martinus Nijhoff/Dr. W. Junk Publishers, The Hague (1984).

47. A. Stemler, Carbonic Anhydrase: Molecular Insights Applied to Photosystem II Research in Thylakoid Membranes, in: "Inorganic Carbon Transport in Aquatic Photosynthetic Organisms," J. Berry and W. Lucas, eds., American Society of Plant Physiologists, Rockwell, MD. In the Press (1984).

48. R. Barr, and R. Melhem, A. L. Lezotte and F. L. Crane, Stimulation of Electron Transport from Photosystem II to Photosystem I in Spinach Chloroplasts, J. Bioenerg. Biomemb. 12:197 (1980).

49. W. F. J. Vermaas and Govindjee, Bicarbonate Effects on Chlorophyll $\underline{a}$ Fluorescence Transients in the Presence and the Absence of Diuron, Biochim. Biophys. Acta 680:202 (1982).

THE ANION AND CALCIUM REQUIREMENT OF THE PHOTOSYNTHETIC WATER SPLITTING COMPLEX

Peter H. Homann[1] and Yorinao Inoue[2]

[1]Institute of Molecular Biophysics, Florida State University, Tallahassee, FL 32306, U.S.A.
[2]Solar Energy Research Group, The Institute of Physical and Chemical Research (RIKEN), Wako-shi, Saitama 351, Japan

INTRODUCTION

Considerable evidence has accumulated which suggests that Cl^- and Ca^{2+} are essential participants in the process of photosynthetic water oxidation. However, without special treatments of membrane preparations from chloroplast thylakoids, neither anions, nor Ca^{2+}, need to be provided to assure water oxidizing activity. Apparently, these ions are very strongly held at the water oxidizing site of the thylakoid membrane. An alternative, albeit less likely, possibility would be that a need for any one of the ions is artifactual and a consequence of a perturbation caused by the "depletion" treatment.

The conditions which create a Cl^- requirement of the water oxidizing enzyme ("water oxidase") have in common that they expose it to an elevated pH and/or cause structural disturbances (as during exposure to elevated temperatures, or during aging) [1,2,3]. In contrast, addition of Ca^{2+} to preparation from chloroplast membranes usually becomes necessary only after removal of the extrinsic 18 and 23 kDa polypeptides from the site of the water oxidizing complex.

This contribution examines some experimental results which have been obtained in studies of the role of Cl^- and Ca^{2+} in photosystem II (PSII) preparations from chloroplast thylakoids. An understanding of the function of these ions obviously is a prerequisite for the elucidation of the intricate mechanism by which water is oxidized in organisms performing oxygenic photosynthesis. Unfortunately, many very fundamental questions will not be answered satisfactorily. It can be said with some certainty, however, that the presence of Cl^- at the water oxidizing site is essential for the orderly accumulation of the oxidizing equivalents needed for the production of an O_2 molecule by abstraction of four electrons from two water molecules. The functions of Ca^{2+}, on the other hand, may be more indirect, and directed at conformational requirements of PSII-associated polypeptides.

It has been suggested [4,5] that Cl^- may serve as a bridging ligand in the putative Mn-cluster of the water oxidase. However, in analogy to the perceived role of Cl^- in other anion requiring enzymes, a less straightforward function of Cl^- is more likely.

MATERIALS AND METHODS

Most of the experiments discussed here were performed with membrane preparations enriched in the water oxidizing photosystem II (PSII). They were obtained from Triton-X100 treated chloroplast thylakoids according to Berthold et al. [6]. The extrinsic 23 and 18 kDa polypeptides of PSII were removed from those membrane particles by a 20 min treatment with 1.6 M NaCl at pH 6.5 ("high [salt] treatment") [7]. Thermoluminescence measurements were performed at RIKEN, The Institute of Physical and Chemical Research (Wako, Japan) with the setup described elsewhere [8].

RESULTS

Cl^- binding at the water oxidizing site

At a medium pH < 7, PSII-enriched thylakoid particles usually are capable of photo-oxidizing water regardless of the presence of Cl^- in the suspension medium. However, after the preparation is exposed to Cl^--free media of moderate alkalinity, an addition of Cl^- or certain other mono-valent anions, becomes necessary for optimal O_2-evolving activity. As shown by Kelley and Izawa [9], the dependence of the rate of water oxidation on Cl^- in such "depleted" preparations can be analyzed by treating Cl^- as if it were a substrate (or perhaps, more appropriately, an essential activator) of the water oxidizing enzyme (Fig. 1). An apparent decrease of the affinity of the enzyme for Cl^- with increasing pH is reflected by such kinetic analysis as an increase of the K_m for Cl^-. Indeed, the dependence of the K_m on $[OH^-]$ is linear, suggesting a competitive inhibition of Cl^- binding by $[OH^-]$. This is shown in Fig. 2, which reveals a $K_I = 10$ nM for OH^-. Since the extrapolated K_m for Cl^- in the absence of OH^- is approximately 70 μM, OH^- appears to have a 10^4 fold competitive advantage over Cl^- at the "binding site". However, the plot can also be construed as being linear with $[H^+]^{-1}$. Such a relation-ship implies [10] that Cl^- binding is contingent upon the protonation of a membrane group with a $pK_a \simeq 6$, bringing the Cl^--water oxidase inter-action in line with concepts developed for the association of Cl^- with other anion activated proteins.

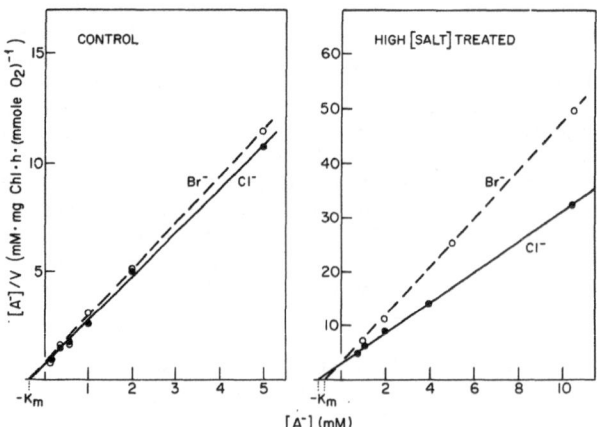

Fig. 1.   Hanes-Woolf plots of the dependence of the rate of photo-synthetic O_2 evolution on the concentration of the activating anion. Measurements on PSII membranes from <u>Phytolacca americana</u> leaves in 400 mM sorbitol, 10 mM Na_2SO_4, 20mM Na-Mes or (for high [salt] treated particles) Ca-Mes pH 6.2; 2 kW/m^2 red light, 25 C.

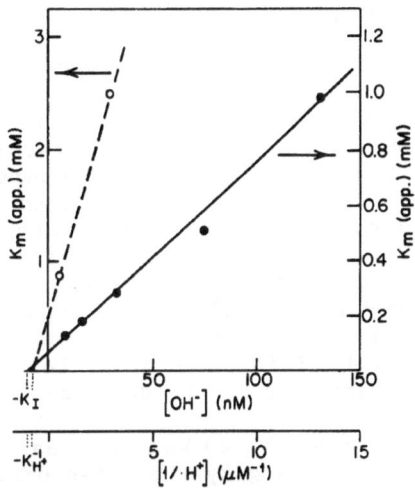

Fig. 2.   Secondary plots of the apparent K_m for Cl^- binding against
          $[OH^-]$ or $1/[H^+]$: Data from determinations of the type shown in
          Fig. 1, but with PSII particles from <u>Spinacea oleracea</u>.
          ● : normal preparations;   ○ : high [salt]-treated
          preparations.

The role of the 18 and 23 kDa extrinsic polypeptides in Cl^- and Ca^{2+} binding

It has become apparent [11] that a removal of the extrinsic 18 and
23 kDa polypeptides from the water oxidizing site in PSII greatly
accelerates the loss of Cl^- and increases its K_m of binding. These
observations suggest that the two polypeptides in some way control the
association of Cl^- with the water oxidase. Their function as diffusion
barrier for Cl^- was evident not only when we measured the rate of Cl^-
release, but also during the reconstitution of Cl^- depleted samples with
added Cl^-. As can be seen from the data of Table 1, the diffusion rate
of Cl^- to its site of action was markedly faster after removal of the 23
and 18 kDa polypeptides. For optimal stability of the polypeptide
deficient membrane particles (see below), the experiments were performed

Table 1.   Reactivation by added Cl^- of Cl^--depleted PSII preparations
           from <u>Phytolacca americana</u>

PSII membranes	Na-Mes pH 6.2[a] (s)	Ca-Mes pH 6.2[a] (s)
normal	12	20
lacking 18 and 23 kDa polypeptide	-	4

[a]Data are approximate halftimes of Cl^- acquisition at 25C from
a medium (400 mM sorbitol, 10 mM Na_2SO_4 and 20 mM buffer as
indicated) containing 1.4 mM Cl^-. The $[Cl^-]$ at the reaction
site was estimated from the measured rate of O_2 evolution.

in Ca^{2+} containing media. Peculiarly, the inclusion of this cation caused, for reasons that are unknown to us, a marked retardation of the Cl^- acquisition by control preparations.

Another consequence of the loss of the two polypeptides turned out to be a significantly altered effectiveness of different anions to substitute for Cl^- as activators of the water oxidizing enzyme (Fig. 1 and Table 2). This surprising discovery underscored the cooperation, either directly or indirectly, of several components of the water oxidizing enzyme complex in establishing the proper function of an anion in photosynthetic O_2 evolution. The almost total loss of activating power seen with I^- could have been due to its ability to become a substitute for water as electron donor to the oxidized reaction center of PSII. Such function of I^- added to the polypeptide depleted preparation could indeed be established. However, unlike other reductants [12], I^- did not cause an irreversible inhibition due to Mn-loss when incubated with polypeptide depleted preparations in the dark (unpubl. results).

Of the two polypeptides, the 23 kDa species has the most profound influence on the K_m of Cl^- binding by the water oxidase [11]. Interestingly, it is this polypeptide which also modulates the action of Ca^{2+}. Just as is the case with Cl^-, it does so by significantly decreasing the demand by the water oxidase for the added ion [13,14].

Table 2. Relative effectiveness of monovalent anions in supporting photosynthetic O_2 evolution in _Phytolacca_ _americana_ PSII particles before and after removal of the extrinsic 18 and 23 kDa polypeptides.

Anion used in reconstitution	normal preparation[a,b]	high [salt] treated preparation[a,c]
none	(12.5)	(2.5)
Cl^-	100	100
Br^-	95	55
NO_3^-	70	50
I^-	45	15
ClO_4^-	40	25
CNS^-	40	15
HCO_2^-	25	15
ClO_3^-	25	n.d.
F^-	15	5
BrO_3^-	15	n.d.
$CH_3CO_2^-$	15	10

Assay media as for Fig. 1, buffered at pH 6.2 with 25 mM Na-Mes or, in the case of polypeptide depleted preparations, Ca-Mes (Ca^{2+} did not affect the anion effectiveness).

[a] values rounded off to nearest 5 (except for control when no anions were added)

[b] 10 mM anion added

[c] 20 mM anion added

The phenomenology of the Ca^{2+} requirement is complex, because there appear to be several types of actions between which a relation has not yet been established convincingly. Among them are an action in cyano-bacterial membranes close to the reaction center of PSII [14], an involvement in the insertion of functional Mn into the water oxidase of chloroplast thylakoids [15], a stabilization of the water oxidase in preparations stripped of the 23 and 18 kDa polypeptides, and the afore-mentioned stimulation of its turnover.

Miyao and Murata [13] have shown that lack of Ca^{2+} lowers both the apparent quantum efficiency and the maximal rate, yet in our own experiments an addition of Ca^{2+} to preparations depleted of the 18 and 23 kDa polypeptides improved mainly the quantum efficiency and had comparatively little effect on the maximal attainable rate (Fig. 3). Ca^{2+} also curtailed a progressing inactivation of the preparation in the course of the measurement, and stabilized it in the dark when Cl^- was withheld. This latter observation may be indicative of some interdependence of the actions of Cl^- and Ca^{2+}. However, a direct cooperation is unlikely because the persistence of considerable O_2 evolving activity after Ca^{2+} depleting treatments makes it uncertain that Ca^{2+} is in fact an essential cofactor for O_2-evolution.

The impairment of photosynthetic water oxidation by Cl^- removal

Some progress has been made recently in attempts to identify the site of the lesion in PSII caused by the removal of functional Cl^-. It was predicted that a failure to oxidize water should find its expression in an inability to successfully store and use oxidizing equivalents on the oxidizing side of the reaction center P_{680} of PSII. In previous independent studies, Itoh et al. [16] and Theg et al. [17] have obtained evidence that in Cl^- depleted thylakoids the photoxidized reaction center P_{680} can regain in succession no more than two electrons. An adequate

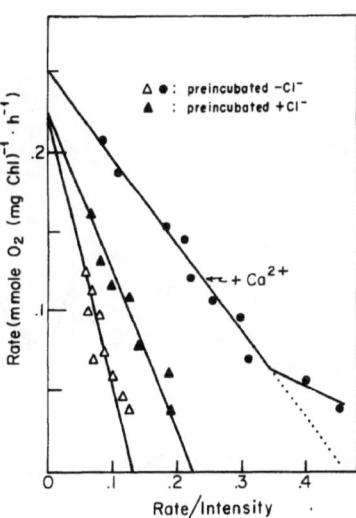

Fig. 3.   Rate vs. rate/light intensity plots for high [salt] washed preparations of _Spinacea oleracea_. Effect of Ca^{2+} on the rate of O_2 evolution, and effect of 20s preincubation in Cl^--free medium. Assay conditions as for Fig. 1, except that 25 mM NaCl were present. (Even though not shown, the rates measured with $+Ca^{2+}$ samples were not dependent on Cl^- presence during the 20s preincubation)

explanation of this result must relate it to the postulated states S_i of the water oxidase with $i = 0,1,2,3,4$ [18]. These states symbolize the extent of accumulation of oxidants, with water oxidation being coupled to the spontaneous conversion of S_4 to S_0. Since in normal dark-adapted membrane preparations 75%, i.e., most, of the water oxidases are in the S_1 state, our results and those of Itoh et al., might mean that in the absence of Cl^- an advance is possible to the S_3, but not to the S_4 state. However, it is not certain [16,17] whether the positive charge left behind by the second electron on the primary donor Z, can be transferred to the water oxidizing S complex.

Unfortunately, methods allowing a direct and unequivocal identification of the prevailing S state are not yet available. However, an indirect look at the oxidation state of the water oxidase is possible through thermoluminescence measurements on PSII preparations. This experimental approach makes use of the fact that the recombination of charges of photoreduced and photoxidized products in PSII may lead to a re-excitation of P_{680} and, thence, other chlorophyll molecules. The free energy of activation of this reversal of the photochemistry depends characteristically on the nature of the electron donor and acceptor [19]. Under normal conditions, the donor is a reduced secondary e^--acceptor of PSII, namely Q_B^-, while the prospective acceptors are either S_2 or S_3 (S_0 and S_1 are stable, and S_4 spontaneously converts to S_0 as it oxidizes water). Thus, from what is known about the primary photochemical events in PSII, it has been predicted [20] that highly luminescent states can be formed from dark adapted thylakoids only after one or two short light flashes have been given, and again after the fifth and sixth flash, and so forth. Since the emission yield, furthermore, is higher after the first and fifth flash than after the second and sixth, a clear oscillation of the luminescence intensity with a period of four and maxima after the first and fifth (and ninth, etc.) flash is typical.

Fig. 4 is a presentation of the thermoluminescence pattern measured with a Cl^- depleted PSII preparation before and after readdition of Cl^-.

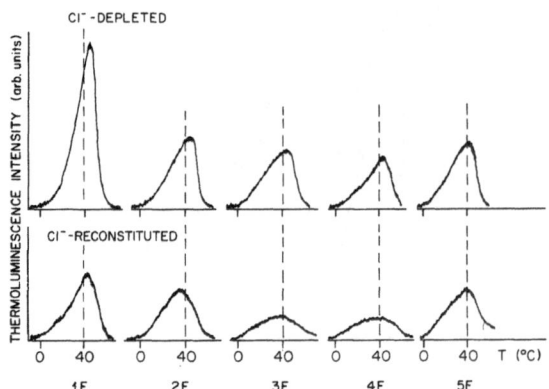

Fig. 4.   Thermoluminescence pattern measured with PSII membranes from Spinacea oleracea thylakoids after depletion of Cl^-, and reconstitution with 25 mM NaCl. Samples were preilluminated at 2 C with 1 to 5 flashes (F), frozen to 77 K, and then rewarmed. Suspension medium as for Fig. 1, but buffered with Na-Mes at pH 6.0.

The samples were illuminated with one to five short light flashes, quickly frozen in liquid N_2, and the light emission was then recorded during the subsequent period of rewarming. The experiment reveals three important effects of Cl^- removal: (1) the luminescence yield after the first flash ($Q_B^--S_2$ backreaction) is unusually high; (2) hardly any changes of the emission intensity occur after the second flash; (3) the peak of the emission occurs at a high temperature regardless of the preillumination regime.

Two comments on these results must be made. First, for the figure we did not choose the "best" experiment because we wish to admit that it proved to be quite difficult to obtain "ideal" responses with respect to the characteristics listed above. This clearly was caused by residual Cl^- in the preparation, either due to incomplete removal or to recontamination. It was not feasible to circumvent these problems by using membranes from which the 18 and 23 kDa polypeptides had been removed because their thermoluminescence pattern was, for yet unexplored reasons, somewhat different than that of normal membranes, especially with respect to the luminescence yield after illumination with one flash.

The second comment concerns the temperature at which the maximal luminescence emissions occurred. These characteristic temperatures are affected by the pH of the suspension medium. The effect of pH is thought to be caused by a complicated interplay between changes of the protonation state of Q_B^- and of the oxidized species in the S-system [20]. The data of Fig. 5 show that in normal preparations the emission temperature was around 35 C except at pH 6 where the emission for the $Q_B^--S_2$ backreaction was elevated above 40 C. After Cl^- depletion, all emissions occurred at such high temperature, regardless of pH and flash number. A corresponding upward shift of the emission temperature as the result of Cl^- deficiency was also seen with DCMU poisoned membranes (Fig. 5). Since in the presence of DCMU the luminescence is elicited by a

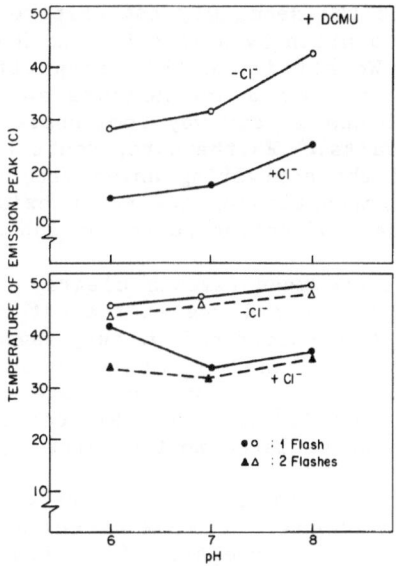

Fig. 5.   pH dependence of temperatures at which peak emission of the thermoluminescence was measured. PSII membranes from <u>Spinacea oleracea</u> in the absence of DCMU, or with 12 µM DCMU present. Medium buffered with Na-Mes (pH 6.0) or Na-Hepes (pH 7.0 and 8.0). Other conditions as for Fig. 4.

backreaction in which no longer Q_B^- but rather Q_A^- is the electron donor, the cause of the altered temperature position of the luminescence peak may be sought at the oxidizing side of PSII, i.e. in the charge-accumulating S-complex.

DISCUSSION

Our general lack of detailed knowledge about the mechanism of photosynthetic water oxidation in PSII precludes a full understanding of the roles played in it by Cl^- and Ca^{2+} ions. However, the data presented in this contribution, together with those available in the literature, allow us to develop some concepts which may guide further explorations.

Specific actions of anions in biological systems are no rare occurrences. In general, the positively charged binding sites for the anion appear to be provided by protonated amino groups. Our measurements of the pH dependence of the activation of the water oxidase by Cl^- have revealed an intimate involvement of functional groups with a pK_a around 6 in the association of Cl^- with PSII. Such pK_a would be unusual for amino groups, but is not unrealistic because a similarly low pK_a caused by a unique microenvironment has been reported for the Cl^--binding arginine of the anion transporter of erythrocytes [21], and for a lysine at the active site of acetoacetate decarboxylase [22].

Since the removal of the 18 and 23 kDa polypeptides changes the apparent binding strength, but not the pK_a of the involved group, we have to assume that the association of the anions with the water oxidizing site is governed by proximity effects of "binding" groups on more than one polypeptide. This cooperativity may also explain the altered anion specificity, and the peculiarly low K_m of Br^- binding, of centers devoid of the 18 and 23 kDa polypeptides.

Our studies do not exclude an involvement of the Mn cluster in anion binding, but make unlikely an association as intimate as the proposed inner sphere bridging ligand function [4,5]. One argument in favor of this mode of binding is the seemingly competitive prevention by Cl^- of the inhibition of O_2 evolution by amines [5] which supposedly bind to the functional Mn [23]. We attribute this competitiveness to indirect causes. For example, we could not measure an effect of NH_3 on Cl^- release from PSII which was in any way commensurate with its inhibitory effect (unpublished results). Furthermore, doubt is cast on a concept of a close association of the activating anion with Mn by our observation that the polypeptide composition of the water oxidizing site determines the relative effectiveness of anions to act as substitutes for Cl^-.

The thermoluminescence measurements clearly showed that the normal course of charge accumulation and use at the water oxidase was impaired after removal of Cl^-. More specifically, they suggested that a sensitive step is the advance to the final S_4 state which would be capable of producing O_2. Hence, the two electrons which are successively available to a photo-oxidized P_{680} of Cl^- depleted PSII centers [16,17], allow two oxidizing charges to be accumulated in the water oxidase.

With respect to the progression of the charge accumulating sequence Cl^- deficiency appears to have the same effect as a removal of the Mn-stabilizing 33 kDa extrinsic polypeptide [24]. However, it is clear from our experiments that the Cl^- involvement is not restricted to the S_3-S_4 transition. Judging from the yield of the thermoluminescence after one flash, and the high thermal energy required for the luminescence emission under all conditions, we must conclude that the S_2 and S_3 states formed in Cl^--deficient enzymes have profoundly altered properties.

Theoretical considerations have established a rather good correlation between emission temperatures, and expected free energies of activation for the generation of excited reaction centers from a back-reaction between reductants and oxidants located close to the center [19]. According to this view, we can postulate that, at least at pH > 6, the oxidant-reductant pairs in Cl^- depleted membranes are separated by smaller redox potential differences than in normal preparations. Since this was true in the absence and in the presence of DCMU, i.e. regardless of the nature of the reacting e^--donor, one may be tempted to ascribe the narrowing of the redox-potential gap to a lowered oxidation potential of the accumulated oxidants at the Cl^--free water oxidase.

Some caution with these interpretations is necessary, however, on account of the unique situation at pH 6 where once flashed plus and minus Cl^- preparations emitted their luminescence at an almost identical temperature in the absence of DCMU, but not in its presence. If the elevation of the emission temperature at pH 6 for once flashed control preparations is indeed attributable to a protonation of Q_B^- as proposed by Rutherford et al. [20] we may have to postulate that Cl^- depletion affects its pK_a. This would imply a hitherto unsuspected effect of Cl^- on the reducing side of PSII (see, however, the contribution by Stemler in this volume).

Alternatively, due to secondary effects of the heating, a clear correlation between the emission temperature and the energy loss during the original process of energy storage may no longer exist above approximately 45 C. This might apply especially to Cl^- depleted preparations which are known to be very heat sensitive [26]. The very high luminescence yield from the recombination reaction after illumination with one flash may be a further expression of an altered mechanism of the backreaction in heat stressed Cl^- deficient membrane particles. It must be mentioned, though, that detrimental effects on PSII usually find their expression in a lower thermoluminescence yield, e.g. those caused by removal of the 18 and 23 kDa polypeptides [24], or by exposure to an elevated pH, especially in Cl^- free media (Vass, pers. comm.; our measurements). In these instances, backreactions probably fail to cause a re-excitation of P_{680}, or the charge storage itself is impaired as appears to be the case with Cl^--deficient PSII particles at pH 7.5 (Itoh, pers. comm.).

Any proposal of a function for Cl^- has to take into account the fact that the activity profile of anion requiring enzymes is, quite generally, shifted to lower pH values when the anion has been removed [26]. Hence, removal of Cl^- from the water oxidase, for example, is in some way equivalent to an increase of the pH at the active site. This may imply a deprotonation of essential enzyme sites, or an inability of certain groups to perform their role as proton acceptors during water oxidation. Hence, we suggest that the unique S_2 and S_3 states formed in the absence of Cl^- are defect with respect to the proton distribution on the water oxidase. This concept is a further development of our previously proposed model [27] and would readily accommodate the mechanism of Cl^- action envisaged by Coleman and Govindjee [28].

Contrary to the situation with respect to the role of Cl^-, there is as yet no evidence which would unequivocally establish an essential function for Ca^{2+} in the mechanism of water oxidation. Hence, Ca^{2+} may simply satisfy a need for specific conformations of the polypeptides close to the active site of the water oxidase. The abundance of glutamic and aspartic acid in the extrinsic polypeptides at the O_2 evolving site [29] would make them particularly susceptible to conformation altering interactions with Ca^{2+} [30]. In intact preparations, the conformation may to a large extent be established by interactions between the poly-peptides themselves. Consequently, the greater need for Ca^{2+} after detachment of especially the 23 kDa species [12,13] may reflect the removal of the natural partner of the remaining 33 kDa polypeptide. The apparent interplay between Cl^- and Ca^{2+} in stabilizing such preparations can perhaps be rationalized by the influence these ions have on the interactions between their respective binding groups on the proteins.

Our view of the role of Ca^{2+} and Cl^- in photosynthetic water oxidation focuses on their interactions with the various proteinaceous components of the enzyme complex. Given our inability to provide mechanistic details, we must admit that we have not made much progress since Gorham and Clendenning [31] suggested more than 30 years ago that anions act in the Hill reaction by "the provision of a favorable electro-kinetic potential at (the involved) colloidal surfaces". In their perceptive article they point out what is even more evident today than it was then, namely that anion requiring enzymes typically use water as a reactant. Indeed, one of them, pancreatic α-amylase, not only requires Cl^-, but Ca^{2+} as well [26]. Progress in the elucidation of the mechanism of any of these enzymes, including the water oxidase, may well depend on a better understanding how the water at their active sites is affected by the functional ions.

CONCLUSIONS

(1) The binding of Cl^- at the water oxidase complex "E" of chloroplast thylakoids is suggested to be contingent upon the protonation of functional groups with a $pK_a \approx 6$ according to the following mechanism:

$$E + H^+ \rightleftharpoons EH^+$$
$$+$$
$$Cl^- \rightleftharpoons EHCl$$

(2) The Cl^--binding sites presumably are provided by amino residues located in a special microenvironment, but other functional groups (e.g. on histidines) cannot be ruled out yet. Cooperativity of binding groups located on different polypeptides at the site of water oxidation is suggested by a dependence of the binding constants, and of the anion selectivity, on the polypeptide composition of the PSII preparations.

(3) When Cl^- has been removed from the site of water oxidation, the normal sequence of light mediated reactions according to the following scheme

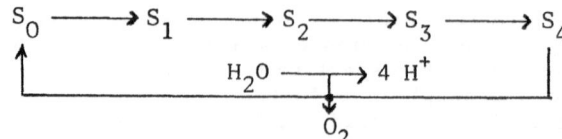

is changed to

$$(S_0 \dashrightarrow S_1) \longrightarrow S_2' \longrightarrow S_3' \nrightarrow .$$

In the defect reaction sequence, starting from the S_1 state which prevails in dark adapted thylakoids, S_2 and S_3 states with unique properties are formed, and a block exists after S_3. It is suggested that Cl^- depletion alters the oxidation potential of these S_2 and S_3 states by changing the pattern of protonation/deprotonation steps during the reaction, and abolishes the ability to generate a state capable of oxidizing water.

(4) Unique conformations of all polypeptides at the site of photosynthetic water oxidation are prerequisite for a most efficient catalysis of the oxidation-reduction and H^+ transfer reactions associated with water oxidation. Ca^{2+} ions aid in the maintenance of these conformations, and can to some extent take over such functions from at least one of the extrinsic polypeptides [13].

(5) The properties of water at the water oxidase, and how they are influenced by Cl^-, Ca^{2+} and the various polypeptides, deserve our attention.

Acknowledgements

These studies were aided by NSF grant PCM-8304416, and grant INT-8407399 of the U.S.-Japan Program of Cooperation in Photoconversion and Photosynthesis, awarded to PHH. PHH wishes to express gratitude to the staff of the Solar Energy Research Group of RIKEN (The Institute of Physical and Chemical Research, Wako-shi, Japan) and its director, Dr. Y. Inoue, for the help and hospitality he enjoyed during his research visit. Dr. I. Vass deserves special thanks for much advice and many stimulating discussions. The assistance of Ms. Laura A. Hagan with some experiments is gratefully acknowledged.

References

1. S. Izawa, R. L. Hind, and G. Hind, The role of chloride in photosynthesis, II, Biochim. Biophys. Acta. 180:388 (1969).
2. C. Critchley, I. C. Bainu, Govindjee and H. S. Gutowsky, The role of chloride in O_2 evolution by thylakoids from salt tolerant higher plants, Biochim. Biophys. Acta. 682:436 (1982).
3. S. M. Theg and P. H. Homann, Light- pH-, and uncoupler dependent association of chloride with chloroplast thylakoids, Biochim. Biophys. Acta 679:221 (1982).
4. J. M. Bové, C. Bové, F. R. Whatley and D. I. Arnon, Chloride requirement for O_2 evolution in photosynthesis, Z. Naturforsch. 18b:863 (1963).
5. P. O. Sandusky and C. F. Yocum, The Chloride requirement for photosynthetic oxygen evolution, Biochim. Biophys. Acta 766:603 (1984).
6. D. A. Berthold, G. T. Babcock and C. F. Yocum, A highly resolved, oxygen-evolving photosystem II preparation from spinach thylakoid membranes, FEBS Lett. 134:231 (1981).
7. M. Miyao and N. Murata, Partial disintegration and reconstitution of the photosynthetic oxygen evolving system, Biochim. Biophys. Acta 725:87 (1983).
8. A. W. Rutherford, A. R. Crofts and Y. Inoue, Thermoluminescence as a probe of photosystem II photochemistry, Biochim. Biophys. Acta 682:457 (1982).
9. P. M. Kelley and S. Izawa, The role of chloride in photosynthesis, I, Biochim. Biophys. Acta 502:198 (1978).
10. I. H. Segel, "Enzyme Kinetics", Wiley and Sons, Inc. New York (1975).
11. A. Imaoka, M. Yanagi, K. Akabori and Y. Toyoshima, Reconstitution of photosynthetic charge accumulation and oxygen evolution in $CaCl_2$ treated PSII particles, FEBS Lett. 176:341 (1984).
12. D. F. Ghanotakis, J. N. Topper and C. F. Yocum, Structural organization of the oxidizing side of photosystem II, Biochim. Biophys. Acta 767:524 (1984).

13. M. Miyao and N. Murata, Calcium ions can be substituted for the 24 kDa polypeptide in photosynthetic oxygen evolution, FEBS Lett. 168:118 (1984)

14. D. W. Becker and J. J. Brand, in "Advances in Photosynthesis Research", Vol. II, C. Sybesma ed., Martinus Nijhoff/Dr. W. Junk, Publ., The Hague (1983).

15. T. Ono and Y. Inoue, Requirement of divalent cations for photo-activation of the latent water oxidation systems in intact chloro-plasts from flashed leaves. Biochim. Biophys. Acta 723:191 (1983).

16. S. Itoh, C. T. Yerkes, H. Koike, H. H. Robinson and A. R. Crofts, Effects of chloride depletion on electron donation from the water oxidizing complex to the photosystem II reaction center as measured by the microsecond rise of chlorophyll fluorescence in isolated pea chloroplasts, Biochim. Biophys. Acta 766:612 (1984).

17. S. M. Theg, P. A. Jursinic and P. H. Homann, Studies on the mechanism of chloride action on photosynthetic water oxidation, Biochim. Biophys. Acta 766:636 (1984).

18. B. Kok, B. Forbush and M. McGloin, Cooperation of charges in photo-synthetic oxygen evolution, I, Photochem. Photobiol. 11:457 (1970).

19. I. Vass, G. Horvath, T. Herczeg and S. Demeter, Photosynthetic energy conservation investigated by thermoluminescence, Biochim. Biophys. Acta 634:140 (1981).

20. A. W. Rutherford, G. Renger, H. Koike and Y. Inoue, Thermo-luminescence as probe of PSII: The redox and protonation states of the secondary acceptor quinone and the O_2 evolving enzyme, Biochim. Biophys. Acta 767:548 (1984).

21. L. Zaki, Anion transport in red blood cells and arginine specific reagents, FEBS Lett. 169:234 (1984).

22. F. C. Kokesh and F. H. Westheimer, A reporter group at the active site of acetoacetate carboxylase, II, J. Am. Chem. Soc. 93:26 (1971).

23. B. R. Velthuys, Binding of the inhibitor NH_3 to the oxygen evolving apparatus of spinach chloroplasts, Biochim. Biophys. Acta 396:392 (1975).

24. T. Ono and Y. Inoue, S-state turnover in the O_2 evolving system of $CaCl_2$ washed Photosystem II particles depleted of three peripheral proteins as measured by thermoluminescence, Biochim. Biophys. Acta 806:331 (1985).

25. W. J. Coleman, I. C. Bainu, H. S. Gutowsky and Govindjee, The effect of chloride and other anions on the thermal inactivation of oxygen evolution in spinach thylakoids, in "Advances in Photosynthesis" Vol. I, I. C. Sybesma, ed., Martinus Nijhoff/Dr. W. Junk Publ. The Hague (1983).

26. M. Dixon and E. C. Webb, "Enzymes", Longman Group Ltd., London.

27. J. D. Johnson, V. R. Pfister and P. H. Homann, Metastable proton pools in thylakoids and their importance for the stability of photo-system II, Biochim. Biophys. Acta 723:256 (1983).

28. W. J. Coleman and Govindjee, The role of chloride in oxygen evolution in "Proceedings of the 16th FEBS Congress", VNU Science Press BV, Utrecht (1985).

29. N. Murata and M. Miyao, Organization of the photosynthetic oxygen evolving systems, in "The Oxygen Evolving System of Photosynthesis", Y. Inoue et al, eds., Academic Press, Tokyo (1983).

30. R. J. P. Williams, Calcium chemistry and its relation to protein binding, in "Calcium Binding Proteins and Calcium-Function", R. H. Wasserman et al, eds., North Holland, New York (1977).

31. P. R. Gorham and K. A. Clendenning, Anionic stimulation of the Hill reaction in isolated chloroplasts, Arch. Biochem. Biophys. 37:199 (1952).

THE ROLE OF WATER-SOLUBLE POLYPEPTIDES AND CALCIUM IN PHOTOSYNTHETIC OXYGEN EVOLUTION

Demetrios F. Ghanotakis and Charles F. Yocum

Division of Biological Sciences and Department of
Chemistry, The University of Michigan, Ann Arbor
MI 48109

INTRODUCTION

The oxidizing side of photosystem II operates at unusually high redox potentials ($2H_2O \rightarrow O_2 + 4H^+ + 4e^-$, $E_0' = + 0.82V$), and this implies the presence of a structural environment which can contribute to the stabilization of the reactive chemical intermediates formed during water oxidation. The photocatalysts of PSII activity are thought to be associated with two hydrophobic proteins of molecular weights of 43 and 47 kDa found in the PSII "core" complex as described by Satoh, et al. (1). This complex, although photochemically active, does not possess the capacity to evolve oxygen. A substantial body of experimental evidence now supports the view that three water-soluble proteins (33, 23 and 17 kDa) form part of the structure of the oxygen evolving apparatus. These water soluble polypeptides are believed to be essential for productive binding of inorganic cofactors such as Mn, Cl^- and Ca^{2+} within the oxygen-evolving complex. The involvement of Mn in water oxidation is well-documented (2-5). The implication of Cl^- as a cofactor for oxygen evolution activity by Warburg and Luttgens (6) and later by Bové, et al. (7) has been confirmed by Izawa and his colleagues (8,9). The presence of the anion stabilizes activity against inactivation by heat, elevated pH and amines (10-13); Cl^- probably binds to a site on, or very near, Mn (13,14). Recent work has established the existence of a high-affinity Ca^{2+} binding site within the oxygen-evolving complex (15-18). Our own observations indicate that the 17 and 23 kDa polypeptides are important for the structural integrity of the oxidizing side of PSII (19,20). The presence of the 17 and 23 kDa species forms an environment which favors tight binding of Ca^{2+} (19) and Cl^- (21,22), and also shields the manganese complex from destruction by bulky reductants (20). Many studies have provided data to indicate that the 33 kDa protein is somehow associated with manganese binding to the PSII complex, although there are reports in the literature which describe conditions for release of the 33 kDa polypeptide which produce minimal effects on Mn binding to the PSII complex (23,24). To further investigate the involvement of calcium in oxygen evolution activity, we have used various lanthanides which are known to substitute for calcium in other calcium-binding proteins isolated from biological systems (25). In this paper, we show that calcium and lanthanum compete for sites on the oxidizing side of photosystem II, and in addition we report conditions which allow release of the 33 kDa protein without extraction of manganese.

MATERIALS AND METHODS

Subchloroplast membranes, free of Photosystem I and having high rates of oxygen evolution, were prepared as in ref. 26. Complete release of the 17 and 23 kDa polypeptides from the PSII complex was carried out by incubation of the preparation in 2 M NaCl (pH 6.0) on ice for 1 hr in the dark. After centrifugation the membranes were washed once with SMN (0.4 M sucrose/50 mM Mes, pH 6.0/15 mM NaCl) and then stored in the same buffer. Chloride depletion of high-salt treated PSII membranes was carried out by two 1 hr-dialysis steps against a medium containing 0.4 M sucrose/50 mM Mes, pH 6.0/10 mM NaF. Oxygen evolution activity was assayed using a Clark-type electrode and the polypeptide content of the membranes was examined by gel electrophoresis. Determination of the Mn content of PSII preparations was carried out by use of EPR spectroscopy as described in ref. 26, but using $HClO_4$ instead of HCl for acidification. EPR spectroscopy was carried out with a Bruker ER-200D spectrometer operated at X-band.

RESULTS

Structural Role of 17 and 23 kDa Polypeptides

Exposure of PSII membranes to 2 M NaCl produces an extensive depletion of two polypeptides (17 and 23 kDa) concomitant with loss of O_2 evolution activity. An EPR study of the decay of flash-induced Z^+ in salt washed PSII membranes allowed us to localize the site of inhibition at the oxidizing side of Photosystem II (27). Åkerlund et al. (28) and Dekker et al. (29) reported an enhanced rate of back reaction, $P_{680}^+ Q_A^- \longrightarrow P_{680}Q_A$ in salt-washed membranes, due to an inhibition between the oxygen evolving complex and Z. Even though rebinding of the 17 and 23 kDa polypeptides to the salt-washed PSII complex restores oxygen evolution activity (19,30), addition of high concentrations (>5 mM) of Ca^{2+} to polypeptide-depleted PSII membranes also reactivates the oxygen evolving complex (Table I). When a dialysis procedure was used to rebind the 17 and 23 kDa proteins the role of Ca^{2+} and the 17, 23 kDa species could be explored in more detail. Dialysis executed under conditions which excluded Ca^{2+}, by use of EGTA, did not reconstitute oxygen evolution activity, even though the two polypeptides did rebind (19). Subsequent addition of Ca^{2+} then reactivated the oxygen evolving complex; a Ca^{2+} titration of the reactivation process showed that the presence of the 17 and 23 kDa polypeptides resulted in a substantially lowered Ca^{2+} requirement for maximal activity. Åkerlund et al. (21) and Akabori et al. (22) have shown that the salt-washed PSII complex also required higher concentrations of Cl^- (>10 mM) compared to the intact PSII system. As shown in Table I PSII membranes depleted of the 17 and 23 kDa polypeptides as well as Cl^- require addition of both Ca^{2+} and Cl^- for high rates of oxygen evolution activity. It is shown in the same Table that even though the cation effect is specific for Ca^{2+}, Cl^- can be replaced by Br^- and NO_3^- (13,14). It is thus apparent that although release of the 23 and 17 kDa polypeptides by salt-washing does not extract the pool of functional Mn, this treatment causes structural changes in the PSII complex which result in a requirement for high, non-physiological, concentrations of Cl^- and Ca^{2+}. The structural role of the 17 and 23 kDa proteins was also demonstrated by our own observations regarding the accessibility of the Mn complex to exogenous reductants; using high salt treated PSII membranes we have shown that in the absence of the 17 and 23 kDa species, exogenous reductants reduce and destroy the Mn-complex (20).

Use of Lanthanides as a Probe for Studying Calcium Binding Sites

As shown in Table II, high concentrations of $CaCl_2$ reactivate oxygen evolution-activity even in the absence of the 17 and 23 kDa polypeptides. When sufficient calcium is present during exposure to a lanthanide, the PSII complex retains activity. However, calcium does not reactivate PSII membranes previously treated with La^{3+}, Tb^{3+} or Ce^{3+} (Table II); this observation indicates that an irreversible inhibition is induced by the lanthanide in the absence of calcium. In order to determine whether the Mn-complex was perturbed, we exposed high-salt treated PSII membranes to $LaCl_3$ and then examined their Mn and polypeptide content as well as oxygen evolution activity in the presence of $CaCl_2$. The results of such an experiment are summarized in Table III; as shown in this Table, exposure to $LaCl_3$ in the absence of $CaCl_2$ results in destruction of the manganese complex with concomitant release of the 33 kDa protein. The presence of

Table I. Effect of Cations and Anions on Oxygen Evolution Activity of Cl^--Depleted High-Salt Treated PSII Membranes

Additions	% Activity[a]
None	4
NaCl (20 mM)	26
$Ca(Ac^-)_2$ (10 mM)	8
$CaCl_2$ (10 mM)	80
$CaBr_2$ (10 mM)	76
$Ca(NO_3)_2$ (10 mM)	46
$MgCl_2$ (10 mM)	28
$SrCl_2$ (10 mM)	32

[a]The PSII complex was illuminated with continuous light in a solution containing 50 mM Mes, pH:6.0, 0.4 M sucrose, the salt indicated above, 10 µg Chl/ml and 400 µM DCBQ (Control activity: 720 µmol O_2·mg Chl^{-1}· hr^{-1}).

Table II. Effect of Lanthanides on Oxygen Evolution Activity in PSII Preparations Depleted of the 17 and 23 kDa Polypeptides

Additions	% Activity[a]
None	24
$CaCl_2$	81
$LaCl_3$ (1 mM) + $CaCl_2$ (20 mM)	77
$LaCl_3$ (1 mM) + $MgCl_2$ (20 mM)	10
$LaCl_3$ (1 mM); $CaCl_2$ (20 mM)[b]	8
$TbCl_3$ (1 mM); $CaCl_2$ (20 mM)[b]	10
$CeCl_3$ (1 mM); $CaCl_2$ (20 mM)[b]	14

[a]High-salt (2M NaCl) treated PSII membranes (10 µg Chl/ml) were incubated for 10 min at 25°C in a solution containing 0.4 M sucrose/15 mM NaCl/50 mM Mes, pH:6.0, along with the salts indicated above. After incubation, the membranes were assayed for oxygen evolution activity with 400 µm DCBQ as an acceptor (Control activity:660 µmoles O_2/hr/mg Chl).
[b]5 min incubation with the lanthanide was followed by addition of $CaCl_2$ and another 5 min incubation.

calcium during treatment with LaCl₃ protects against manganese release and the system retains its oxygen evolving capacity when calcium is present during assay. Thus, the data presented in Table III explain the irreversible inhibition by lanthanides, when calcium was not included in the inactivation incubation medium, observed in Table II. A titration of the inhibitory effect of lanthanum in the presence of various concentrations of $CaCl_2$ is shown in Fig. 1A; Fig. 1B shows that the slope of the lines in 1A varies linearly with the inverse of the $CaCl_2$ concentration. These results indicate that lanthanum and calcium compete for the same site(s); the estimated constants are $K_I \sim 50$ µM for lanthanum and $K_M = 0.6$ mM for calcium in preparations of the PSII complex which have been depleted of the 17 and 23 kDa polypeptides.

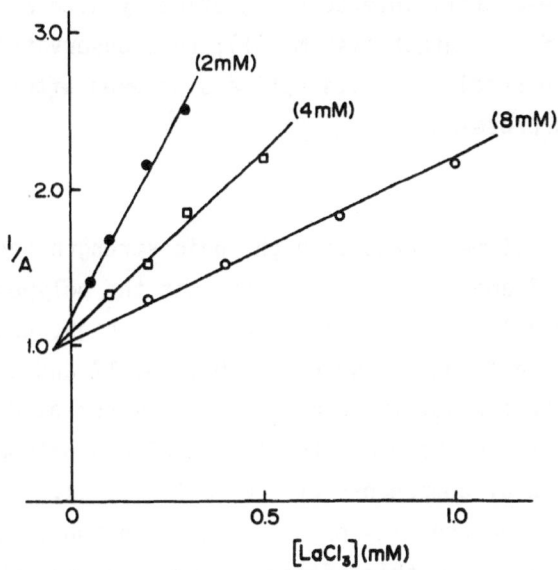

Fig. 1A. Inhibitory effect of LaCl₃ in the presence of 2 (●), 4 (□) or 8 (○) mM $CaCl_2$ in high-salt treated PSII membranes. Experimental conditions as in Table I.

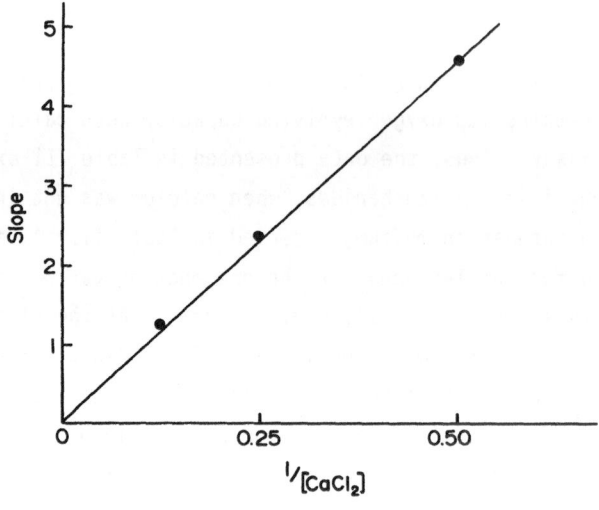

Fig. 1B. Slope replot from Fig. 1A.

The data presented above suggest that replacement of calcium by a lanthanide is followed by destruction of the Mn-complex and concomitant release of the 33 kDa polypeptide. Since Cl^- has been shown to protect against manganese release under conditions which perturb binding of the 33 kDa species (23,24), we repeated the dialysis against $LaCl_3$ in the presence of 2 M NaCl. As shown in Table III and Fig. 2 this treatment releases the 33 kDa polypeptide, but not Mn; this modified PSII complex shows no oxygen evolution activity even when assayed in the presence of high concentrations of $CaCl_2$. A study of the microwave power saturation of Z^+, in the polypeptide-depleted PSII complex which retains Mn has shown that manganese still interacts magnetically with Z^+ in a manner similar to that of the intact system (31); this observation suggests that manganese has been retained in its native site even after removal of the 17, 23 and 33 kDa proteins.

DISCUSSION

Exposure of PSII membranes to high ionic strength results in release of two proteins, 17 and 23 kDa. The fact that the polypeptide-depleted PSII complex is highly active in the presence of high, non-physiological, concentrations of Ca^{2+} and Cl^- suggests that the 17 and 23 kDa species are not involved in the catalytic mechanism of water oxidation. Under physiological conditions however, the 17 and 23 kDa polypeptides are significant structural components of the PSII complex. It has been shown that the 17 and 23 kDa proteins promote tight binding of both Ca^{2+} and Cl^- (19,21,22); speculating on this effect, we suggest that the 17 and 23 kDa species affect the structure of putative Ca^{2+} and Cl^- receptor

Table III.   Effect of Lanthanum on Activity, Mn Content and Binding of the 33 kDa Polypeptide in High-Salt Treated PSII Membranes.

Additions During Treatment[a]	Mn Content	33 kDa Bound	% Activity[b] (+ 10 mM $CaCl_2$)
30 mM $CaCl_2$	96	+	78
1 mM $LaCl_3$	28	-	8
1 mM $LaCl_3$ + 30 mM $CaCl_2$	94	+	76
1 mM LaCl + 2 M NaCl	96	-	4[c]

[a]High-Salt treated PSII membranes (2.5 mg Chl/ml) were dialyzed for 6 hrs at 4°C against a medium containing 0.4 M sucrose/50 mM Mes, pH:6.0 and the salts indicated above.   After dialysis the membranes were collected by centrifugation; washed once with a medium containing 0.4 M sucrose/50 mM Mes, pH:6.0/100 mM NaCl, and finally were resuspended in the same medium.
[b]Control Activity: 650 μmoles O_2/hr/mg Chl.
[c]An increased concentration of NaCl (100 mM) during assay had very little effect on oxygen evolution activity.

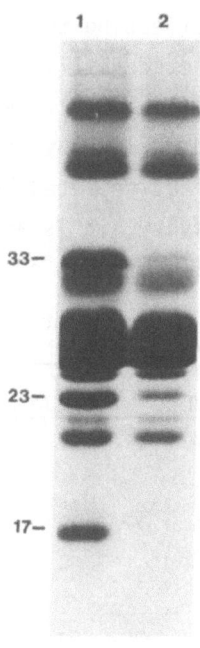

Fig. 2.   Gel electrophoresis patterns of PSII membranes.   a) Control membranes; b) PSII membranes treated with 2M NaCl + 1 mM $LaCl_3$.

protein(s) favoring tight binding of both inorganic ions; alternatively, they form part of a structural domain which concentrates Ca^{2+} and Cl^-. A structural role for the 17 and 23 kDa polypeptides has been further demonstrated by our own observations regarding the accessibility of the Mn-complex to exogenous compounds. We have shown that the two proteins form a structural barrier which prevents bulky reductants from reducing and destroying the manganese complex (20). The presence of a structural barrier at the oxidizing side of PSII, which selectively allows only molecules of a certain size to penetrate and react with the OEC, has been also demonstrated by Radmer and Ollinger (32). The third water-soluble protein, which is a component of the oxygen-evolving complex, is the 33 kDa species. Research in several laboratories, using PSII-enriched membranes, has shown that the presence of the 33 kDa protein can be correlated with binding and stability of functional manganese within the oxygen-evolving complex. Of considerable interest is the apparent correlation between extraction of Mn and the 33 kDa protein. Even though treatments such as exposure to high concentrations of Tris buffer (or to NH_2OH, Urea, or high pH) which release Mn also release the 33 kDa protein (see Table IV and ref. 33), there are conditions which selectively extract either Mn or the 33 kDa species. As shown in Table IV, exposure of the (17,23 kDa)-depleted PSII complex to a reductant releases most of the manganese while leaving the 33 kDa polypeptide membrane-bound. Release of the 33 kDa protein without affecting the Mn-complex is also possible by exposure of (17,23 kDa)-depleted PSII membranes to $LaCl_3$ in the presence of high concentrations of NaCl (Table IV). A Mn-retaining extraction of the 33 kDa species was reported earlier by Ono and Inoue (24) and by Miyao and Murata (23). Since the 33 kDa-depleted Mn-retaining PSII complex shows low oxygen evolution activity, it is apparent that the 33 kDa protein is somehow involved in productive binding of functional manganese.

Table IV. Effect of Various Treatments on the 17, 23, 33 kDa and Manganese Content of PSII Membranes.

Treatment	33kDa	23kDa	17kDa	Mn Content (%)
None	+	+	+	100
0.8 M Tris, pH:8.0	-	-	-	8
2 M NaCl, pH:6.0	+	-	-	96
i) 2 M NaCl, pH 6.0[a] ii) 5 mM HQ	+	-	-	15
2 M NaCl + 1 mM $LaCl_3$, pH:6.0	-	-	-	98

[a]Treatment of the PSII complex with 2 M NaCl was followed by a wash with SMN and subsequent exposure to the reductant.

Cations like the lanthanides, which have ionic radii close to that of calcium, effectively substitute for the latter even though in most cases the resulting proteins have no physiological activity (25). In this communication we have used lanthanides as a probe of the calcium binding site(s) associated with the oxidizing side of photosystem II. As we show here, substitution of calcium by a lanthanide results in destruction of the Mn-complex with concomitant release of the water-soluble 33 kDa polypeptide; this observation suggests that calcium is acting close to manganese. It is possible that the oxygen evolving complex has a Concanavalin-A type organization with calcium and manganese interacting closely with one another. In such a model calcium would be important for an active conformation of the Mn-complex; replacement of calcium by a lanthanide, which has the same ionic radius but different charge and coordination properties, would cause drastic conformational changes in the Mn-complex which would result in its destruction. Our observations demonstrate that the 17, 23 and 33 kDa polypeptides are organized so as to provide a structure which stabilizes and also shields the Mn-complex from reaction with exogenous reductants. Based on the close interaction between calcium and manganese demonstrated in the present work, we would propose the model shown in Fig. 3. This model is still tentative, and further work is required in order to localize precisely the binding sites for manganese, calcium and chloride.

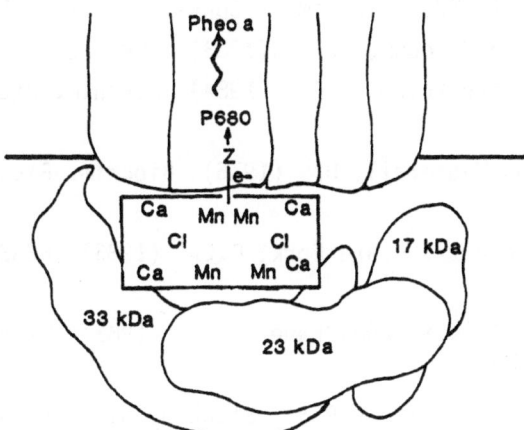

Fig. 3. Structural organization of the oxidizing side of PSII. A possible arrangement of the water-soluble polypeptides involved, as well as the inorganic species important for photosynthetic water oxidation are shown.

ACKNOWLEDGEMENT

The research described in this communication was supported by a grant (PCM82-14240) from the National Science Foundation. We thank Prof. G. T. Babcock for helpful discussions.

REFERENCES

1. Satoh, K., Nakatani, H.Y., Steinbach, K.E., Watson, J. and Arntzen, C.J. (1983) Biochim. Biophys. Acta 724, 142-150.

2. Radmer, R. and Cheniae, G.M. (1977) in Primary Processes in Photosynthesis (Barber, J., ed.) Elsevier/North Holland Biomedical Press, Amsterdam, pp. 303-348.

3. Amesz, J. (1983) Biochim. Biophys. Acta 725, 1-12.

4. Cheniae, G.M. (1980) in Methods in Enzymology (San Pietro, A., ed.) Academic press, New York, Vol. 69, pp. 349-363.

5. Andreasson, L.E., Hansson, O. and Vanngard, T. (1983) Chemica Scripta 21, 71-74.

6. Warburg, O. and Luttgens, W. (1944) Naturwissenchaften 32, 301.

7. Bovè, J.M., Bovè, C., Whatley, F.R. and Arnon, D.I. (1963) Z. Naturforch. 186, 683-688.

8. Hind, G., Nakatani, H.Y. and Izawa, S. (1969) Biochim. Biophys. Acta 172, 277-289.

9. Kelly, P.M. and Izawa, S. (1978) Biochim. Biophys. Acta 502, 198-210.

10. Theg, S. and Homann, P. (1982) Biochim. Biophys. Acta 679, 221-234.

11. Coleman, W.J., Baianu, I.C., Gutowsky, H.S. and Govindjee (1984) in Advances in Photosynthesis Research (Sybesma, C., ed.) M. Nijhoff/Dr. W. Junk, The Hague, Vol. I, pp. 283-286.

12. Critchley, C. (1983) Biochim. Biophys. Acta 724, 1-5.

13. Sandusky, P.O. and Yocum, C.F. (1983) FEBS Lett. 162, 339-343.

14. Sandusky, P.O. and Yocum, C.F. (1984) Biochim. Biophys. Acta 766, 603-611.

15. Picconi, R. and Mauzerall, D. (1976) Biochim. Biophys. Acta 423, 605-609.

16. Brand, J.J., Mohanty, P.and Fork, D.C. (1983) FEBS Lett. 155, 120-124.

17. Barr, R., Troxel, K.S. and Crane, F.L. (1983) Plant Physiol. 72, 309-315.

18. Ono, T.A. and Inoue, Y. (1983) Biochim. Biophys. Acta 723, 191-201.

19. Ghanotakis, D.F., Topper, J.N., Babcock, G.T. and Yocum, C.F. (1984) FEBS Lett. 170, 169-173.

20. Ghanotakis, D.F., Topper, J.N. and Yocum, C.F. (1984) Biochim. Biophys. Acta 767, 524-531.
21. Andersson, B., Critchley, C., Ryrie, I.J. Jansson, C., Larsson, C., and Anderson, J.M. (1984) FEBS Lett. 168, 113-117.
22. Akabori, K., Imaoka, A. and Toyoshima, Y. (1984) FEBS Lett. 173, 36-40.
23. Miyao, M. and Murata, N. (1984) FEBS Lett. 170, 350-354.
24. Ono, T.-A. and Inoue, Y. (1984) FEBS Lett. 168, 281-286.
25. Kretsinger, R.H. and Nelson, D.J. (1976) Coord. Chem. Rev. 18, 29-124.
26. Ghanotakis, D.F., Babcock, G.T. and Yocum, C.F. 91984) Biochim. Biophys. Acta 765, 388-398.
27. Ghanotakis, D.F., Babcock, G.T. and Yocum, C.F. (1984) FEBS Lett. 167, 127-130.
28. Åkerlund, H.-E., Brettel, K. and Witt, H.T. (1984) Biochim. Biophys. Acta 765, 7-11.
29. Dekker, J.P., Ghanotakis, D.F., Plijter, J.J., Van Gorkam, H.J. and Babcock, G.T. (1984) Biochim. Biophys. Acta 767, 515-523.
30. Åkerlund, H.-E., Renger, G., Weiss, W. and Hagemann, R. (1984) Biochim. Biophys. Acta 765, 1-6.
31. Ghanotakis, D.F., Babcock, G.T. and Yocum, C.F. (1985) Biochim. Biophys. Acta, in press.
32. Radmer, R. and Ollinger, O. (1983) FEBS Lett, 152, 39-43.
33. Yamamoto, Y., Doi, M., Tamura, N. and Nishimura, M. (1981) FEBS Lett. 133, 265-268.

MODULATION OF THE HILL REACTION RATES BY IONS

INTERACTING WITH THE OUTER SURFACE OF CYANOBACTERIAL THYLAKOIDS

G. Sotiropoulou and G. C. Papageorgiou

National Research Center Demokritos
Department of Biology
Athens, 153 10 Greece

INTRODUCTION

Polypeptide-complexed ions (Mn^{2+}, Ca^{2+}, Cl^-) are indispensible components of the hydrophilic domain of the photosystem II complex of higher plants and probably of cyanobacteria (reviewed by Ghanotakis and Yocum, ref. 1). In addition, many studies suggest that diffusible cations exert a controlling influence on photoinduced electron transport in oxygenic plant thylakoids. Particularly prominent is the role of cations in cyanobacteria.

Early research on thylakoid fragments from these prokaryotes stressed the Mg^{2+} requirement for efficient Hill reactions.[2-4] Subsequently, in a systematic study on membranes obtained by osmolysis of Phormidium luridum spheroplasts, Binder et al[5] noted a much stronger stimulation of the FeCN-supported O2 evolution by $CaCl_2$ (x53) than by $MgCl_2$ (x18), or KCl (x6). Maximal stimulation required approximately 0.02 M divalent metal chloride and 0.2-0.3 M monovalent metal chloride. Similar observations in Anacystis nidulans thylakoid fragments, obtained by French press disruption, led Piccioni and Mauzerall[6] and Brand[7] to advocate a specific role for the Ca^{2+} ion in cyanobacteria. Brand[7] further suggested that Ca^{2+} is effective only when trapped inside the resealed thylakoid vesicle, in view of the fact that optimized activities were possible only when Ca^{2+} was present in the cell breakage medium. As reported by DeRoo and Yocum[8] in the case of sonicated Spirulina platensis vesicles, by Wavare and Mohanty[9] in the case of Synechococcus cedrorum spheroplasts, and by us in the case of Anacystis nidulans permeaplasts,[10-11] cations are ranked in increasing valence order (trivalent > divalent > monovalent) with respect to their ability to stimulate the FeCN Hill reaction. Finally, it has been generally observed[2-11] that at higher cation concentrations the stimulation of electron transport reverts to an inhibition.

It is quite likely that membrane fragments prepared by osmotic or mechanical cell rupture consist of mixed populations of properly and improperly sealed vesicles, as well as of imperfectly sealed vesicles. In

---

Abbreviations: Chl, chlorophyll; DBMIB, dibromothymoquinone; DCMU, 1,1-dimethyl-3-(3,5-dichlorophenyl)-urea; DPC, 1,5-diphenylcarbazide; FeCN, potassium ferricyanide; Hepes, N-2-hydroxyethyl piperazine-N'-2-ethane-sulfonic acid; MV, methylviologen.

such preparations, it is not possible to define membrane orientation. On the other hand, osmoresistant, ion-permeabilized cells (permeaplasts), having ion-impermeable thylakoids, afford the experimenter the possibility to examine ionic interactions occuring strictly on the outer thylakoid surface. In the present work, we have employed Anacystis nidulans permeaplasts in order to investigate the effects of medium electrolytes on the photoinduced electron transport of thylakoids.

MATERIALS AND METHODS

Anacystis nidulans was cultured photoautotrophically as in ref. 12. Permeaplasts were prepared by partial digestion of the cell wall peptidoglycan as in ref. 13. Permeaplast stocks were maintained in 0.5 M sorbitol, 0.05 M Hepes.NaOH, pH 7.5, at 0.45 mg Chl a/ml. Assays were performed on permeaplasts resuspended in 0.5 M sorbitol, 0.005 M Hepes.NaOH, pH 7.5

Oxygen evolution was measured polarographically, under saturating (550 W/m^2), heat-filtered polychromatic illumination, as described in ref. 12.The reaction mixtures contained permeaplasts equivalent to 15 µg Chl a/ml, and depending on the particular assay the following: FeCN, 1 mM; DCIP, 34 µM; Na-ascorbate, 2 mM; methylviologen, 100 µM; and DCMU, 20 µM. DCIP photoreduction was measured by difference absorption spectrophotometry, with a Hitachi Modell 557 spectrophotometer. Saturating actinic light (162 W/m2; red glass filter C.S. 2-64) was directed to the sample at a right angle to the measuring beam via a flexible light guide. The reference cuvette was completely shielded from the actinic light, and the photomultiplier was protected by two guard filters (C.S. 4-96 and interference filter with λ max = 580 nm and $\Delta\lambda$ = 12 nm). Samples, in assay buffer, contained permeaplasts equivalent to 5 µg Chl a/ml, 10 µM DCIP, and depending on the assay 0.5 mM DPC and 10 µM DCMU. An absorbance coefficient of 19.8/mM.cm at 580 nm and pH 7.5 was used to convert absorption readings to DCIP concentrations.[14]

Incident light intensities were measured with a Lambda Instruments Model LI-185A photometer. Chl a was assayed in methanolic extracts as in ref. 15. All experimental procedures were carried out at room temperature.

RESULTS

We have shown[13] that permeaplasts prepared by short (15-20 min) treatment of Anacystis with lysozyme in hypotonic medium (0.05 M Hepes. NaOH, pH 7.5) contain virtually intact, ion-impermeable and fully functional thylakoids. Thylakoid impermeability to H+ can be inferred from data such as those displayed in Table 1, which show that permeaplasts are capable of photosynthetic control. In addition, we have shown elsewhere that permeaplast thylakoids are impermeable to K$^+$ ions.[16] The phycobilisomes, on the other hand, are detached from the thylakoid surface and are dissociated to phycobiliprotein oligomers.[13] This evidences communication of the low ionic strength conditions of the medium to the cytoplasm, hence cell envelope permeabilization. Electrolytes present in the suspension medium pass freely to the cytoplasm, where they interact with the outer surface of thylakoids. The dismantling, however, of the phycobilisome light-harvesting system shifts light saturation to higher actinic light intensities.

Results will be discussed with reference to the electron transport sequence of Scheme I. Fig. 1 displays the variation of the rate of O2 evolution in the presence of FeCN as a function of the valence and the concentration of cations in the suspension medium. Rates increase up to a maximum and then decrease, resulting in the bell-shaped curves shown. Cations are ranked in increasing valence order with regard to their ability

Table 1. Effect of phosphorylation cofactors and of uncoupler (NH$_4$Cl) on the rate of FeCN-supported photosynthetic O$_2$ evolution by Anacystis nidulans permeaplasts, measured under nonsaturating and saturating actinic illumination. In parentheses, relative rate values.

	Electron transport rate, μequiv/mg Chl a·h	
Additions	30 W/m^2	550 W/m^2
None	57 (100)	128 (100)
ADP, phosphate, MgCl$_2$	75 (132)	198 (155)
" " " , NH$_4$Cl	131 (230)	323 (252)

either to promote, or to inhibit the FeCN Hill reaction, implicating electrostatic repulsion and screening as rate determining factors. These results resemble closely results obtained with cell-free cyanobacterial thylakoid fragments [5,8] and spheroplasts. [9]

The Gouy-Chapman theory requires the space function of the electrostatic potential in a membrane-aqueous phase system to depend only on the bulk concentration and the ion valence of the electrolyte present.[17] Accordingly, cations of the same valence should be indistinguishable with regard to their charge screening ability and its consequences on the photoreduction of FeCN. However, as Fig. 2 shows, monovalent cations deviate from this expectation.

Wax and Lockau [18] reported that the plastoquinone antagonist DBMIB fails to inhibit fully the reduction of FeCN by cyanobacterial thylkoids, while it obliterates the photoreduction of NADP$^+$ (see Scheme I). This implies that part of FeCN is reduced ahead of the DBMIB inhibition site. The same is true for Anacystis permeaplasts (Fig. 3). Approx. 50% of the activity survives at maximal inhibition (i.e at 1 μM DBMIB).

Figure 4 shows the consequences of increasing concentrations of MgCl$_2$ and CaCl$_2$ on the FeCN reduction rate, either in the absence, or in the presence of 2.5 μM DBMIB. From the ordinate intercepts, we may infer fewer (by 30-40 %) thylakoid surface sites, both for Ca^{2+} and Mg^{2+} in the presence of DBMIB, and fewer sites for Mg^{2+} relative to Ca^{2+}. Finally, the dissociation constants calculated from the ordinate intercepts and the slopes are in the range of 0.45 to 0.55 mM for both cations.

While electrostatic screening of the negative surface charge by cations should promote the photoreduction of anionic acceptors like FeCN, it should have the opposite effect on the photoreduction of cationic acceptors like methylviologen. Fig. 5 shows, however, a qualitatively similar response of methylviologen mediated O$_2$ uptake by permeaplasts, with ascorbate reduced DCIP as the source of electrons (DCIPH2 ⟶ MV; see Scheme I), as for the FeCN Hill reaction.

In contrast to our results, Wavare and Mohanty [9] found no stimulation, but only inhibition of the DCIPH$_2$ ⟶ MV reaction in S. cedrorum spheroplasts by divalent and trivalent metal cations. A possible explanation is that long incubations with lysozyme, which is a prerequisite for the

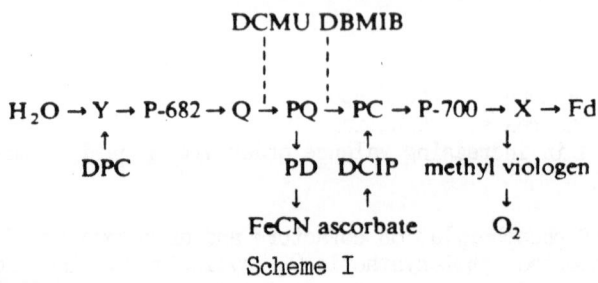

Scheme I

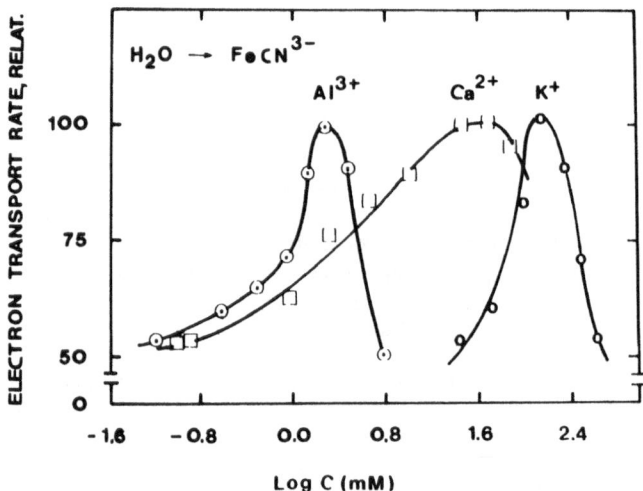

Fig. 1. Dependence of the FeCN supported photosynthetic O_2 evolution by Anacystis permeaplasts on the valence and the concentration of cations present in the suspension medium. Maximal rates in μequiv. / mgChl/h: $AlCl_3$ 248; $CaCl_2$, 497; KCl, 194.

the conversion of Synechococcus species to spheroplasts, causes light-induced, DCMU-insensitive $O2$ uptake, via an endogenous autoxidizable intermediate (first reported by Honeycutt and Krogmann; ref. 18). This activity is indeed suppressed by cations [20]. Permeaplasts, on the other hand, prepared by short treatments with lysozyme, do not develop the endogenous autoxidizable system, and thus they are free of such complications. In fact, this is one of the advan- tages of using permeaplasts instead of spheroplasts.

Figure 6 examines the effect of monovalent cations on the $DCIPH_2$ ⟶ MV reaction by Anacystis permeaplasts. Here in contrast to the FeCN Hill reaction (Fig. 2), curves peak at approximately the same concentration, but their shapes and absolute rate maxima differ. The nonmetal cholinate causes the least stimulation (only about 10%). One logical hypothesis is that cholinate does not bind specifically to membrane ligands while metal cations do. Thus, metal cations suppress both the negative surface charge and the potential, while cholinate suppresses only the latter.

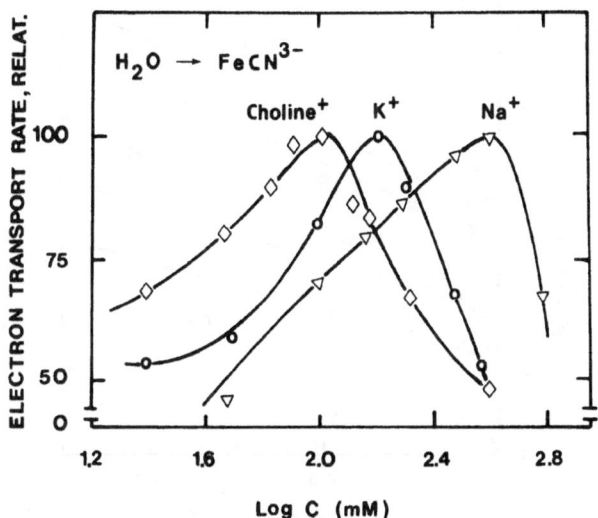

Fig. 2. Dependence of the rate of FeCN-supported photosynthetic O_2 evolution by Anacystis permeaplasts on the concentration of monovalent cations in the suspension medium. Maximal rates: Choline-Cl, 303; KCl, 497; NaCl, 177.

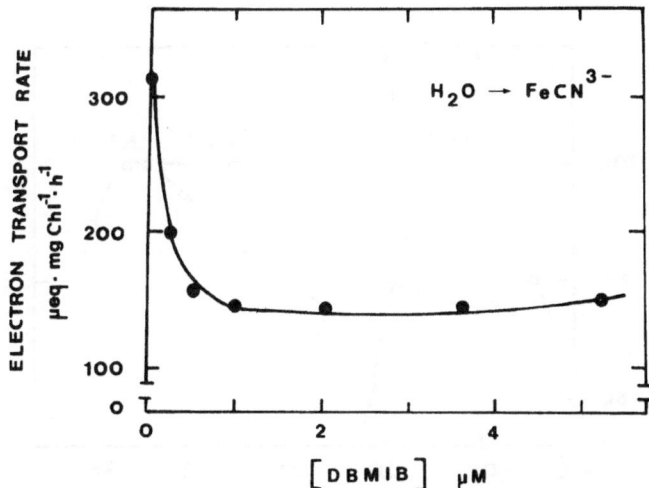

Fig. 3. Effect of DBMIB on the FeCN-supported photosynthetic O_2 evolution by Anacystis permeaplasts.

In the experiment of Fig 7, we examined the effect of several metal cations on the $H_2O \rightarrow FeCN$ reaction, on a constant background concentration of 150 mM choline-Cl, in order to suppress the negative surface potential. KCl exerts litle effect under these conditions, while prominent stimulations are caused by $MgCl_2$ and particularly by $CaCl_2$. We would like to ascribe this to specific binding of the divalent caations to ligands on the outer thylakoid surface. Calculated dissociation constants are 0.45 mM for Ca^{2+} and 0.96 for Mg^{2+}.

The stimulation of electron transfer reactions between membrane embedded intermediates and medium oxidoreducible medium solutes by cations at a lower concentration range, followed by inhibition at a higher

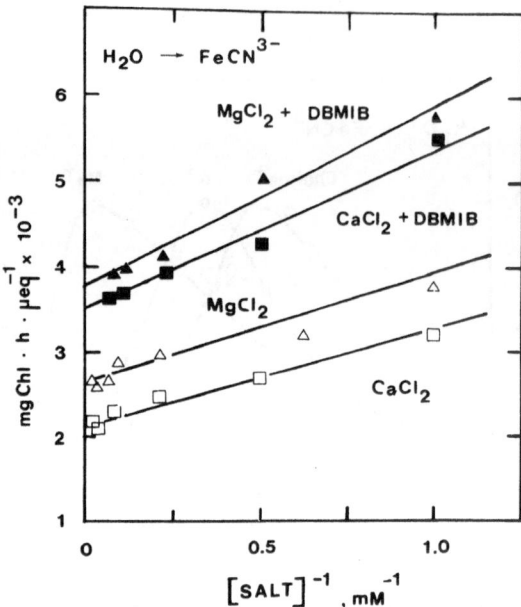

Fig. 4. Lineweaver-Burk plots of the FeCN-supported photosynthetic O_2 evolution by Anacystis permeaplasts as a function of the concentrations of $MgCl_2$ and $CaCl_2$ in the suspension in medium, in the absence and in the presence of 2.5 μM DBMIB.

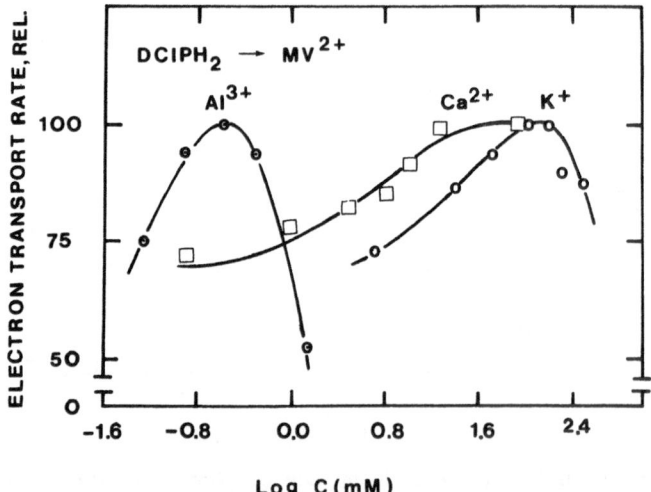

Fig. 5. Dependence of the rate of photosynthetic electron transport across photosystem I ($DCIPH_2 \rightarrow MV$) by Anacystis permeaplasts on the valence and the concentration of cations in the suspension medium. Maximal rates: $AlCl_3$, 521; $CaCl_2$, 475; KCl, 486.

concentration range, appears to be a typical response observed with neutral, anionic and cationic solutes. The characteristic valence hierarchy obeyed, and the concentrations required for stimulation and inhibition, implicate electrostatic interactions between membranes and aqueous phase cations. Another deactivation factor is possibly the chaotropic effect of the anions, also a function of the bulk electrolyte concentration. We examined this possibility in the experiment of Fig. 8, where we studied the effect of several K-salts and of choline-Cl on the

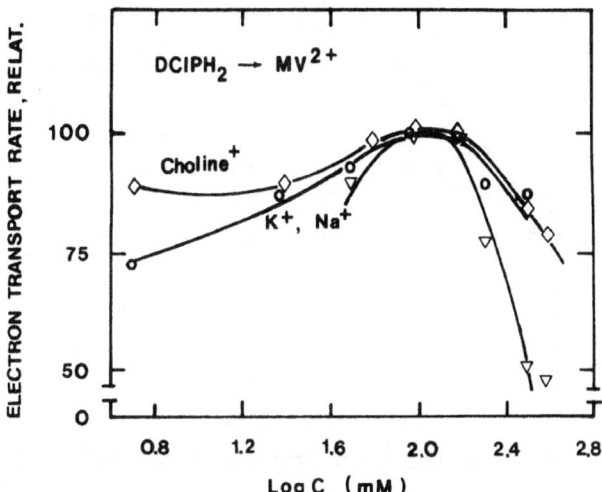

Fig. 6. Dependence of the rate of photosynthetic electron transport across photosystem I (DCIPH₂ → MV) by Anacystis permeaplasts on the concentration of monovalent cations. Maximal rates: NaCl, 679; KCl, 486; choline-Cl, 316.

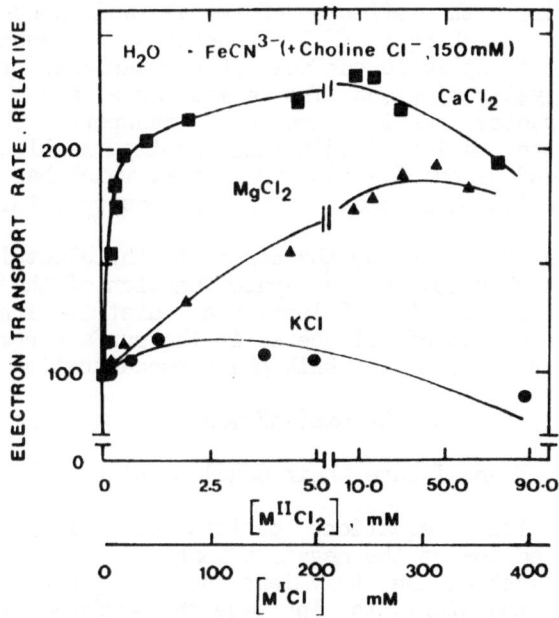

Fig. 7. Effect of metal chlorides on the FeCN Hill reaction of Anacystis permeaplasts, measured on a background concentration of 150 mM choline-Cl.

photoreduction of DCIP (H2O → DCIP) by Anacystis permeaplasts. The anion ranking is in the order sulfate< chloride< nitrate, with regard to their efficiency in promoting deactivation, which corresponds to their ranking as chaotrops.

Inhibition at a lower concentration of cholinate, compared to metal chlorides, may possibly indicate a structure stabilizing effect as a result of cation binding to the membrane.

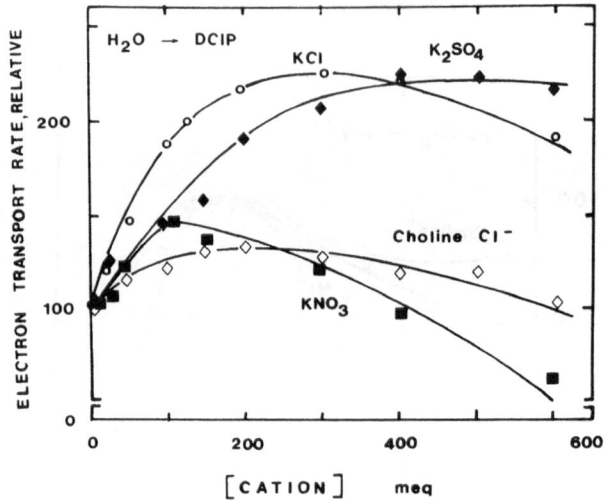

Fig. 8. Effects of anions on the deactivation of the H2O → DCIP reaction.

DISCUSSION

We have studied the effects of metal and nonmetal cations on photoinduced electron exchanges (Hill reactions) between oxidoreducible solutes and the outer thylakoid surface in permeaplasts of the unicellular cyanobacterium Anacystis nidulans. The permeaplasts are characterized by a fully functional photosynthetic electron transport chain and high photosynthetic control ratios (indicating ion-impermeable thylakoids) but are devoid of the endogenous O_2-reducing system, which has been detected in cell-free cyanobacterial thylakoids[19] and in spheroplasts[20].

The distribution of ions along the normal to the electrically charged membrane surface is governed by the space function of the electrostatic potential, in accordance to Boltzmann's distribution law. The concentration Cs of an ion of valence z in the surface region where the potential is Ψo, is related to the bulk phase concentration Cb as follows

$$Cs = Cb \exp(-zF\Psi_0/RT)$$

where the symbols R, F and T have their usual meaning.

According to this equation, cations of higher valence are preferentially pulled toward the negative membrane surface, while anions of higher valence, such as the trivalent FeCN are preferentially pushed away. The only way, therefore, to increase the surface concentration of FeCN is to lower the membrane surface potential, which can be achieved both by counterion screening and neutralization of the surface charge. Cations were found to stimulate, in general, such electron exchanges at a low range of concentrations, and to inhibit them at higher concentrations. The valence and concentration dependence suggests the involvement of electrostatic interactions between aqueous phase ions and the negative electric charge of the thylakoid surface. Our results deviate, however, from the expectations of the Gouy-Chapman[17] theory in several respects, and implicate both membrane charge screening by cations accumulating at the membrane-water interface and membrane charge neutralization by cations binding to anionic ligands.

These deviations include: (1) The inhibition of the Hill reactions at

310

higher cation concentrations. A possible cause is the suppression of electrostatic repulsion between membrane subunits, and the enhancement of attractive forces due to hydrophobic interactions, as a result of the partial neutralization of the surface charge by bound cations. The involvement of hydrophobic interactions is suggested by the dependence of the inhibition of the Hill reaction rates at higher electrolyte concentration on the chaotropic character of the anion. (2) Rate stimulation and inhibition are observed not only with respect to the anionic oxidant $FeCN^{3-}$, but also with respect to the cationic MV^{2+} and the electroneutral DCIP. Secondary structural changes, caused by the lowering of the thylakoid surface charge by the bound cations, may contribute to these phenomena. (3) Considerable differences exist among cations of the same valence with regard to the concentration requirement for rate stimulation and the maximal enhancement achieved. Notable is the rate enhancement by Ca^{2+}, observed both in the absence and in the presence of a high background concentration of monovalent nonbinding cation (cholinate). The dissociation constants for Ca^{2+} binding to the outer thylakoid surface are in the mM range. Accordingly, the only high affinity site for Ca2+ appears to be the one detected in the water-splitting complex of photosystem II.

REFERENCES

1. D. F. Ghanotakis and C. F. Yocum, Polypeptides of photosystem II and their role in oxygen evolution, Photosynth. Res. (in press, 1985).
2. W. W. Fredricks and A. T. Jagendorf, A soluble component of the Hill reaction in Anacystis nidulans, Arch. Biochem. Biophys. 104:39 (1964).
3. W. A. Susor and D. W. Krogmann, Hill activity in cell-free preparation of a blue-green alga, Biochim.Biophys. Acta 88:11 (1964).
4. D. I. Arnon, B. D. McSwain, H. Y. Tsujimoto and W. Wada, Photochemical activity and components of membrane preparations from blue-green algae. I. Coexistence of two photosystems in relation to chlorophyll a and removal of phycocyanin, Biochim. Biophys. Acta 357:231 (1974).
5. A. Binder, E. Tel-Or and M. Avron, Photosynthetic activities of membrane preparations of the blue-green alga Phormidium luridum, Eur. J. Biochem. 67:187 (1976).
6. R. G. Piccioni and D. C. Mauzerall, Calcium and photosynthetic oxygen evolution in cyanobacteria, Biochim. Biophys. Acta 504:384 (1978).
7. J. J. Brand, The effect of Ca^{2+} on oxygen evolution in membrane preparations from Anacystis nidulans, FEBS Letters 103:114 (1979).
8. C. L. S. DeRoo and Yocum D. F., Cation-induced, inhibitor-resistant photosystem II reactions in cyanobacterial membranes, Biochem. Biophys. Res. Commun. 100:1025 (1981).
9. R. A. Wavare and P. K. Mohanty, Cations stimulate electron transport associated with photosystem II and inhibit electron flow linked with photosystem I in spheroplasts of the cyanobacterium Synechococcus cedrorum, Photobiochem. Photobiophys. 6:189 (1983).
10. G. C. Papageorgiou, K. Kalosaka, T. Lagoyanni and G. Sotiropoulou, The role of cations in the photoinduced electron transport of cyanobacteria, in: "New Methods in Membrane Research and Biological Energy Transduction", L. Packer, ed., Plenum Press, New York (in press, 1985).
11. G. C. Papageorgiou, G. Sotiropoulou, T. Lagoyanni and K. Kalosaka, Electrolyte control of photosynthetic electron transport in cyanobacteria, in: "Creation and deactivation of excited states of biological molecules", D. Frackowiak, ed., Polish Academy of Sciences, Poznan (1985).
12. G. C. Papageorgiou, Photosynthetic activity of diimidoester modified

cells, permeaplasts and cell-free membrane fragments of the blue-green alga _Anacystis nidulans, Biochim. Biophys. Acta_ 461:379 (1977).

13. G. C. Papageorgiou and T. Lagoyanni, Photosynthetic properties of rapidly permeabilized cells of the cyanobacterium _Anacystis nidulans, Biochim. Biophys. Acta_ 807:230 (1985).

14. J. M. Armstrong, The molar extinction coefficient 2,6-dichlorophenol indophenol, _Biochim. Biophys. Acta_ 86:194 (1964).

15. G. McKinney, Absorption of light by chlorophyll solutions, _J. Biol. Chem._ 140:315 (1941).

16. K. Kalosaka, G. Sotiropoulou and G. C. Papageorgiou, Retardation of electron donation to photosystem I in aged cyanobacteria and its reversal by metal cations, _Biochim. Biopphys. Acta_ (1985, in press).

17. S. McLaughlin, Electrostatic potentials in membrane-solution interfaces, _in_: "Current Topics in Membranes and Transport," F. Bronner and A. Kleinzeller, eds., Academic Press, New York (1977).

18. E. Wax and W. Lockau, Stoichiometric photophosphorylation in thylakoids from the blue-green alga _Anabaena variabilis, Z. Naturforsch_ 35c:98 (1980).

19. R. C. Honeycutt and D. W. Krogmann, Alight-dependent oxygen reducing system from _Anabaena variabilis, Biochim. Biophys. Acta_ 197:267 (1970).

20. G. Sotiropoulou, T. Lagoyanni and G. C. Papageorgiou, Effects of Ca^{2+} ions on the light-induced electron transport of Anacystis nidulans pemeaplasts and spheroplasts, _in_: "Advances in Photosynthesis Research", vol. 2, C. Sybesma, ed., M. Nijhoff/Dr. W. Junk, publishers, The Hague (1984).

IMMOBILIZED CHLOROPLAST MEMBRANES :

EFFECTS OF CATIONS AND ANTICHAOTROPES ANIONS

Brigitte Thomasset, Jean-Noël Barbotin and Daniel Thomas

Laboratoire de Technologie Enzymatique, UA 523 du CNRS
Université de Technologie de Compiègne, B.P. 233
60206 Compiegne Cedex, France

INTRODUCTION

Biophotolysis of water, photohydrogen production and ATP regeneration can be used for the direct bioconversion of solar energy and have been studied by numerous authors[1]. Isolated chloroplast membranes are able to perform the photolysis of water, but the stability of photosystems over a long period of time is a crucial limitation for technological applications.

Joussaume and Bourdu[2] have emphasized the importance of ionic and protective agents on the functional properties of isolated thylakoids. In fact, the stabilization of thylakoids can be obtained by using high concentrations of ions[3], serum albumin[4], glutaraldehyde[5] and by several immobilization methods[6]. In this respect, the immobilization avoids extraction and purification phases and allows multistep reactions in bioconversion processes.

In this report, investigations are presented to determine the optimal conditions of mono- and divalent cation concentrations in the resuspending medium of thylakoids. Then, the thylakoids are immobilized in an insoluble matrix to obtain a best functional stability (ie. continuous use at 24°C under saturating illumination). The artificial membrane is considered as a weak ionic exchanger which introduces interactions between the electron transport chain and the polyelectrolyte environment. In order to study the response of thylakoids to environmental changes, we examine the influence of salt concentrations on kinetic parameters, morphological aspects, spectroscopic observations and on the stability of immobilized thylakoids.

MATERIALS AND METHODS

Thylakoid preparation

Thylakoids were isolated from lettuce leaves after osmotic shock according to the method of Epel and Neuman[7] and resuspended in a solution at pH 7.4 containing 50 mM Hepes, 330 mM sorbitol, 1 mM $MgCl_2$, 0.1 mM $MnCl_2$, 0.5 mM EDTA and 0.15 mM potassium phosphate buffer[8] ("sorbitol buffer", SB). Chlorophyll was assayed by the method of Mac Kinney[9] and by the method described by Laval-Martin et al.[10]. In this latter case, the pigments were extracted in acetone/water solution (90/10, v/v) and spectrophotometrically determined using a kinetic method of controlled pheophitinization.

Immobilization methods

As previously described by Thomas and Broun[11], cocrosslinking was performed by freezing at −20°C over 2 hours with 58.7 mg/ml bovine serum albumin (fraction V from Sigma chemical Co.) and 3.3 mg/ml glutaraldehyde (from Polysciences) in 50 mM sodium phosphate buffer pH 7.1 containing thylakoids corresponding to 450 µg/ml of chlorophyll. After thawing, an insoluble proteic matrix was obtained.

Reaction media

The reaction media were "sorbitol buffer" (SB) as described above, or 20 mM Hepes containing either 0.75 M potassium citrate ("potassium citrate buffer", PCB) that is the concentration previously described to obtain the maximal oxygen production[12]. Photosynthetic samples were illuminated at 24°C under saturating light (800 W/m^2) produced by a 50 W quartz/iodine lamp and filtered with a red filter ($\lambda > 580$ nm). Potassium ferricyanide (3 mM) was used as electron acceptor and NH_4Cl (5 mM) as the uncoupling agent.

Ferricyanide reduction

Immobilized thylakoids were tested either in a batch reactor or in a continuous stirred tank reactor[13] (C.S.T.R.). For both systems, the volume of the reactor was 15 ml and 100 µg of chlorophyll were used to reduce the ferricyanide. The reduction was recorded continuously at 420 nm. For batch assays, the activities were determined by measuring the initial rate of the photosynthetic reaction. In CSTR assays, the conversion ratio depends on the flow regime, the reaction rates and the size of the reactor[14] and was calculated at the steady state as follows :

$$\frac{(\text{reduced ferricyanide})}{(\text{reduced ferricyanide}) + (\text{untransformed ferricyanide})} \times 100$$

For immobilized thylakoids, the optimal parameters were determined by Cocquempot and Thomas[13]. A flow rate of 12 ml per hour was employed.

Fluorescence Measurements

Fluorescence emission spectra were recorded at 77 K using a laboratory-built apparatus described by Sironval et al.[15]

Photoacoustic experiments

A laboratory-built two beam photoacoustic (PA) spectrometer, previously described[16], was employed to record the photoacoustic spectra of immobilized thylakoids. According to Cahen et al.[17], the spectra can be normalized at 440 nm to facilitate the comparison of the different samples.

Electron microscopy

The specimens were fixed and dehydrated according to the method of Barbotin and Thomasset[18].

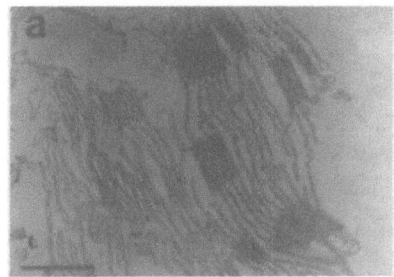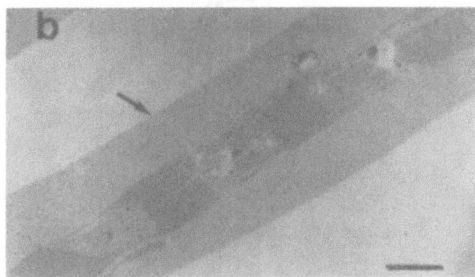

Figure 1. Ultrathin sections of native thylakoids (a) and of immobilized
thylakoids in an albumin-glutaraldehyde matrix (b). The arrow
indicates the matrix. Bars indicate 0.5 μm.

RESULTS AND DISCUSSION

Influence of immobilization on the functional stability of the electron
transport chain

Several authors[3,19-20] have pointed out that mono- and divalent cations
concurrently affect the excitation energy distribution from PSII to PSI,
and the stacking of thylakoid membranes. The highest activity yield
(i.e. fractional retention of activity after immobilization) was obtained
when 0.15 mM K^+, 1 mM Mg^{2+} and 0.1 mM Mn^{2+} were present in the resuspending media[8]. In such conditions, the thylakoid integrity is not modified
by the immobilization process. Moreover, the liquid-solid phase separations
which occurs during the insolubilization process, reinforces the degree of
stacking of the chloroplast membranes (Fig.1).

Furthermore, the albumin-glutaraldehyde matrix efficiently protects
the thylakoids against the photoinactivation. The functional stability
(measured at 24°C, under saturating light, in the presence of ferricyanide)
of immobilized thylakoids is considerably increased (Fig. 2).

In these studies, the fluorescence emission (Fig. 3a) and the photoacoustic spectroscopy (Fig. 3b) are used to provide informations about the
energy transfer efficiency and the non-radiative processes. Thus, correlations between the activities and the spectral changes have been obtained[8].

Diffusional constraints and exchange ionic properties of the matrix

The binding of thylakoids into an ion-exchange matrix allows a stabilization of the photosynthetic activities, but introduce interactions
between the catalytic reaction and the polyelectrolytic environment which
generally affect the activity. In order to study the response of thylakoids
to environmental changes, we have studied the influence of high concentrations
of salt on the kinetic parameters, on the initial activities and on the
functional stability of immobilized thylakoids.

Determination of apparent kinetic parameters. The initial rate data
from a batch reactor, was used to evaluate the kinetic parameter of the
system. In our case, the kinetic model relating electron acceptor transport
as a function of electron acceptor concentration can be described by a

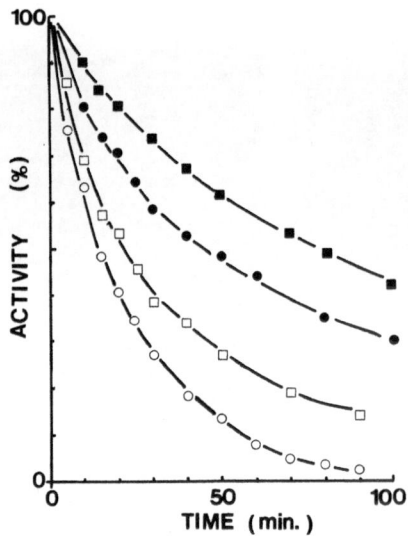

Figure 2. Ferricyanide reduction in function of continuous illumination of native (open symbols) or immobilized (full symbols) thylakoids tested in SB (○-●), PCB (□--■).

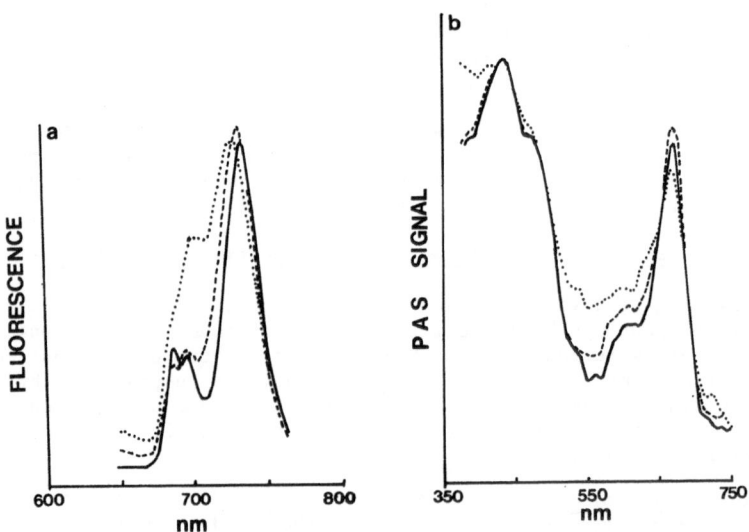

Figure 3. Low-temperature (77 K) fluorescence emission spectra (a) and photoacoustic spectra (b) of immobilized thylakoids (——) and after continuous illumination of 1 hr. 30 min. (---) and 4 hr. 30 min. (···).

Table I. Apparent kinetic parameters for immobilized thylakoids.

	"Sorbitol buffer"	"Potassium citrate buffer"
Kp ($\pm$ SEM)* mM	1.21 $\pm$ 0.094	0.6 $\pm$ 0.085
Vm ($\pm$ SEM)* mM ferrocyanide/ hr./100 µg chlorophyll	1.84 $\pm$ 0.086	1.63 $\pm$ 0.085

* SEM = Standard Errors of the means.

rectangular hyperbolic function similar to the Michaelis-Menten equation[21]. The apparent affinity for ferricyanide increases when high concentrations of citrate are present in the reaction medium (Table I).

Influence of high concentrations of salts on the electron transport. The artificial matrix, used in this study, can be considered as a weak ion-exchanger and the distribution of the solute between the matrix surface and the external solution is partially governed by a Donnan equilibrium[22]. For immobilized thylakoids, high concentrations of citrate highly increase the ferricyanide reduction and allow a better transformation of the substrate in the CSTR than a low ionic strength medium (Tables II b and III, time 0).

Table II. Photosynthetic activity of native thylakoids (a) and immobilized thylakoids (b) after preincubation periods (24°C, in the dark, and without ferricyanide).

	Pre-incubation Period (min.)	Activity mM ferrocyanide/mg chlorophyll/hr.	
		"Sorbitol buffer"	"Potassium citrate buffer"
a) Native Thylakoids	0	36.25	38
	30	23.6	27.8
	60	20.1	27
	120	14.1	24.75
b) Immobilized Thylakoids	0	15	32.25
	30	11.7	29.9
	60	10.9	27.5
	120	10.8	25.75

Table III. Maximum production of ferrocyanide and conversion ratio of
immobilized thylakoids in function of pre-incubation time (at
24°C, in the dark and in the absence of ferricyanide).
The production is expressed in mM ferrocyanide/100 µg chloro-
phyll/hr.

Time of pre-incubation (min.)	"Sorbitol buffer"		"Potassium citrate buffer"	
	Production *	Conversion ration (%)	Production *	Conversion ratio (%)
0	0.33	11.3	0.575	19.15
30	0.297	9.9	0.395	13.3
120	0.254	8.5	0.354	12

An increase of the ionic strength in the external media can explained
this phenomenon (Fig.4).
So, the local ion concentration seems to be strongly different from the
external concentrations. To increase the local concentration of ions,
equilibration of the matrix with the electrolyte solutions are performed
at 24°C in the dark and in the absence of ferricyanide. The apparent
activity recorded after 2 hours of preincubation in the presence of 0.75 M
citrate are higher than the initial activity of the native thylakoids after
the same treatment (Table II a, b). Moreover, the maximal production
obtained after 2 hours at 24°C in "potassium citrate buffer" is higher than
the initial production of immobilized thylakoids without pre-incubation

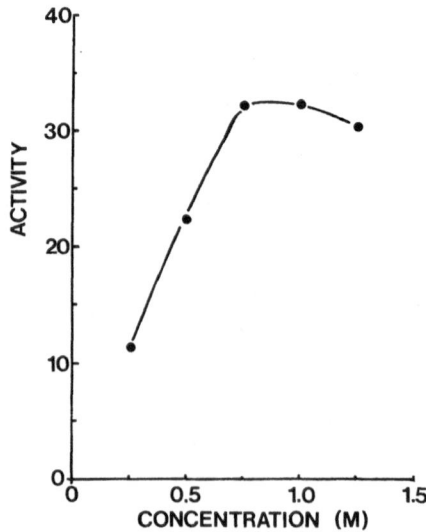

Figure 4. Influence of different concentrations of citrate on the ferricyanide reduction for immobilized thylakoids. The activity is
expressed in mM ferrocyanide/mg chlorophyll/hr.

Table IV. Ferrocyanide production in the CSTR over 8 hours of continuous
functional work of immobilized thylakoids.

	Ferrocyanide Production (mM)	
Time of pre-incubation	"Sorbitol buffer"	"Citrate buffer"
0	2.8	4.45
2 hours	2.02	5.1

period (Table III). The denaturation observed for the immobilized systems
is less important than that of the native systems. The immobilization
process highly increases the stability of thylakoids[23], and local high
concentrations of antichaotropic anions (e.g. citrate) are known to
stabilize the chlorophyll-protein complexes against denaturation[24]. The
use of high concentrations of citrate in the external medium, allows to
increase the functional stability of native and immobilized thylakoids
(Fig. 2). After pre-incubation in the presence of citrate, the total
ferrocyanide production in a CSTR over 8 hours of continuous functional
work is highly increased (Table IV) and a protective effect against denaturation is observed (Fig. 5). In fact, there is a balance between the
reaction rate, the salt diffusion and the denaturation process.

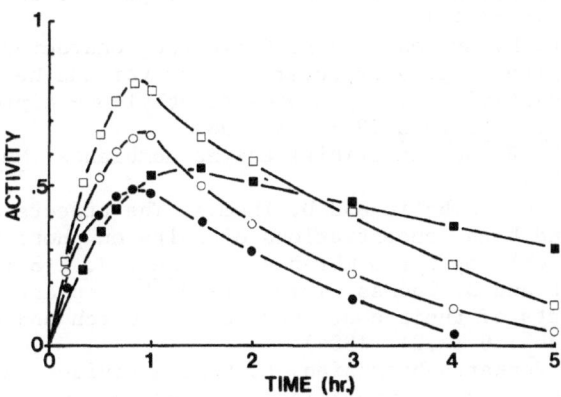

Figure 5 : Ferricyanide reduction as a function of time in a CSTR under
saturating light for immobilized thylakoids.
Without pre-incubation period : SB (○ - ○), PCB (□ - □)
media, and after 2 hours of pre-incubation : SB (● - ●) and
PCB (■ - ■).
The activity is expressed in mM ferrocyanide/100 μg chlorophyll/
hr.

CONCLUSION

These studies show that it is necessary to rigorously define the re-suspended media of biological organelles before performing their immobil-ization. However, the immobilization process considerably increases the stability of labile photosystems, but does not prevent a partial denatur-ation during pre-incubation periods, or continuous illumination. But, the photoinactivation, which occurs throughout the long term use in a CSTR, can be limited by using high concentration of citrate. A compromise should be found between the initial activity, the denaturation process and the highest transformation.

REFERENCES

1. J.R. Benemann, K. Miyamoto and P.C. Hallenbeck, Bioengineering aspects of biophotolysis, Enzyme Microb. Technol. 2 : 103 (1980).
2. M. Joussaume and R. Bourdu, Retention of properties by chloroplasts upon storage outside their original cellular medium, Biol. Cell 47 : 251 (1983)
3. J. Barber, Influence of surface charges on thylakoid structure and function, Ann. Rev. Plant Physiol. 33 : 261 (1982)
4. G. Kulandaivelu and D.O. Hall, Stabilization of the photosynthetic activities of isolated spinach chloroplasts during prolonged ageing, Z. Naturforsch 31 : 452 (1976)
5. G. Papageorgiou, Stabilization of chloroplasts and subchloroplast particles, Methods Enzymol. 69 : 613 (1980).
6. A. Tanaka and S. Fukui, Immobilized organelles in : "Immobilized cells and organelles", Mattiasson, ed., CRC Press, (1983).
7. B.L. Epel and J. Neuman, The mechanism of the oxidation of ascorbate and Mg^{2+} by chloroplasts. The role of the radical superoxide, Biochim. Biophys. Acta 325 : 520 (1973)
8. B. Thomasset, T. Thomasset, A. Vejux, J. Jeanfils, J.N. Barbotin and D. Thomas, Immobilized thylakoids in a cross-linked albumin matrix : effects of cations studied by electron microscopy, fluorescence emission, photoacoustic spectroscopy, and kinetic measurements, Plant Physiol. 70 : 714 (1982).
9. G. Mac Kinney, Absorption of light by chlorophyll solutions, J. Biol. Chem. 140 : 315 (1941).
10. D. Laval-Martin, D. Grizeau and R. Calvayrac, Characterization of resistant Euglena : greater tolerance for various herbicides and increased sensitivity of thylakoids to ethyl- s - dipropylthiocarba-mate. Plant Sci. Letters 29 : 155 (1983).
11. D. Thomas and G. Broun, Artificial enzyme membranes. Methods Enzymol. 44 : 901 (1976)
12. B. Thomasset, J.N. Barbotin and D. Thomas. The effects of oxygen solubility and high concentrations of salts on photosynthetic electron transport of chloroplast membranes, Biochem. J. 218 : 539 (1984)
13. M.F. Cocquempot and D. Thomas, Immobilized chloroplast membranes : kinetic aspects of their continuous use in batch and chemostat, Enzyme Microb. Technol. 6 : 321 (1984)
14. W.R. Vieth, K. Venkatasubramanian, A. Constantinides and B. Davidson, Design and analysis of immobilized enzymes and immobilized microbial cells, Appl. Biochem. Bioeng. 1 : 221 (1976).
15. C. Sironval, M. Brouers, J.M. Michel and Y. Kuiper, The reduction of photochlorophyllide into chlorophyllide I. The kinetics of the P 657-647 P 688-676 phototransformation, Photosynthetica 2 : 268 (1968)
16. B. Thomasset, T. Thomasset, J.N. Barbotin and A.M. Vejux, Photoacoustic spectroscopy of active immobilized chloroplast membranes, Appl. Optics 21 : 124 (1982)

17. D. Cahen, S. Malkin and E.I. Lerner, Photoacoustic spectroscopy of chloroplast membranes : listening to photosynthesis, FEBS Lett. 91 : 339 (1978)

18. J.N. Barbotin and B. Thomasset, Immobilized organelles and whole cells into protein foam structures : scanning and transmission electron microscopic observations, Biochimie 62 : 359 (1980)

19. S. Izawa and N.E. Good, Effect of salts and electron transport on the conformation of isolated chloroplasts II : electron microscopy, Plant Physiol. 41 : 552 (1966)

20. D. Wong, Govindjee and H. Merkelo, Effects of bulk pH and of monovalent and divalent cations on chlorophyll a fluorescence and electron transport in pea thylakoids, Biochim. Biophys. Acta, 592 : 546 (1980)

21. J.M. Howell, W.R. Vieth, Biophotolytic membranes : simplified kinetic model of photosynthetic electron transport, J. Mol. Cat. 16 : 245 (1982).

22. A. Friboulet and D. Thomas, Electrical excitability of artificial enzyme membranes. I. ion-exchange properties of synthetic properties films, Biophys. Chem. 16 : 139 (1982)

23. M.F. Cocquempot, B. Thomasset, J.N. Barbotin, G. Gellf and D. Thomas, Comparative stabilization of biological photosystems by several immobilization procedures, 2. storage and functional stability of immobilized procedures, Eur. J. Appl. Microbiol. Biotechnol. 11 : 193 (1981)

24. B.A. Zilinskas and R.E. Glick, Monovalent intermolecular forces in phycobilisomes of "Porphyridium cruentum", Plant Physiol. 68 : 447 (1981)

ANIONIC INHIBITION OF PS II: INSIGHTS

DERIVED FROM CARBONIC ANHYDRASE

Alan Stemler
Département de Biologie, Service de Radioagronomie
CEN Cadarache, BP n°1 - 13108 Saint-Paul-Lez-Durance CEDEX
France
Permanent address
Botany Dept.
University of California
Davis, CA 95616 U.S.A.

Paul Jursinic
Northern Regional Research Center
Agricultural Research Service
U.S. Dept. of Agriculture
Peoria, IL 61604 U.S.A.

Judith B. Murphy
Botany Dept.
University of California
Davis, CA 95616 U.S.A.

INTRODUCTION

At high concentrations, a number of monovalent anions inhibit photosystem II (PSII)[1]. Often this inhibition can be overcome by added bicarbonate[1]. Sinclair[2] has suggested that several anions inhibit PSII by replacing chloride ions on the oxygen-evolving mechanism. Thus both the Cl^- and HCO_3^- binding sites on PSII appear to be targets for inhibition by other anions. This inhibition bears striking resemblance to similar effects on the enzyme carbonic anhydrase. Indeed all of the classic inhibitors of this enzyme, monovalent anions, acetazolamide, and imidazole also inhibit PSII[3,4]. So close is the resemblance between the response of PSII and carbonic anhydrase to anions and other inhibitors, that we investigated the possibility that the PSII complex actually has carbonic anhydrase activity. We used a mass spectrometer to measure the rate of oxygen exchange between doubly labelled $C^{18}O_2$ and $H_2^{16}O$ in the presence and absence of osmotically shocked, washed maize thylakoids. Our results indicate that such thylakoids do demonstrate carbonic anhydrase activity. Moreover, the exchange activity was definitely greater in the dark than when the experiment was done in room light. PSII particles also showed carbonic anhydrase activity. These results and others suggest that some anion effects on the PSII complex (including that of bicarbonate) can be understood in terms of a specific interaction with a polypeptide that has carbonic anhydrase activity. Since a great deal is known about the binding of anions to carbonic anhydrase, this knowledge can be profitably applied to PSII.

MATERIALS AND METHODS

Maize thylakoids were used in these studies. They were obtained as previously described[1]. Hill reaction rates were measured with a Clark-type oxygen monitor under saturating white light. Flash yields of oxygen were measured as in reference [3]. The methods for measuring $H^{14}CO_3^-$ binding to thylakoids were described in detail previously[5].

Carbonic anhydrase activity was measured as the catalysed rate of isotopic exchange between $C^{18}O_2$ and $H_2^{16}O$. Experiments were carried out with a MAT Atlas CH_4 mass spectrometer. A temperature-controlled glass reaction cell was connected to the spectrometer by means of a vacuum line. A polypropylene membrane located at the bottom of the magnetically-stirred cell allowed dissolved gases to be introduced into the line. Molecular masses 44 ($C^{16}O^{16}O$), 46 ($C^{18}O^{16}O$), and 48 ($C^{18}O^{18}O$) were monitored alternately with sweep time of about 30 s. CO_2/HCO_3^- labelled with ^{18}O was made by dissolving a saturating amount of unlabelled $NaHCO_3$ in $H_2^{18}O$ (>97 percent ^{18}O) and allowing the solution to equilibrate for several days. For each measurement, a small amount (8 µl) of this solution was injected into the reaction cell which contained 8 ml of reaction mixture.

RESULTS AND DISCUSSION

Inhibition of PSII by anions and reversal on inhibition be added HCO$_3^-$

Hill reaction rates are influenced by the presence of various anions. This is shown in Table I. F^-, HCO_2^-, NO_2^-, OAc, and NO_3^- all inhibit the Hill reaction. For these anions, HCO_3^- reverses the inhibition. All of these anions inhibit carbonic anhydrase to some degree (for a review see Pocker and Sarkanen, [6]) the carbonic anhydrase inhibitor, gold cyanide, $Au(CN)_2^-$, is also a relatively powerful inhibitor of the Hill reaction, as in another linear anion, N_3^-. In the case of $Au(CN)_2^-$, inhibition is only partially reversed by HCO_3^-. For N_3^-, no reversal was detected. On carbonic anhydrase also $Au(CN)_2^-$ does not share the same binding site as other anions[6]. More detailed discussion can be found in reference 4.

Table I

Inhibition of Hill reaction by various monovalent anions in the presence or absence of bicarbonate. See ref.1 for details.

Anion	Conc.	Rate O_2 mg^{-1}Chl hr-1 − HCO$_3$ % of control	+ HCO$_3$ % of control	$\dfrac{+ HCO_3^-}{- HCO_3^-}$
Control	–	100	104	1.04
F^-	100	44	85	1.95
formate	100	15	76	5.0
NO_2^-	20	20	73	3.64
acetate	100	61	93	1.52
NO_3^-	100	22	94	4.31
$Au(CN)_2^-$	5	33	52	1.55
N_3^-	20	20	20	1.00

Competitive inhibition of $H^{14}CO_3^-$ binding to thylakoid membranes by various anions

F^- and NO_2^- are competitive inhibitors of HCO_3^- binding to the PSII complex. This is shown in Fig. I where double reciprocal plots give $H^{14}CO_3^-$ binding parameters in thylakoids in the presence of various added anions. HCO_2^-, NO_3^- and OAc (not shown) also competitively inhibit $H^{14}CO_3^-$ binding. $Au(CN)_2^-$ (and also NO_3^-, not shown), the two most powerful anionic inhibitors of the Hill reaction (Table I) have only weak effects on HCO_3^- binding. It appears that these two anions do not inhibit at the same site as other anions. Thus the inhibitory effects of anions on $H^{14}CO_3^-$ binding correlate with their HCO_3^- - reversible effects on Hill activity (Table I). Details can be found in ref.1.

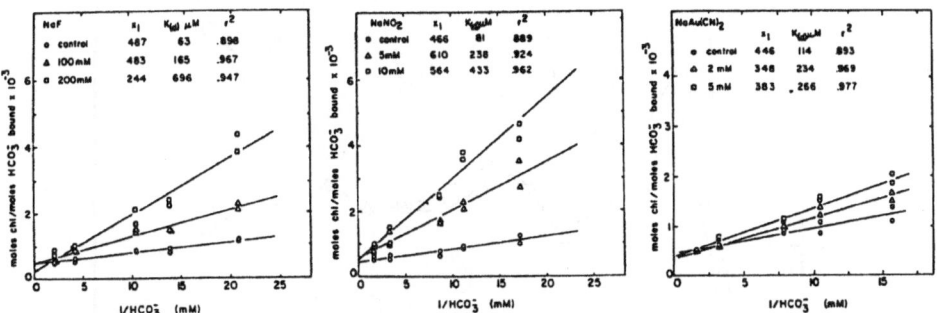

Fig. 1. Binding parameters of $H^{14}CO_3^-$ to PSII in the presence or absence of various inhibitory anions. X_1 is the number of chlorophyll molecules per HCO_3^- binding site. K_d is the dissociation constant for HCO_3^-. r^2 is the correlation coefficient from linear regression analysis. Details in ref. 1.

Inhibition of flash-yields of oxygen in chloroplasts given acetazolamide and imidazole

Acetazolamide is thought to be a specific inhibitor of carbonic anhydrase. Indeed the response of a system to his reagent is generally taken to be diagnostic of the presence of this enzyme. Fig.2 shows that oxygen yields induced by saturating light flashes are strongly inhibited by 1 mM acetazolamide. Imidazole also inhibits carbonic anhydrase as well as PSII (Fig. 2). In both cases, inhibition is pH dependent. Only the basic form on imidazole is effective. Details can be found in ref. 3.

Inhibition of $H^{14}CO_3^-$ binding to PSII by atrazine at a low-affinity herbicide binding site

Atrazine, at concentrations of 1 μM or greater, non-competitively inhibits $H^{14}CO_3^-$ binding to PSII. This is shown in Fig.3. Maximum inhibition occurs at a concentration of 5 μm when only half of the $H^{14}CO_3^-$ binding sites are eliminated. Even a 10-fold higher concentration of atrazine fails to eliminate more than half of the sites. This indicates heterogeneity of HCO_3^- binding sites. It now appears that there are two HCO_3^- binding sites on the PSII complex[1],[7], only one is influenced by atrazine. A parallel phenomenon is seen on carbonic anhydrase. The enzyme contains two binding sites for acetate[8] and also for non-substrate HCO_3^-[9]. Sulfonamide eliminates binding of anions to one site on carbonic anhydrase but not to the other site. Additional information is given in reference 7.

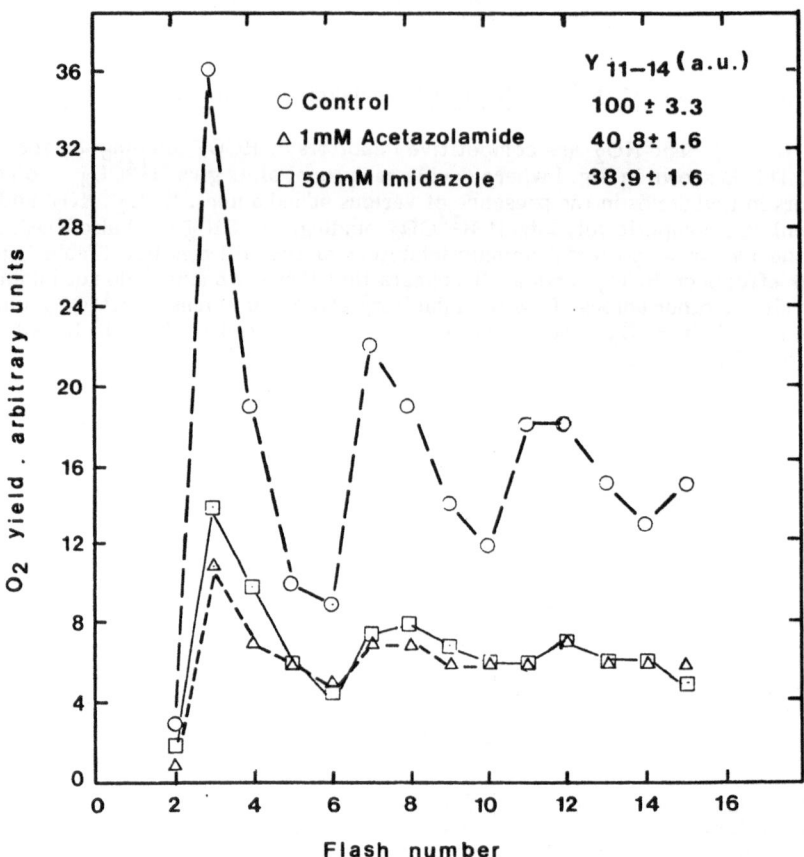

Fig. 2. Effects of acetazolamide and imidazole on flash yields of oxygen. Thylakoids were dark-adapted for 10 min on a bare-platinum electrode, then given a series of saturating light flashes spaces 1S apart. Details in ref. 3.

Pretreatment of thylakoids with bromoacetate eliminates $H^{14}CO_3^-$ binding sites

Bromoacetate is a chemical modifier of carbonic anhydrase (for a review, see 10). Initially it binds ionically at the anion binding sites on the enzyme. After a time, the binding becomes covalent as the bromoacetate reacts with a specific histidine near the active site of the enzyme. This inhibits carbonic anhydrase non-competitively and irreversibly. We treated thylakoids with bromoacetate to see if a similar effect could be observed on PSII. Initially, bromoacetate acted as a competitive inhibitor of $H^{14}CO_3^-$ binding, like other anions shown in Fig. 1. However, when we pretreated the chloroplasts overnight and in room light with bromoacetate, the binding of $H^{14}CO_3^-$ was then inhibited non-competitively (Fig.4) compared to controls pretreated with acetate. We interpret this to mean that bromoacetate binds covalently to the anion binding site on PSII just as it does on carbonic anhydrase. Once bound, the carboxymethyl ligand prevents the binding of other anions such as bicarbonate. This work demonstrates that experiments suggested by the carbonic anhydrase literature can have predictable results when done on PSII. See reference 4 for additional discussion.

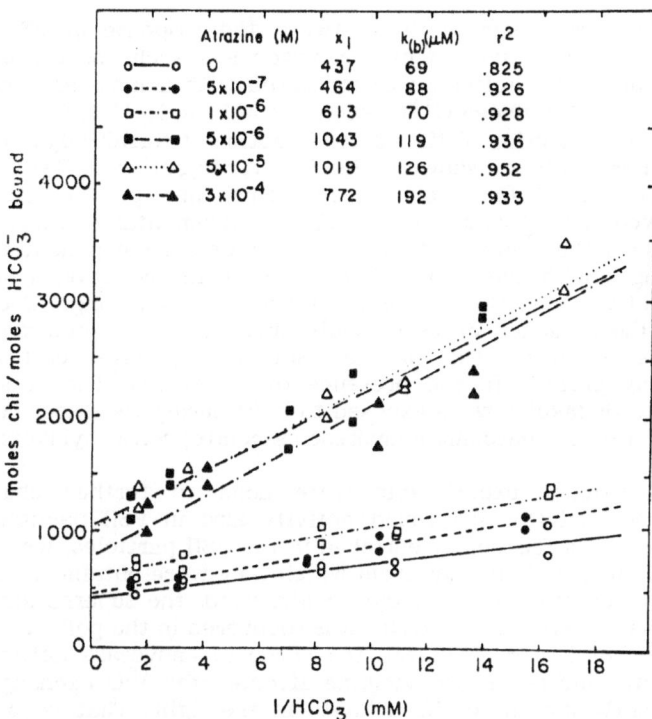

Fig. 3. HCO_3^- binding parameters in thylakoids given various concentrations of atrazine. See reference 7 for experimental details.

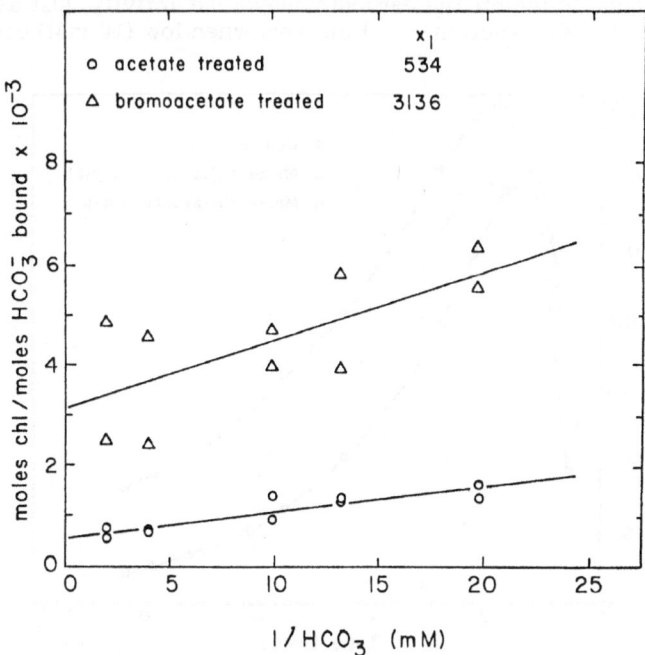

Fig. 4. Inhibition of $H^{14}CO_3^-$-binding to PSII by pretreatment with bromoacetate. Thylakoids were pretreated overnight at room temperature in either acetate or bromoacetate. The chloroplasts were washed to remove unreacted anion and then given $H^{14}CO_3^-$.

Carbonic anhydrase activity associated with washed thylakoid membranes and PSII particles

The strong similarities between the response of PSII and carbonic anhydrase to a number of treatments prompted a search for carbonic anhydrase activity in thylakoids. The activity was measured with a mass spectrometer as the rate of exchange of ^{18}O between $C^{18}O^{18}O$ (mass 48) and $H_2^{16}O$. The amount of mass 48 was monitored as a function of time after injection of $NaHC^{18}O_3^-$. The results of representative experiments are shown in Fig.5. After injection of $NaHC^{18}O_3^-$, the mass 48 signal increased for about 2 min. This time period reflects chemical equilibration between HCO_3^- and CO_2 as well as instrumental response time. After a maximum is reached, the concentration of mass 48 declines exponentially as the ^{18}O in $C^{18}O_2$ exchanges with the ^{16}O in the water of the reaction medium. The top redrawn trace in Fig.5 shows the rate of spontaneous ^{18}O exchange. The bottom trace was obtained in the presence of osmotically-shocked and washed maize thylakoids kept in complete darkness. The rate of isotopic exchange (carbonic anhydrase activity) is clearly greater in the presence of thylakoids. The middle trace was observed when the thylakoids were exposed to light during the experiment. Light has an inhibitory effect on the carbonic anhydrase associated with thylakoids.

Additional experiments were done to further characterize the carbonic anhydrase activity. We found activity also in PSII particles made from maize. The activity was frequently enriched in the PSII particles, when compared on a chlorophyll basis to the thylakoids from which they were obtained. When thylakoids were suspended in reaction mixture, then centrifuged, the supernatant contained no carbonic anhydrase activity. The activity was recovered in the pellet. Acetazolamide, at a concentration of 1 mM, inhibited the carbonic anhydrase activity to a great extent. Neither atrazine nor hydroxylamine affected the ^{18}O exchange in the dark, but both significantly increased the activity in the light. That is, both these PSII modifiers partially reversed the inhibitory effects of light on the carbonic anhydrase in thylakoids. The anionic inhibitors of PSII, formate and 10 mM NO_3^- inhibited thylakoid carbonic anhydrase. Interesting effects were observed with Cl^- at a concentration of 100 mM, Cl^- was found to inhibit the carbonic anhydrase activity, just as high concentrations can inhibit the Hill reaction[11]. However, when low (10 mM) concentrations

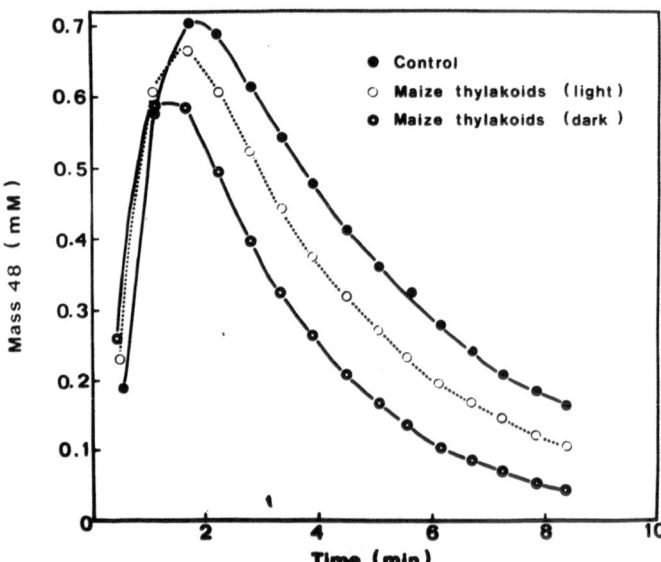

Fig.5. Decay of mass 48 signal following isotopic exchange between $C^{18}O^{18}O$ and $H_2^{16}O$. The top trace represents spontaneous isotopic exchange while the bottom two traces show exchange catalyzed by the presence of carbonic anhydrase in maize thylakoids.

were given to "Cl$^-$-depleted" thylakoids, a marked increase in carbonic anhydrase activity was observed. Br$^-$, and a low (1 mM) concentration of NO$_3^-$ also stimulated the carbonic anhydrase activity while F$^-$ and SO$_4^=$ had no effect. Thus a "Cl$^=$ effect", typical of PSII electron transport[12] was also observed on thylakoid carbonic anhydrase. These observations, and others, will be discussed in a detailed manuscript now in preparation.

CONCLUDING REMARKS

A number of phenomena associated with PSII in thylakoid membranes occur also in carbonic anhydrase. Most notably are the stimulatory effects of low concentrations of certain anions (Cl$^-$, Br$^-$, NO$_3^-$) and the inhibitory effect of high concentrations of Cl$^-$, HCO$_2^-$, NO$_3^-$, F$^-$ and others. All of the classic inhibitors of carbonic anhydrase, including acetazolamide and imidazole, also inhibit PSII. We suggest that the reason why these and other strong similarities exist between the two systems is that one of the polypeptides of the PSII complex may actually have carbonic anhydrase activity. This possibility is supported by direct measurements of ^{18}O exchange between C^{18}O$_2$ and H$_2$^{16}O catalysed by thylakoid membranes and PSII particles. We propose that the response of PSII to anions (including bicarbonate) can be understood in terms of direct binding of the anions to a polypeptide intrinsic to PSII that has carbonic anhydrase activity. In this view, the "bicarbonate-effect" on PSII is not a phenomenon unique to photosynthetic electron transport, but simply indicative of the presence of a common enzyme, carbonic anhydrase. The most significant advantage to this view is that the carbonic anhydrase literature is rich in detailed studies of anion interactions. This knowledge can be applied to PSII to increase our present understanding and for predicting the outcome of future work.

ACKNOWLEDGEMENTS

The work reported here was supported in part by a National Science Foundation (U.S.A.) Grant PCM 80 04 075 to A.S.

REFERENCES

1. A. STEMLER and J.B. MURPHY., Bicarbonate-reversible and irreversible inhibition of photosystem II by monovalent anions, Plant Physiol. 77 : 974 (1985)

2. J. SINCLAIR. The influence of anions on oxygen evolution by isolated spinach chloroplasts. Biochim. Biophys. Acta 764 : 247 (1984).

3. A. STEMLER and P. JURSINIC, The effects of carbonic anhydrase inhibitors formate, bicarbonate, acetazolamide, and imidazole on photosystem II in maize chloroplasts, Arch. Biochem. Biophys. 221 : 227 (1983).

4. A. STEMLER, Carbonic anhydrase : molecular insights applied to photosystem II research in thylakoid membranes. in : Inorganic Carbon Transport in Aquatic Photosynthetic Organisms. J. Berry, W. Lucas, eds., American Society of Plant Physiologists, Rockeville, MD, in press.

5. A. STEMLER and J.B. MURPHY, Determination of the binding constant of H^{14}CO$_3^-$ to the photosystem II complex in maize chloroplasts : effects of inhibitors and light. Photochem. Photobiol. 38 : 701 (1983).

6. Y. POCKER and S. SARKANEN, Carbonic anhydrase : structure, catalytic versatility, and inhibition. Adv. Enzymol. 47 : 149 (1978).

7. A. STEMLER and J. MURPHY, Inhibition of HCO$_3^-$ binding to photosystem II by atrazine at a low-affinity herbicide binding site. Plant Physiol. 76 : 179 (1985).

8. A. LANIR and G. NAVON, NMR studies of the two binding sites of acetate ions to manganese (II) carbonic anhydrase. Biochim. Biophys. Acta 341 : 75 (1974).

9. P.L. YEAGLE, C.H. LOCHMULLER, R.W. HENKENS ^{13}C nuclear magnetic resonance studies on the mechanism of action of carbonic anhydrase. Proc. Natl. Acad. Sci. U.S.A., 72 : 454 (1975).

10. S. LINDSKOG, Carbonic anhydrase, in : Advan. in Inorganic Biochem, Vol.4, G.L. Eickhorn, L.G. Marzilli, eds, Elsevier Biomedical (1982).

11. N.E. GOOD, Carbon dioxide and the Hill reaction. Plant. Physiol. 38 : 298 (1963).

12. P.M. KELLEY and S. IZAWA, The role of chloride ion in photosystem II : Effects of chloride ion on photosystem II electron transport and on hydroxylamine inhibition. Biochim. Biophys. Acta 502 : 198 (1978).

INDEX